10 0312012 5

D0302916

ROM THE LIBRAR

DATE DUE FOR RETURN

UNIVERSITY LIBRARY

3 NOV 2010

OWL JUB 05

This book may be recalled before the above date.

Introductory Techniques for 3-D Computer Vision

Emanuele Trucco

Heriot-Watt University,
Edinburgh, UK

Alessandro Verri

Università di Genova,
Genova, Italy

JUBILEE
CAMPUS
LRC

Prentice Hall
Prentice Hall
Upper Saddle River, New Jersey 07458

Library of Congress Cataloging-in-Publication Data

Trucco, Emanuele
   Introductory Techniques for 3-D Computer Vision,
   Emanuele Trucco and Alessandro Verri
      p. cm.
   Includes bibliographical references and index.
   ISBN: 0-13-261108-2
   CIP data available

Acquisitions editor: **TOM ROBBINS**
Editor-in-chief: **MARCIA HORTON**
Production editor: **IRWIN ZUCKER**
Managing editor: **BAYANI MENDOZA DE LEON**
Director of production and manufacturing: **DAVID W. RICCARDI**
Cover director: **JAYNE CONTE**
Manufacturing buyer: **JULIA MEEHAN**
Editorial assistant: **NANCY GARCIA**
Composition: **WINDFALL SOFTWARE**, using ZzT_EX

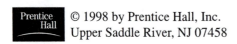

© 1998 by Prentice Hall, Inc.
Upper Saddle River, NJ 07458

I 0031201 25

All rights reserved. No part of this book may be reproduced, in any form or by any means, without permission in writing from the publisher.

The author and publisher of this book have used their best efforts in preparing this book. These efforts include the development, research, and testing of the theories and programs to determine their effectiveness. The author and publisher make no warranty of any kind, expressed or implied, with regard to these programs or the documentation contained in this book. The author and publisher shall not be liable in any event for incidental or consequential damages in connection with, or arising out of, the furnishing, performance, or use of these programs.

Printed in the United States of America

10 9 8 7 6 5

ISBN 0-13-261108-2

Prentice-Hall International (UK) Limited, London
Prentice-Hall of Australia Pty. Limited, Sydney
Prentice-Hall Canada, Inc., Toronto
Prentice-Hall Hispanoamericana, S.A., Mexico
Prentice-Hall of India Private Limited, New Delhi
Pearson Education Asia Pte. Ltd., Singapore
Prentice-Hall of Japan, Inc., Tokyo
Editora Prentice-Hall do Brazil, Ltda., Rio de Janeiro

To my mother

Emanuele Trucco

To my parents

Alessandro Verri

. . . nunc agere incipiam tibi, quod vementer ad has res
attinet, esse ea quae rerum *simulacra* vocamus.

(. . . you shall now see me begin to deal with what is of high importance to the
subject, and to show that there exists what we call *images* of things.)

Lucretius, *De Rerum Natura*, 4.24–44

Contents

Foreword

Until recently, computer vision was regarded as a field of research still in its infancy, not yet mature and stable enough to be considered part of a standard curriculum in computer science. As a consequence, most books on computer vision became obsolete as soon as they were published. No book thus far has ever managed to provide a comprehensive overview of the field since even the good ones focus on a narrow subarea, typically the author's research endeavor.

With Trucco and Verri, the situation has finally changed. Their book promises to be the first true textbook of computer vision, the first to show that computer vision is now a mature discipline with solid foundations. Among connoisseurs, the authors are well known as careful and critical experts in the field. (I am proud to have figured in the career of one of them: Alessandro Verri worked with me at MIT for a short year, and it was a joy to work with him.)

Over the years, I have been asked many times by new graduate students or colleagues what to read in order to learn about computer vision. Until now, my answer was that I could not recommend any single book. As a substitute, I would suggest an ever-changing list of existing books together with a small collection of specific papers. From now on, however, my answer is clear: *Introductory Techniques for 3-D Computer Vision* is the text to read.

I personally believe that *Introductory Techniques for 3-D Computer Vision* will be the standard textbook for graduate and undergraduate courses on computer vision in years to come. It is an almost perfect combination of theory and practice. It provides a complete introduction to computer vision, effectively giving the basic background for practitioners and future researchers in the field.

Trucco and Verri have written a textbook that is exemplary in its clarity of exposition and in its intentions. Despite the initial warning ("Fra il dire e il fare c'è di mezzo il mare"[1]), the objectives stated in the preface are indeed achieved. The book

---

[1] Between words and deeds there is the sea.

not only places a correctly balanced emphasis on theory and practice but also provides needed material about typically neglected but important topics such as measurements, calibration, SVD, robust estimation, and numerical differentiation.

Computer vision is just now maturing from an almost esoteric corner of research to a key discipline in computer science. In the last couple of years, the first billion-dollar computer vision companies have emerged, a phenomenon no doubt facilitated by the irrational exuberance of the stock market. We will undoubtedly see many more commercial applications of computer vision in the near future, ranging from industrial inspection and measurements to security database search, surveillance, multimedia and computer interfaces. This is a transition that other fields in engineering, such as signal processing and computer graphics, underwent long ago. Trucco and Verri's timely book is the first to represent the discipline of computer vision in its new, mature state, as the industries and applications of computer vision grow and mature as well. As it reaches adulthood, computer vision is still far from being a solved problem. The most exciting developments, discoveries and applications lie ahead of us. Though a similar statement can be made about most areas of computer science, it is true for computer vision in a much deeper sense than, say, for databases or graphics. After all, understanding the principles of vision has implications far beyond engineering, since visual perception is one of the key modules of human intelligence. Ultimately, understanding the problem of vision is likely to help us understand the brain. For this reason, I am sure that a long and successful series of new editions will follow this book, with updates most likely to come in the chapters dedicated to object recognition and in new hot topics such as adaptation and learning.

Introductory Techniques for 3-D Computer Vision is much more than a good textbook: It is the first book to mark the coming of age of our own discipline, computer vision.

Tomaso Poggio
Cambridge, MA
Brain Sciences Department and Artificial Intelligence Laboratory
Massachussetts Institute of Technology

Preface: About this Book

> Here, take this book and peruse it well.
>
> Christopher Marlowe, *Doctor Faustus*, II.i

> Fra il dire e il fare c'e' di mezzo il mare[1]
>
> Italian proverb

What this Book is and is Not

This book *is* meant to be:

- an applied introduction to the problems and solutions of modern computer vision.
- a practical textbook, teaching how to develop and implement algorithms to representative problems.
- a structured, easy-to-follow textbook, in which each chapter concentrates on a specific problem and solves it building on previous results, and all chapters form a logical progression.
- a collection of selected, well-tested methods (theory and algorithms), aiming to balance difficulty and applicability.
- a starting point to understand and investigate the literature of computer vision, including conferences, journals, and Internet sites.
- a self-teaching tool for research students, academics, and professional scientists.

This book *is not* meant to be:

- an all-embracing book on computer vision and image processing.
- a book reporting research results that only specialists can appreciate: It is meant for teaching.
- an exhaustive or historical review of methods and algorithms proposed for each problem.

The choice of topics has been guided by our feeling of practitioners. There is no implication whatsoever that what is left out is unimportant. A selection has been

---

[1] Between words and deeds there is the sea.

imposed by space limits and the intention of explaining both theory and algorithms to the level of detail necessary to make implementation really possible.

What are the Objectives of this Book?

- To introduce the fundamental problems of computer vision.
- To enable the reader to implement solutions for reasonably complex problems.
- To develop two parallel tracks, showing how fundamental problems are solved using both intensity and range images, two most popular types of images in today's computer vision community.
- To enable the reader to make sense of the literature of computer vision.

What is the Reader Expected to Know?

This book has been written for people interested in programming solutions to computer vision problems. The best way of reading it is to try out the algorithms on a computer. We assume that the reader is able to translate our pseudocode into computer programs, and therefore that he or she is familiar with a language suitable for numerical computations (for instance C, Fortran). We also expect that the reader has access to popular numerical libraries like the Numerical Recipes[2] or Meschach, or to high-level languages for developing numerical software, like MATLAB, Mathematica or Scilab.

The whole book is non-language specific. We have endeavored to present all the necessary vision-specific information, so that the reader only needs some competence in a programming language.

Although some of the mathematics may appear complex at first glance, the whole book revolves around basic calculus, linear algebra (including least squares, eigenvectors, and singular value decomposition), and the fundamentals of analytic and projective geometry.

Who can Benefit from this Book?

- Students of university courses on computer vision, typically final-year undergraduates or postgraduates of degrees like Computer Science, Engineering, Mathematics, and Physics. Most of the knowledge required to read this book should be part of their normal background.
- Researchers looking for a modern presentation of computer vision, as well as a collection of practical algorithms covering the main problems of the discipline.
- Teachers and students of professional training courses.
- Industry scientists and academics interested in learning the fundamentals and the practical aspects of computer vision.

---

[2] For information on this and the other packages mentioned here, see Chapter 1.

How is this Book Organized?

Each chapter is opened by a summary of its contents, and concluded by a self-check list, review questions, a concise guide to further readings, as well as exercises and suggestions for computer projects.

For each problem analyzed, we give

1. a problem statement, defining the objective to be achieved.
2. a theoretical treatment of the problem.
3. one or two algorithms in pseudocode.
4. hints on the practical applicability of the algorithms.

A few mathematical concepts are crucial to the understanding of solutions and algorithms, but not necessarily known to everybody. To make the book reasonably self-contained, we have included an appendix with several brief sections reviewing background topics. We tried to gear the appendix to the level of detail necessary to understand the discussions of the main text, in the attempt to avoid just a mere sprinkle of vague reminders.

We made an effort to keep the tone informal throughout, hopefully without relaxing too much the mathematical rigour.

The graphics have been designed to facilitate quick identification of important material. Problem statements, important definitions, and algorithms are enclosed in rules; hints and comments of practical relevance, including coding suggestions, appear in a different point size, and are highlighted by a pointer (☞).

Finally, we have included in Chapter 1 information on the computer vision community, including pointers to Internet vision sites (software, images, and documents), and a list of the main publications, electronic newsletters and conferences.

Suggestions for Instructors

The material in this text should be enough for two semesters at the senior undergraduate level, assuming three hours per week. Ultimately, this depends on the students' background, the desired level of detail, the choice of topics, and how much time is allocated to project work. Instructors may want to review some of the material in the appendix in the first few lectures of the course.

In case only one semester is available, we suggest two selections of topics.

- *Stereo and Motion* Chapters 1 to 6 (image acquisition, noise attenuation, feature extraction and calibration), then Chapters 7 (stereopsis) and 8 (motion analysis).
- *Object Recognition* Chapters 1 to 6, then Chapters 10 (object recognition) and 11 (object location).

Ideally, the students should be assigned projects to implement and test at least some of the algorithms. It is up to the instructor to decide which ones, depending on how the course is structured, what existing software is available to students, and which parts of the book one wants to cover.

So Why Another Book on Computer Vision?

We like to think of this textbook, first and foremost, as a practical guide to the solutions of problems characteristic of today's computer vision community. As this book is meant for both students and practitioners, we have tried to give a reasonably complete theoretical treatment of each problem while emphasising practical solutions. We have tried to state algorithms as clearly as possible, and to lay out the material in a graphically appealing manner, in a logical progression.

It seems to us that there is a shortage of such textbooks on computer vision. There are books surveying large numbers of topics and techniques, often large and expensive, sometimes vague in many places because of the amount of material included; books very detailed on theory, but lacking on algorithms and practical advice; books meant for the specialist, reporting advanced results in specific research or application areas, but of little use to students; and books which are nearly completely out of date. Moreover, and not infrequently in computer vision, the style and contents of research articles makes it difficult (sometimes close to impossible) to reimplement the algorithms reported. When working on such articles for this book, we have tried to explain the theory in what seemed to us a more understandable manner, and to add details necessary for implementation. Of course, we take full and sole responsibility for our interpretation.

We hope our book fills a gap, and satisfies a real demand. Whether or not we have succeeded is for you, the reader, to decide, and we would be delighted to hear your comments. Above all, we hope you enjoy reading this book and find it useful.

Acknowledgments

We are indebted to a number of persons who contributed in various ways to the making of this book.

We thank Dave Braunegg, Bob Fisher, Andrea Fusiello, Massimiliano Pontil, Claudio Uras, and Larry Wolff for their precious comments, which allowed us to remove several flaws from preliminary drafts. Thanks also to Massimiliano Aonzo, Adele Lorusso, Jean-Francois Lots, Alessandro Migliorini, Adriano Pascoletti, Piero Parodi, and Maurizio Pilu for their careful proofreading.

Many people kindly contributed various material which has been incorporated in the book; in the hope of mentioning them all, we want to thank Tiziana Aicardi, Bill Austin, Brian Calder, Stuart Clarke, Bob Fisher, Andrea Fusiello, Christian Frühling, Alois Goller, Dave Lane, Gerald McGunnigle, Stephen McKenna, Alessandro Migliorini, Majid Mirmehdi, David Murray, Francesca Odone, Maurizio Pilu, Costas Plakas, Ioseba Tena Ruiz, John Selkirk, Marco Straforini, Manickam Umasuthan, and Andy Wallace.

Thanks to Marco Campani, Marco Cappello, Bruno Caprile, Enrico De Micheli, Andrea Fusiello, Federico Girosi, Francesco Isgrò, Greg Michaelson, Pasquale Ottonello, and Vito Roberto for many useful discussions.

Our thanks to Chris Glennie and Jackie Harbor of Prentice-Hall UK, the former for taking us through the early stages of this adventure, the latter for following up with remarkably light-hearted patience the development of this book, which was peppered by our consistent infringing of deadlines. Thanks to Irwin Zucker and Tom Robbins of Prentice Hall in the U.S. for taking the book through its very final stage.

Finally, very special thanks to Clare, Daniela, Emanuele, Emily, Francesca and Lorenzo, who put up with two absent fathers and husbands for many a month, for their support and love. Fortunately for us, maybe unfortunately for them, we are back.

Emanuele Trucco
Dept. of Computing and
  Electrical Engineering
Heriot-Watt University
Riccarton
Edinburgh EH9 2JH
UK
mtc@cee.hw.ac.uk

Alessandro Verri
Dip. di Informatica e
  Scienze dell'Informazione[3]
Università di Genova
Via Dodecaneso 35
16146 Genova
Italia
verri@ge.infm.it

---

[3] This book was written while the author was with the Department of Physics at the University of Genova.

1

Introduction

"All set?"
"Ready when you are."
Big Trouble in Little China

1.1 What is Computer Vision?

This is the first, inescapable question of this book. Since it is very difficult to produce an uncontroversial definition of such a multifaceted discipline as computer vision, let us ask more precise questions. Which problems are we attempting to tackle? And how do we plan to solve them? Answering these questions will limit and define the scope of this book, and, in doing so, motivate our definition of computer vision.

The Problems of Computer Vision. The target problem of this book is *computing properties of the 3-D world from one or more digital images*. The properties that interest us are mainly *geometric* (for instance, shape and position of solid objects) and *dynamic* (for instance, object velocities). Most of the solutions we present assume that a considerable amount of *image processing* has already taken place; that is, new images have been computed from the original ones, or some image parts have been identified to make explicit the information necessary to the target computation.

The Tools of Computer Vision. As the name suggests, computer vision involves *computers interpreting images*. Therefore, the tools needed by a computer vision system include hardware for acquiring and storing digital images in a computer, processing the images, and communicating results to users or other automated systems. This is a book about the *algorithms* of computer vision; it contains very little material about hardware, but hopefully enough to realize where digital images come from. This does *not* mean that algorithms and software are the only important aspect of a vision system. On the contrary, in some applications, one can choose the hardware and can engineer the scene to facilitate the task of the vision system; for instance, by controlling the illumination, using high-resolution cameras, or constraining the pose and location of the objects. In many situations, however, one has little or no control over the scene. For instance, in the case of outdoors surveillance or autonomous navigation in unknown environments, appropriate algorithms are the key to success.

We are now ready to define the scope of computer vision targeted by this book: *a set of computational techniques aimed at estimating or making explicit the geometric and dynamic properties of the 3-D world from digital images.*

1.2   The Many Faces of Computer Vision

An exhaustive list of all the topics covered by the term "computer vision" is difficult to collate, because the field is vast, multidisciplinary, and in continuous expansion: new, exciting applications appear all the time. So there is more to computer vision than this book can cover, and we complement our definition in the previous section with a quick overview of the main research and application areas, and some related disciplines.

1.2.1   Related Disciplines

Computer vision has been evolving as a multidisciplinary subject for about thirty years. Its contours blend into those of artificial intelligence, robotics, signal processing, pattern recognition, control theory, psychology, neuroscience, and other fields. Two consequences of the rapid growth and young age of the field of computer vision have been that

- the objectives, tools, and people of the computer vision community overlap those of several other disciplines.
- the definition and scope of computer vision are still matters of discussion, so that all definitions should be taken with a grain of salt.

You are likely to come across terms like *image analysis*, *scene analysis*, and *image understanding*, that in this book we simply regard as synonyms for computer vision. Some other terms, however, denote disciplines closely related but not identical to computer vision. Here are the principal ones:

Image Processing.   Image processing is a vast research area. For our purposes, it differs from computer vision in that it concerns *image properties* and *image-to-image transformations*, whereas the main target of computer vision is the 3-D world. As most computer vision algorithms require some preliminary image processing, the overlap between the two disciplines is significant. Examples of image processing include *enhancement* (computing an image of better quality than the original one), *compression* (devising compact representations for digital images, typically for transmission purposes), *restoration* (eliminating the effect of known degradations), and *feature extraction* (locating special image elements like contours, or textured areas). A practical way to understand the difference between representative problems of image processing and computer vision is to compare the contents of Chapters 3, 4, and 5 with those of Chapters 6 to 11.

Pattern Recognition.   For a long time, pattern recognition has produced techniques for *recognizing and classifying objects using digital images.* Many methods developed in the past worked well with 2-D objects or 3-D objects presented in constrained poses, but were unsuitable for the general 3-D world. This triggered much of the research which led to today's field of computer vision. This book does not cover classic pattern recognition, although some of its methods creep up here and there. The *International Association for Pattern Recognition* (IAPR) gathers many researchers and users interested in the field, and maintains a comprehensive WWW site (`http://peipa.essex.ac.uk/iapr/`).

Photogrammetry.   Photogrammetry is concerned with *obtaining reliable and accurate measurements from noncontact imaging.* This discipline overlaps less with computer vision than image processing and pattern recognition. The main differences are that photogrammetry pursues higher levels of accuracies than computer vision, and not all of computer vision is related to measuring. Taking a look at photogrammetric methods before designing a vision system carrying out measurements is always a good idea. The *International Society of Photogrammetry and Remote Sensing* is the international organization promoting the advancement of photogrammetry. It maintains a very comprehensive Internet site (`http://www.p.igp.ethz.ch/isprs/isprs.html`) including archives and activities, and publishes the *Journal of Photogrammetry and Remote Sensing*.

1.2.2   Research and Application Areas

For the purposes of this section, *research areas* refer to topics addressed by a significant number of computer vision publications (a visible indicator of research), and *application areas* refer to domains in which computer vision methods are used, possibly in conjunction with other technologies, to solve real-world problems. The following lists and the accompanying figures should give you the flavor of the variety and scope of computer vision; further applications are illustrated in the book. The lists are meant to be suggestive, not exhaustive; most of the terms that may be unclear now will be explained later in the book.

Examples of Research Areas

> Image feature detection
>
> Contour representation
>
> Feature-based segmentation
>
> Range image analysis
>
> Shape modelling and representation
>
> Shape reconstruction from single-image cues (*shape from X*)

Stereo vision

Motion analysis

Color vision

Active and purposive vision

Invariants

Uncalibrated and self-calibrating systems

Object detection

3-D object recognition

3-D object location

High-performance and real-time architectures

Examples of Application Areas

Industrial inspection and quality control

Reverse engineering

Surveillance and security

Face recognition

Gesture recognition

Road monitoring

Autonomous vehicles (land, underwater, space vehicles)

Hand-eye robotics systems

Space and applications

Military applications

Medical image analysis (e.g., MRI, CT, X-rays, and sonar scan)

Image databases

Virtual reality, telepresence, and telerobotics

1.3 Exploring the Computer Vision World

This section provides a starting set of pointers to the multifaceted world of computer vision. In all the following lists, items appear in no particular order.

1.3.1 Conferences, Journals, and Books

Conferences. The following international conferences cover the most significant advancements on the topics central to this book. Printed proceedings are available for all conferences, and details appear regularly on the Internet.

International Conference on Computer Vision (ICCV)

International Conference on Computer Vision and Pattern Recognition (CVPR)

European Conference on Computer Vision (ECCV)

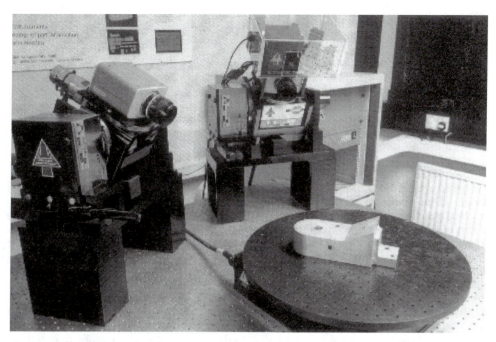

Figure 1.1 A prototype of 3-D inspection cell. The cell includes two types of depth sensors, a laser scanner, and a Moiré fringe system (see Chapter 2), which locate the object in space and perform measurements. Notice the turntable for optimal, automatic object positioning.

International Conference on Image Processing (ICIP)

International Conference on Pattern Recognition (ICPR)

Several national conferences and international workshops are organized on annual or biennial basis. A complete list would be too long, so none of these are mentioned for fairness.

Journals.   The following technical journals cover the most significant advancements on the field. They can be found in the libraries of any university hosting research on computer vision or image processing.

International Journal of Computer Vision

IEEE Transactions on Pattern Analysis and Machine Intelligence

Computer Vision and Image Understanding

Machine Vision and its Applications

Image and Vision Computing Journal

Journal of the Optical Society of America A

Pattern Recognition

Figure 1.2 Left: automatic recognition of road bridges in aerial infrared images (courtesy of Majid Mirmehdi, University of Surrey; Crown copyright reproduced with the permission of the Controller of Her Majesty's Stationery Office). Right: an example of automatic face detection, particularly important for surveillance and security systems. The face regions selected can be subsequently compared with a database of faces for identification (courtesy of Stephen McKenna, Queen Mary and Westfield College, London).

Pattern Recognition Letters

IEEE Transactions on Image Processing

IEEE Transactions on Systems, Man and Cybernetics

IEE Proceedings: Vision, Image and Signal Processing

Biological Cybernetics

Neural Computation

Artificial Intelligence

Books. So many books on computer vision and related fields have been published that it seems futile to produce long lists unless each entry is accompanied by a comment. Since including a complete, commented list here would take too much space, we leave the task of introducing books in specific, technical contexts to the following chapters.

1.3.2 Internet

As the Internet undergoes continuous, ebullient transformation, this information is likely to age faster than the rest of this book, and we can only guarantee that the

Figure 1.3 Computer vision and autonomous road navigation: some images from a sequence acquired from a moving car, and the estimated motion field (optical flow, discussed in Chapter 8) computed by a motion analysis program, indicating the relative motion of world and camera.

list below is correct at the time of printing. Further Internet sites, related to specific problems, are given in the relevant chapters of this book.

- The **Computer Vision Home Page**, `http://www.cs.cmu.edu/~cil/vision.html` and the **Pilot European Image Processing Archive** home page, `http://peipa.essex.ac.uk`, contain links to test images, demos, archives, research

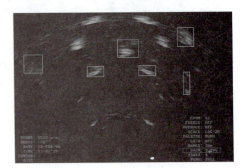

Figure 1.4　Computer vision is becoming increasingly important for remotely operated and autonomous subsea vehicles (ROV/AUVs), like the one shown above, ANGUS, built by the Ocean Systems Laboratory of Heriot-Watt University. As with many ROV/AUVs, ANGUS carries video and sonar sensors (see Chapter 2). Bottom left: an example of underwater sonar image. The white areas are the returns from a diver and the poles of a pier, imaged from above. Bottom right: the result of automatic search for objects of interest (courtesy of Dave Lane, Heriot-Watt University).

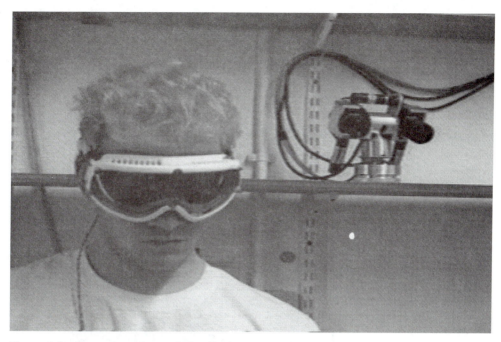

Figure 1.5 Computer vision and virtual telepresence: the movements of the operator's head are tracked by a vision system (not shown) and copied in real time by the head-eye platform (or *stereo head*) on the right (courtesy of David W. Murray, University of Oxford).

groups, research publications, teaching material, frequently asked questions, and plenty of pointers to other interesting sites.

- The **Annotated Computer Vision Bibliography** is an excellent, well-organized source of on-line published papers and reports, as well as announcements of conferences and journals at `http://iris.usc.edu/Vision-Notes/bibliography/contents.html`. You can search the contents by keyword, author, journal, conference, paper title, and other ways.

- Very comprehensive bibliographies on image analysis, pattern recognition and computer vision are produced every year by Azriel Rosenfeld at the University of Maryland (`ftp://ftp.teleos.com/VISION-LIST-ARCHIVE/ROSENFELD-BIBLIOGRAPHIES`). A WWW version can be accessed through the Annotated Computer Vision Bibliographies site.

- **CVonline** is a collection of hypertext summaries of methods and applications of computer vision, recently established by the University of Edinburgh (`http://www.dai.ed.ac.uk/daidb/staff/personal_pages/rbf/CVonline/CVentry.htm`).

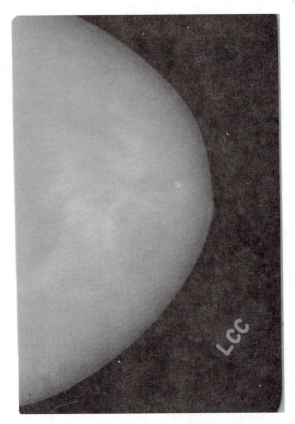

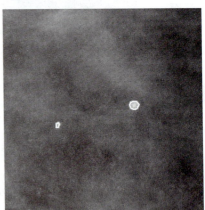

Figure 1.6 An example of medical application of computer vision: computer-assisted diagnoses from mammographic images. Top: X-ray image of a female breast, digitized from a conventional X-ray photography. Bottom: close-up and automatic identification of suspect nodules (courtesy of Stuart Clarke and Brian Calder, Heriot-Watt University, and Matthew Freedman, Georgetown University Medical School, Washington DC).

- **The Vision List** and **The Pixel** are free electronic bulletins circulating news and requests, and hosting technical debates. To subscribe, email `pixel@essex.ac.uk` and `Vision-List-Request@teleos.com`. Both have ftp and WWW archives of useful material.

1.3.3   Some Hints on Math Software

This section gives pointers to numerical computation packages widely used in computer vision, which we found useful. Notice that this list reflects only our experience; no comparison whatsoever with other packages is implied.

- **Numerical Recipes** is a book and software package very popular in the vision community. The source code, in C, FORTRAN and Pascal, is published by Cambridge University Press together with the companion book by Press, Teukolsky, Vettering, and Flannery *Numerical Recipes in C/FORTRAN/Pascal*. The book is an excellent introduction to the practicalities of numerical computation. There is also a *Numerical Recipes: Example Book* illustrating how to call the library routines.

- **Meschach** is a public-domain numerical library of C routines for linear algebra, developed by David E. Stewart and Zbigniew Leyk of the Australian National University, Canberra. For information and how to obtain a copy, see Web page at `http://www.netlib.no/netlib/c/meschach/readme`.

- **MATLAB** is a software environment for fast prototyping of numerical programs, with its own language, interpreter, libraries (called *toolbox*) and visualization tools, commercialized by the U.S. company *The MathWorks*. It is designed to be easy to use, and runs on several platforms, including UNIX and DOS machines. Matlab is described in several recent books, and there is a large community of users. Plenty of information on software, books, bulletins training and so on available at *The MathWorks'* WWW site, `http://www.mathworks.com/`, or contact The MathWorks Inc., 24 Prime Park Way, Natick, MA 01760, USA.

- **Mathematica** is another software environment for mathematical applications, with a large community of users. The standard reference book is Stephen Wolfram's *Mathematica*. Plenty of information on software, books, and bulletins available at Wolfram Research's WWW site, `http://www.wri.com/`.

- **Scilab** is a public-domain scientific software package for numerical computing developed by INRIA (France). It includes linear algebra, control, signal processing, graphics and animation. You can access Scilab from `http://www-rocq.inria.fr/scilab/`, or contact `Scilab@inria.fr`.

1.4   The Road Ahead

This book is organized in two logical parts. The first part (Chapters 2 to 5) deals with the image acquisition and processing methods (noise attenuation, feature extraction, line and curve detection) necessary to produce the input data expected by subsequent algorithms. The primary purpose of this first part is not to give an exhaustive treatment of image processing, but to make the book self-contained by suggesting image processing

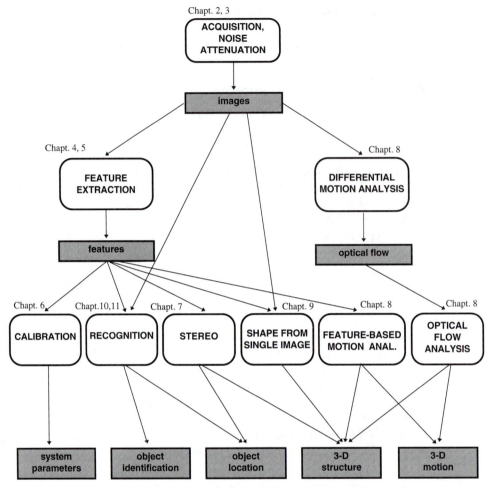

Figure 1.7 The book at a glance: method classes (white boxes), results (grey boxes), their interdependence, and where to find the various topics in this book.

methods commonly found in (in some cases, characteristic of) computer vision. The second part of the book (Chapters 6 to 11) deals with the computer vision problems (stereo, motion analysis, object identification, object location) that we have identified as our targets.

The structure of the book is captured by Figure 1.7, which shows the methods presented, their interdependence, the intermediate and target results, and which chapter contains which topic. Our job begins with the acquisition of one image, or a stereo pair, or a whole video sequence. Before being fed to any algorithms, the images are preprocessed to attenuate the noise introduced by the acquisition process. The target information (structure, motion, location and identity of 3-D objects, and system parameters) are shown at the bottom of Figure 1.7. The diagram suggests that in most cases the same information can be computed in more than one way.

One well-known class of methods rely on the identification of special image elements, called *image features*. Examples of such methods are:

- *calibration*, which determines the value of internal and external parameters of the vision system;
- *stereo analysis*, which exploits the difference between two images to compute the structure (shape) of 3-D objects, and their location in space;
- *recognition*, which determines the objects' identity and location;
- *feature-based motion analysis*, which exploits the finite changes induced in an image sequence by the relative motion of world and camera to estimate 3-D structure and motion; and
- some *shape from single image* methods, which estimate 3-D structure from the information contained in one image only.

Another class of methods computes the target information from the images *directly*. Of these, this book includes:

- one *shape from single image* method, which estimates 3-D structure from the shading of a single image, and
- *optical flow methods*, a class of motion analysis methods which regards an image sequence as a close approximation of a continuous, time-varying signal.

We are now ready to begin our investigation into the theory and algorithms of computer vision.

2

Digital Snapshots

Verweile doch! Du bist so schön![1]

Goethe, *Faust*

This chapter deals with digital images and their relation to the physical world. We learn the principles of image formation, define the two main types of images in this book (*intensity* and *range images*), and discuss how to acquire and store them in a computer.

Chapter Overview

Section 2.2 considers the basic optical, radiometric, and geometric principles underlying the formation of intensity images.

Section 2.3 brings the computer into the picture, laying out the special nature of digital images, their acquisition, and some mathematical models of intensity cameras.

Section 2.4 discusses the fundamental mathematical models of intensity cameras and their parameters.

Section 2.5 introduces range images and describes a class of range sensors based on intensity cameras, so that we can use what we learn about intensity imaging.

What You Need to Know to Understand this Chapter

- Sampling theorem (Appendix, section A.3).
- Rotation matrices (Appendix, section A.9).

---

[1] Stop! You are so beautiful!

15

2.1 Introduction

This chapter deals with the main ingredients of computer vision: *digital images*. We concentrate on two types of images frequently used in computer vision:

intensity images, the familiar, photographlike images encoding light intensities, acquired by television cameras;

range images, encoding shape and distance, acquired by special sensors like sonars or laser scanners.

Intensity images measure the amount of light impinging on a photosensitive device; range images estimate directly the 3-D structure of the viewed scene through a variety of techniques. Throughout the book, we will develop algorithms for both types of images.[2]

It is important to stress immediately that *any digital image, irrespective of its type, is a 2-D array (matrix) of numbers*. Figure 2.1 illustrates this fact for the case of intensity images. Depending on the nature of the image, the numbers may represent light intensities, distances, or other physical quantities. This fact has two fundamental consequences.

- The exact relationship of a digital image to the physical world (i.e., its nature of range or intensity image) is determined by the acquisition process, which depends on the sensor used.

- Any information contained in images (e.g., shape, measurements, or object identity) must ultimately be extracted (computed) from 2-D numerical arrays, in which it is encoded.

In this chapter, we investigate the origin of the numbers forming a digital image; the rest of the book is devoted to computational techniques that make *explicit* some of the information contained *implicitly* in these numbers.

2.2 Intensity Images

We start by introducing the main concepts behind intensity image formation.

2.2.1 Main Concepts

In the visual systems of many animals, including man, the process of image formation begins with the light rays coming from the outside world and impinging on the photoreceptors in the retina. A simple look at any ordinary photograph suggests the variety of physical parameters playing a role in image formation. Here is an incomplete list:

Optical parameters of the lens characterize the sensor's optics. They include:

- lens type,
- focal length,

---

[2] Although, as we shall see, some algorithms make sense for intensity images only.

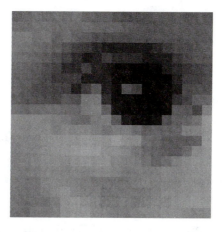

```
117 125 133 127 130 130 133 121 116 115 100  91  93  94   99 103 112 105 109 106
134 133 138 138 132 134 130 133 128 123 121 113 106 102   99 106 113 109 109 113
146 147 138 140 125 134 124 115 102  96  93  94  99  96   99 100 103 110 109 110
144 141 136 130 120 108  88  74  53  37  31  37  35  39   53  79  93 100 109 116
139 136 129 119 102  85  58  31  41  77  51  53  53  33   37  41  69  94 105 108
132 127 117 102  87  57  49  77  42  28  17  15  13  13   17  41  53  69  88 100
124 120 108  94  72  74  72  31  35  31  15  13  15  11   15  13  46  75  83  96
125 115 102  93  88  82  42  79 113  41  19 100  82  11   11  17  31  91  99 100
124 116 109  99  91 113  99 140 144  57  20  20  15  11   15  17  63  87 119 124
136 133 133 135 138 133 132 144 150 120  24  17  15  15   17  20 115 113  88 150
158 157 157 154 149 145 133 127 146 150 116  35  20  19   28 105 124 128 141 171
155 154 156 155 146 155 154 154 147 139 148 150 138 120  128 129 130 151 156 165
150 151 154 162 166 167 169 174 172 167 177 166 164 140  134 120 121 120 127 172
145 149 151 157 165 169 173 179 176 166 166 157 145 136  129 124 120 136 163 168
144 148 153 160 159 158 165 172 165 169 157 151 149 141  130 140 151 162 169 167
144 141 147 155 154 149 156 151 157 157 151 144 147 147  149 159 158 159 166 165
139 140 140 150 153 151 150 146 140 139 138 140 145 151  149 156 156 162 162 161
136 134 138 146 156 164 153 146 145 136 139 139 140 141  149 157 159 161 169 166
136 133 136 135 144 159 168 159 151 142 141 145 139 146  153 156 164 167 172 168
133 129 140 142 146 159 167 165 154 151 146 141 147 154  156 160 161 157 153 154
```

Figure 2.1   Digital images are 2-D arrays of numbers: a 20×20 grey-level image of an eye (pixels have been enlarged for display) and the corresponding 2-D array.

- field of view,
- angular apertures.

Photometric parameters appear in models of the light energy reaching the sensor after being reflected from the objects in the scene. They include:

- type, intensity, and direction of illumination,
- reflectance properties of the viewed surfaces,
- effects of the sensor's structure on the amount of light reaching the photoreceptors.

Geometric parameters determine the image position on which a 3-D point is projected. They include:

- type of projections,
- position and orientation of camera in space,
- perspective distortions introduced by the imaging process.

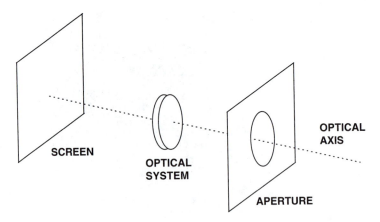

Figure 2.2   The basic elements of an imaging device.

All the above plays a role in *any* intensity imaging device, be it a photographic camera, camcorder, or computer-based system. However, further parameters are needed to characterize digital images and their acquisition systems. These include:

- the physical properties of the photosensitive matrix of the viewing camera,
- the discrete nature of the photoreceptors,
- the quantization of the intensity scale.

We will now review the optical, radiometric, and geometric aspects of image formation.

2.2.2 Basic Optics

We first need to establish a few fundamental notions of optics. As for many natural visual systems, the process of image formation in computer vision begins with the light rays which enter the camera through an *angular aperture* (or *pupil*), and hit a screen or *image plane* (Figure 2.2), the camera's photosensitive device which registers light intensities. Notice that most of these rays are the result of the reflections of the rays emitted by the light sources and hitting object surfaces.

Image Focusing.   Any *single* point of a scene reflects light coming from possibly many directions, so that many rays reflected by the same point may enter the camera. In order to obtain sharp images, all rays coming from a single scene point, P, must converge onto a single point on the image plane, p, the *image of* P. If this happen, we say that the image of P is *in focus*; if not, the image is spread over a circle. *Focusing* all rays from a scene point onto a single image point can be achieved in two ways:

1. Reducing the camera's aperture to a point, called a *pinhole*. This means that only one ray from any given point can enter the camera, and creates a one-to-one correspondence between visible points, rays, and image points. This results in very

sharp, undistorted images of objects at different distances from the camera (see Project 2.1).

2. Introducing an *optical system* composed of lenses, apertures, and other elements, explicitly designed to make all rays coming from the same 3-D point converge onto a single image point.

An obvious disadvantage of a pinhole aperture is its *exposure time;* that is, how long the image plane is allowed to receive light. Any photosensitive device (camera film, electronic sensors) needs a minimum amount of light to register a legible image. As a pinhole allows very little light into the camera per time unit, the exposure time necessary to form the image is too long (typically several seconds) to be of practical use.[3] Optical systems, instead, can be adjusted to work under a wide range of illumination conditions and exposure times (the exposure time being controlled by a *shutter*).

☞   Intuitively, an optical system can be regarded as a device that aims at producing the same image obtained by a pinhole aperture, but by means of a much larger aperture and a shorter exposure time. Moreover, an optical system enhances the light gathering power.

Thin Lenses.   Standard optical systems are quite sophisticated, but we can learn the basic ideas from the simplest optical system, the *thin lens.* The optical behavior of a thin lens (Figure 2.3) is characterized by two elements: an axis, called *optical axis*, going through the lens center, O, and perpendicular to the plane; and two special points, F_l and F_r, called *left* and *right focus*, placed on the optical axis, on the opposite sides of the lens, and at the same distance from O. This distance, called the *focal length* of the lens, is usually indicated by f.

By construction, a thin lens deflects all rays parallel to the optical axis and coming from one side onto the focus on the other side, as described by two *basic properties*.

Thin Lens: Basic Properties

1. Any ray entering the lens parallel to the axis on one side goes through the focus on the other side.

2. Any ray entering the lens from the focus on one side emerges parallel to the axis on the other side.

The Fundamental Equation of Thin Lenses.   Our next task is to derive the *fundamental equation of thin lenses* from the basic properties 1 and 2. Consider a point P, not too far from the optical axis, and let $Z + f$ be the distance of P from the lens along the optical axis (Figure 2.4). By assumption, a thin lens focuses all the rays from P onto the same point, the image point p. Therefore, we can locate p by intersecting only two known rays, and we do not have to worry about tracing the path of any other.

---

[3] The exposure time is, roughly, inversely proportional to the square of the aperture diameter, which in turn is proportional to the amount of light that enters the imaging system.

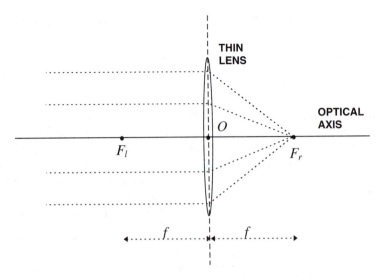

Figure 2.3   Geometric optics of a thin lens (a perpendicular view to the plane approximating the lens).

Note that by applying property 1 to the ray PQ and property 2 to the ray PR, PQ and PR are deflected to intersect at a certain point on the other side of the thin lens. But since the lens focuses all rays coming from P onto the same point, PQ and PR must intersect at p! From Figure 2.4 and using the two pairs of similar triangles $< PF_lS >$ and $< ROF_l >$ and $< psF_r >$ and $< QOF_r >$, we obtain immediately

$$Zz = f^2. \tag{2.1}$$

Setting $\hat{Z} = Z + f$ and $\hat{z} = z + f$, (2.1) reduces to our target equation.

The Fundamental Equation of Thin Lenses

$$\frac{1}{\hat{Z}} + \frac{1}{\hat{z}} = \frac{1}{f}. \tag{2.2}$$

☞   The ray going through the lens center, O, named the *principal ray*, goes through p undeflected.

Field of View.   One last observation about optics. Let d be the *effective diameter of the lens*, identifying the portion of the lens actually reachable by light rays.

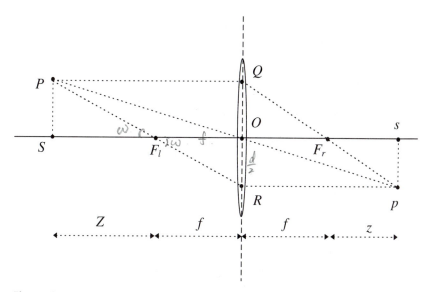

Figure 2.4 Imaging by a thin lens. Notice that, in general, a real lens has two different focal lengths, because the curvatures of its two surfaces may be different. The situation depicted here is a special case, but it is sufficient for our purposes. See the Further Readings at the end of this chapter for more on optics.

☞   We call d the *effective* diameter to emphasize the difference between d and the *physical* diameter of the lens. The aperture may prevent light rays from reaching the peripheral points of the lens, so that d is usually smaller than the physical diameter of the lens.

The effective lens diameter and the focal length determine the *field of view* of the lens, which is *an angular measure of the portion of 3-D space actually seen by the camera*. It is customary to define the field of view, w, as half of the angle subtended by the lens diameter as seen from the focus:

$$\tan w = \frac{d}{2f}. \tag{2.3}$$

This is the minimum amount of optics needed for our purposes. Optical models of real imaging devices are a great deal more complicated than our treatment of thin (and ideal) lenses; problems and phenomena not considered here include *spherical aberration* (defocusing of nonparaxial rays), *chromatic aberration* (different defocusing of rays of different colors), and focusing objects at different distances from the camera.[4]

---

[4] The fundamental equation of thin lenses implies that scene points at different distances from the lens come in focus at different image distances. The optical lens systems of real cameras are designed so that all points within a given range of distances are imaged on or close to the image plane, and therefore acceptably in focus. This range is called the *depth of field* of the camera.

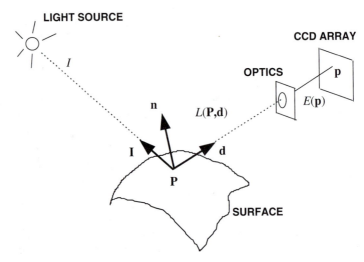

Figure 2.5   Illustration of the basic radiometric concepts.

The Further Readings section at the end of this chapter tell where to find more about optics.

2.2.3   Basic Radiometry

Radiometry is the essential part of image formation concerned with the relation among the amounts of light energy emitted from light sources, reflected from surfaces, and registered by sensors. We shall use radiometric concepts to pursue two objectives:

1. modelling how much of the illuminating light is reflected by object surfaces;
2. modelling how much of the reflected light actually reaches the image plane of the camera.

   Definitions.   We begin with some definitions, illustrated in Figure 2.5 and summarized as follows:

Image Irradiance and Scene Radiance

The *image irradiance* is the power of the light, per unit area and at each point **p** of the image plane.

   The *scene radiance* is the power of the light, per unit area, ideally emitted by each point **P** of a surface in 3-D space in a given direction **d**.

☞   *Ideally* refers to the fact that the surface in the definition of scene radiance might be the illuminated surface of an object, the radiating surface of a light source, or even a fictitious surface. The term *scene radiance* denotes the total radiance emitted by a point;

sometimes *radiance* refers to the energy radiated from a surface (emitted or reflected), whereas *irradiance* refers to the energy incident on a surface.

Surface Reflectance and Lambertian Model.   A model of the way in which a surface reflects incident light is called a *surface reflectance model*. A well-known one is the *Lambertian model*, which assumes that each surface point appears equally bright from all viewing directions. This approximates well the behavior of rough, nonspecular surfaces, as well as various materials like matte paint and paper. If we represent the direction and amount of incident light by a vector **I**, the scene radiance of an ideal Lambertian surface, *L*, is simply proportional to the dot product between **I** and the unit normal to the surface, **n**:

$$L = \rho \mathbf{I}^\top \mathbf{n} \tag{2.4}$$

with $\rho > 0$ a constant called the surface's *albedo*, which is typical of the surface's material. We also assume that $\mathbf{I}^\top \mathbf{n}$ is *positive*; that is, the surface faces the light source. This is a necessary condition for the ray of light to reach **P**. If this condition is not met, the scene radiance should be set equal to 0.

We will use the Lambertian model in several parts of this book; for example, while analyzing image sequences (Chapter 8) and computing shape from shading (Chapter 9). Intuitively, the Lambertian model is based on the exact cancellation of two factors. Neglecting constant terms, the amount of light reaching *any* surface is always proportional to the cosine of the angle between the illuminant and the surface normal **n** (that is, the effective area of the surface as seen from the illuminant direction). According to the model, a Lambertian surface reflects light in a given direction **d** proportionally to the cosine of θ, the angle between **d** and **n**. But since the surface's area seen from the direction **d** is inversely proportional to $\cos\theta$, the two $\cos\theta$ factors cancel out and do not appear in (2.4).

Linking Surface Radiance and Image Irradiance.   Our next task is to link the amounts of light reflected by the surfaces, *L*, and registered by the imaging sensor, *E*.

Assumptions and Problem Statement

Given a thin lens of diameter *d* and focal length *f*, an object at distance *Z* from the lens, and an image plane at distance *Z'* from the lens, with *f*, *Z*, and *Z'* as in (2.1), find the relation between image irradiance and scene radiance.

In order to derive this fundamental relation, we need to recall the geometric notion of *solid angle*. The solid angle of a cone of directions is the area cut out by the cone on the unit sphere centered in the cone's vertex. Therefore, the solid angle $\delta\omega$ subtended by a small, planar patch of area δA at distance *r* from the origin (Figure 2.6) is

$$\delta\omega = \frac{\delta A \cos\psi}{r^2} \tag{2.5}$$

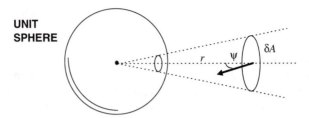

Figure 2.6   The definition of solid angle.

with ψ the angle between the normal to δA and the ray that points from the origin to δA. The factor $\cos \psi$ ensures the proper foreshortening of the area δA as seen from the origin.

We now write the image irradiance at an image point, **p**, as the ratio between δP, the power of light over a small image patch, and δI, the area of the small image patch:

$$E = \frac{\delta P}{\delta I}. \tag{2.6}$$

If δO is the area of a small surface patch around **P**, L the scene radiance at **P** in the direction toward the lens, $\Delta \Omega$ the solid angle subtended by the lens, and θ the angle between the normal to the viewed surface at **P** and the principal ray (Figure 2.7), the power δP is given by $\delta O L \Delta \Omega$ (the total power emitted in the direction of the lens) multiplied by $\cos \theta$ (the foreshortening of the area δO as seen from the lens):

$$\delta P = \delta O L \Delta \Omega \cos \theta. \tag{2.7}$$

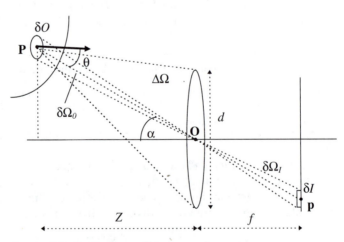

Figure 2.7   Radiometry of the image formation process.

Combining (2.6) and (2.7), we find

$$E = L \Delta\Omega \cos\theta \frac{\delta O}{\delta I}. \tag{2.8}$$

We still need to evaluate $\Delta\Omega$ and $\delta O/\delta I$. For the solid angle $\Delta\Omega$ (Figure 2.7), (2.5) with $\delta A = \pi d^2/4$ (lens area), $\psi = \alpha$ (angle between the principal ray and the optical axis), and $r = Z/\cos\alpha$ (distance of P from the lens center) becomes

$$\Delta\Omega = \frac{\pi}{4} d^2 \frac{\cos^3\alpha}{Z^2}. \tag{2.9}$$

For the solid angle $\delta\Omega_I$, subtended by a small image patch of area δI (see Figure 2.7), (2.5) with $\delta A = \delta I$, $\psi = \alpha$, and $r = f/\cos\alpha$ gives

$$\delta\Omega_I = \frac{\delta I \cos\alpha}{(f/\cos\alpha)^2}. \tag{2.10}$$

Similarly, for the solid angle $\delta\Omega_O$ subtended by the patch δO on the object side, we have

$$\delta\Omega_O = \frac{\delta O \cos\theta}{(Z/\cos\alpha)^2}. \tag{2.11}$$

It is clear from Figure 2.7 that $\delta\Omega_I = \delta\Omega_O$; hence, their ratio is 1, so that dividing (2.11) by (2.10) we obtain

$$\frac{\delta O}{\delta I} = \frac{\cos\alpha}{\cos\theta} \left(\frac{Z}{f}\right)^2 \tag{2.12}$$

Ignoring energy losses within the system, and plugging (2.9) and (2.12) into (2.8), we finally obtain the desired relation between E and L.

The Fundamental Equation of Radiometric Image Formation

$$E(\mathbf{p}) = L(\mathbf{P}) \frac{\pi}{4} \left(\frac{d}{f}\right)^2 \cos^4\alpha. \tag{2.13}$$

Equation (2.13) says that *the illumination of the image at* $\mathbf{p}$ *decreases as the fourth power of the cosine of the angle formed by the principal ray through* $\mathbf{p}$ *with the optical axis*. In the case of small angular aperture, this effect can be neglected; therefore, the image irradiance can be regarded as uniformly proportional to the scene radiance over the *whole* image plane.

The nonuniform illumination predicted by (2.13) is hard to notice in ordinary images, because the major component of brightness changes is usually due to the spatial gradient of the image irradiance. You can try a simple experiment to verify the effect predicted by (2.13) by acquiring an image of a Lambertian surface illuminated by diffuse light (see Exercise 2.2).

☞   We make no substantial distinction among *image irradiance, intensity,* and *brightness.*
Be warned, however, that a distinction does exist; although it is not relevant for our
purposes. The *intensity* is the grey level recorded by an image and is linked to *irradiance*
by a monotonic relation depending on the sensor. *Brightness* often indicates the subjective
human perception of the image intensity.

The fundamental equation of radiometric image formation also shows that the
quantity f/d, called the *F-number*, influences how much light is collected by the camera:
the smaller the F-number, the larger the fraction of L which reaches the image plane.
The F-number is one of the characteristics of the optics. As shown by (2.13), image
irradiance is inversely proportional to the square of the F-number (see footnote 3 in
this chapter).

2.2.4 Geometric Image Formation

We now turn to the geometric aspect of image formation. The aim is to link the position
of scene points with that of their corresponding image points. To do this, we need to
model the *geometric projection* performed by the sensor.

The Perspective Camera.   The most common geometric model of an intensity
camera is the *perspective* or *pinhole* model (Figure 2.8). The model consists of a plane
π, the *image plane*, and a 3-D point **O**, the *center* or *focus of projection*. The distance
between π and **O** is the *focal length*. The line through **O** and perpendicular to π is the
optical axis,[5] and **o**, the intersection between π and the optical axis, is named *principal
point* or *image center*. As shown in Figure 2.8, **p**, the image of **P**, is the point at which the
straight line through **P** and **O** intersects the image plane π. Consider the 3-D reference
frame in which **O** is the origin and the plane π is orthogonal to the Z axis, and let
$\mathbf{P} = [X, Y, Z]^\top$ and $\mathbf{p} = [x, y, z]^\top$. This reference frame, called the *camera frame*, has
fundamental importance in computer vision. We now will write the basic equations of
perspective projections in the camera frame.

Perspective Camera: Fundamental Equations

In the camera frame, we have

$$x = f\frac{X}{Z}$$

$$y = f\frac{Y}{Z} \tag{2.14}$$

☞   In the camera frame, the third component of an image point is always equal to the focal
length (as the equation of the plane π is $z = f$). For this reason, we will often write
$\mathbf{p} = [x, y]^\top$ instead of $\mathbf{p} = [x, y, f]^\top$.

---

[5] You should link these definitions of focal length and optical axis with those in section 2.2.2.

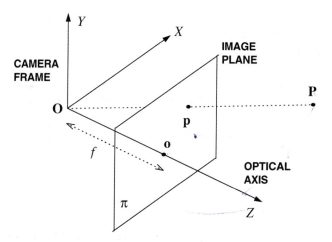

Figure 2.8 The perspective camera model.

Note that (2.14) are nonlinear because of the factor $1/Z$, and do not preserve either distances between points (not even up to a common scaling factor), or angles between lines. However, they do map lines into lines (see Exercise 2.3).

The Weak-Perspective Camera. A classical approximation that turns (2.14) into linear equations is the *weak-perspective* camera model. This model requires that the relative distance along the optical axis, δz, of any two scene points (that is, the scene's depth) is much smaller than the average distance, $\bar{Z}$, of the points from the viewing camera. In this case, for each scene point, **P**, we can write

$$x = f\frac{X}{Z} \approx \frac{f}{\bar{Z}}X$$

$$y = f\frac{Y}{Z} \approx \frac{f}{\bar{Z}}Y \tag{2.15}$$

☞   Indicatively, the weak-perspective approximation becomes viable for $\delta z < \bar{Z}/20$ approximately.

These equations, (2.15), describe a sequence of two transformations: an *orthographic projection*, in which world points are projected along rays parallel to the optical axis,[6] that is,

$$x = X$$

$$y = Y,$$

---

[6] Analytically, the orthographic projection is the limit of the perspective projection for $f \to \infty$. For $f \to \infty$, we have $Z \to \infty$ and thus $f)Z \to 1$.

followed by *isotropic scaling* by the factor $f/\bar{Z}$. Section 2.4 shows that this and other camera models can also be derived in a compact matrix notation. Meanwhile, it is time for a summary.

The Perspective Camera Model

In the *perspective camera model* (nonlinear), the coordinates (x, y) of a point $\mathbf{p}$, image of the 3-D point $\mathbf{P} = [X, Y, Z]^\top$, are given by

$$x = f\frac{X}{Z}$$

$$y = f\frac{Y}{Z}$$

The Weak-Perspective Camera Model

If the average depth of the scene, $\bar{Z}$, is much larger than the relative distance between any two scene points along the optical axis, the *weak-perspective camera model* (linear) holds:

$$x = \frac{f}{\bar{Z}}X$$

$$y = \frac{f}{\bar{Z}}Y.$$

All equations are written in the camera reference frame.

2.3 Acquiring Digital Images

In this section, we now discuss the aspects of image acquisition that are special to *digital* images, namely:

- the essential structure of a typical image acquisition system
- the representation of digital images in a computer
- practical information on spatial sampling and camera noise

2.3.1 Basic Facts

How do we acquire a digital image into a computer? A digital image acquisition system consists of three hardware components: a *viewing camera*, typically a *CCD* (*Charged Coupled Device*) camera, a *frame grabber*, and a *host computer*, on which processing takes place (Figure 2.9).[7]

---

[7] This is the standard configuration, but not the only possibility. For instance, several manufacturers commercialize *smart cameras*, which can acquire images *and* perform a certain amount of image processing.

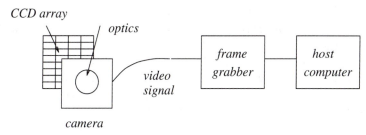

CCD array

optics

frame
grabber

host
computer

video
signal

camera

Figure 2.9   Essential components of a digital image acquisition system.

The input to the camera is, as we know, the incoming light, which enters the camera's lens and hits the image plane. In a CCD camera, the physical image plane is the *CCD array*, a $n \times m$ rectangular grid of photosensors, each sensitive to light intensity. Each photosensor can be regarded as a tiny, rectangular black box which converts light energy into a voltage. The output of the CCD array is usually a continuous electric signal, the *video signal*, which we can regard as generated by scanning the photosensors in the CCD array in a given order (e.g., line by line) and reading out their voltages. The video signal is sent to an electronic device called a *frame grabber*, where it is digitized into a 2-D, rectangular array of $N \times M$ integer values and stored in a memory buffer. At this point, the image can be conveniently represented by a $N \times M$ matrix, E, whose entries are called *pixels* (an acronym for *pic*ture *el*ements), with N and M being two fixed integers expressing the image size, in pixels, along each direction. Finally, the matrix E is transferred to a host computer for processing.

For the purposes of the following chapters, *the starting point of computer vision is the digitized image*, E. Here are the main assumptions we make about E.

Digital Images: Representation

A digital image is represented by a numerical matrix, E, with N rows and M columns.

$E(i, j)$ denotes the image value (image brightness) at pixel (i, j) (i-th row and j-th column), and encodes the intensity recorded by the photosensors of the CCD array contributing to that pixel.

$E(i, j)$ is an integer in the range $[0, 255]$.

This last statement about the range of $E(i, j)$ means that the brightness of an image point can be represented by one byte, or 256 grey levels (typically 0 is black, 255 white). This is an adequate resolution for ordinary, *monochromatic* (or *grey-level*) images and is suitable for many vision tasks. *Color images* require three monochromatic *component images* (red, green, blue) and therefore three numbers. Throughout this book, we shall always refer to grey-level images.

If we assume that the chain of sampling and filtering procedures performed by camera and frame buffer does not distort the video signal, the image stored in the frame buffer is a faithful digitization of the image captured by the CCD array. However, *the number of elements along each side of the CCD arrays is usually different from the dimensions, in pixels, of the frame buffer.* Therefore, the position of the same point on the image plane will be different if measured in CCD elements or image pixels; more precisely, measuring positions from the upper left corner, the relation between the position (x_{im}, y_{im}) (in pixels) in the frame buffer image and the position (x_{CCD}, y_{CCD}) (in CCD elements) on the CCD array is given by

$$x_{im} = \frac{n}{N} x_{CCD}$$

$$y_{im} = \frac{m}{M} y_{CCD}. \tag{2.16}$$

Note that n/N and m/M in (2.16) are not the only parameters responsible for a different scaling of the image with respect to the CCD array along the horizontal and vertical direction; the different ratio of horizontal and vertical sizes of the CCD array has exactly the same effect. This is illustrated in Figure 2.10. The image stored in the computer memory consists of a squared grid of $N \times N$ pixels (Figure 2.10(a)). By inspection, it is easy to see that a grid of $n \times n$ CCD elements with an aspect ratio of n/m between the horizontal and vertical CCD element size (Figure 2.10(b)) produces exactly the same distorsion of an $m \times n$ CCD array with squared elements (Figure 2.10(c)).

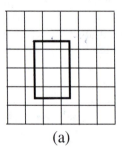

(a)

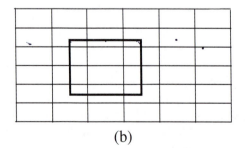

(b)

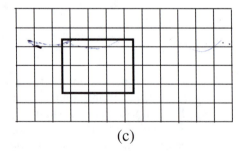

(c)

Figure 2.10 The same distortion of a given pattern on the CCD array (a) is produced by a $n \times n$ grid of rectangular elements of aspect ratio n/m (b), and by a $m \times n$ grid of squared elements (c).

In summary, it is convenient to assume that the CCD elements are always in one-to-one correspondence with the image pixels and to introduce *effective horizontal and vertical sizes* to account for the possible different scaling along the horizontal and vertical direction. The effective sizes of the CCD elements are our first examples of *camera parameters*, which are the subject of section 2.4.

2.3.2 Spatial Sampling

The spatial quantization of images originates at the very early stage of the image formation process, as the photoreceptors of a CCD sensor are organized in a rectangular array of photosensitive elements packed closely together. For simplicity, we assume that the distance d between adjacent CCD elements (specified by the camera manufacturer) is the same in the horizontal and vertical directions. We know from the *sampling theorem* that d determines the highest spatial frequency, v_c, that can be captured by the system, according to the relation

$$v_c = \frac{1}{2d}$$

How does this characteristic frequency compare with the spatial frequency spectrum of images? A classical result of the diffraction theory of aberrations states that *the imaging process can be expressed in terms of a linear low-pass filtering of the spatial frequencies of the visual signal*. (For more information about the diffraction theory of aberrations, see the Further Readings.) In particular, if a is the linear size of the angular aperture of the optics (e.g., the diameter of a circular aperture), λ the wavelength of light, and f the focal length, spatial frequencies larger than

$$v_c' = \frac{a}{\lambda f}$$

do not contribute to the spatial spectrum of the image (that is, they are filtered out).

In a typical image acquisition system, the spatial frequency v_c is nearly one order of magnitude smaller than v_c'. Therefore, since the viewed pattern may well contain spatial frequencies larger than v_c, *we expect aliasing*. You can convince yourself of the reality of spatial aliasing by taking images of a pattern of equally spaced thin black lines on a white background (see Exercise 2.6) at increasing distance from the camera. As predicted by the sampling theorem, if n is the number of the CCD elements in the horizontal direction, the camera cannot *see* more than n' vertical lines (with n' somewhat less than $n/2$, say $n' \sim n/3$). Until the number of lines within the field of view remains smaller than n', all the lines are correctly imaged and resolved. Once the limit is reached, if the distance of the pattern is increased further, but *before* blurring effects take over, the number of imaged lines *decreases* as the distance of the pattern *increases*!

☞   The main reason why spatial aliasing is often neglected is that the amplitude (that is, the information content) of high-frequency components of ordinary images is usually, though by no means always, very small.

2.3.3 Acquisition Noise and How to Estimate It

Let us briefly touch upon the problem of *noise* introduced by the imaging system and how it is estimated. The effect of noise is, essentially, that image values are not those expected, as these are corrupted during the various stages of image acquisition. As a consequence, the pixel values of two images of the same scene taken by the same camera and in the same light conditions are never *exactly* the same (try it!). Such fluctuations will introduce errors in the results of calculations based on pixel values; it is therefore important to estimate the magnitude of the noise.

The main objective of this section is *to suggest a simple characterization of image noise*, which can be used by the algorithms of following chapters. Noise attenuation, in particular, is the subject of Chapter 3.

An obvious way to proceed is to regard noisy variations as random variables, and try to characterize their statistical behavior. To do this, we acquire a sequence of images of the same scene, in the same acquisition conditions, and compute the pointwise average of the image brightness over all the images. The same sequence can also be used to estimate the *signal-to-noise ratio* of the acquisition system, as follows.[8]

Algorithm EST_NOISE

We are given n images of the same scene, $E_0, E_1, \ldots, E_{n-1}$, which we assume square ($N \times N$) for simplicity.

For each $i, j = 0, \ldots, N - 1$, let

$$\overline{E(i, j)} = \frac{1}{n} \sum_{k=0}^{n-1} E_k(i, j)$$

$$\sigma(i, j) = \left(\frac{1}{n-1} \sum_{k=0}^{n-1} (\overline{E(i, j)} - E_k(i, j))^2 \right)^{\frac{1}{2}} \tag{2.17}$$

The quantity $\sigma(i, j)$ is an estimate of the standard deviation of the acquisition noise at each pixel. The average of $\sigma(i, j)$ over the image is an estimate of the average noise, while $\max_{i, j \in [0, \ldots, N-1]} \{\sigma(i, j)\}$ an estimate of the worst case acquisition noise.

☞   Notice that the beat frequency of some fluorescent room lights may skew the results of EST_NOISE.

Figure 2.11 shows the noise estimates relative to a particular acquisition system. A static camera was pointed at a picture posted on the wall. A sequence of $n = 100$ images was then acquired. The graphs in Figure 2.11 reproduce the average plus and minus

---

[8] The signal-to-noise ratio is usually expressed in *decibel* (dB), and is defined as 10 times the logarithm in base 10 of the ratio of two powers (in our case, of signal and noise). For example, a signal-to-noise ratio of 100 corresponds to $10\log_{10}100 = 20dB$.

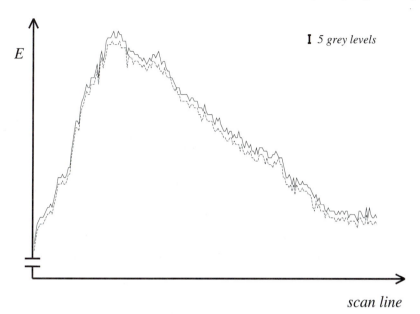

Figure 2.11   Estimated acquisition noise. Graphs of the average image brightness, plus (solid line) and minus (dotted line) the estimated standard deviation, over a sequence of images of the same scene along the same horizontal scan line. The image brightness ranges from 73 to 211 grey levels.

the standard deviation of the image brightness (pixel values) over the entire sequence, along an horizontal scanline (image row). Notice that the standard deviation is almost independent of the average, typically less than 2 and never larger than 2.5 grey values. This corresponds to an average signal-to-noise ratio of nearly one hundred.

Another cause of noise, which is important when a vision system is used for fine measurements, is that *pixel values are not completely independent of each other:* some *cross-talking* occurs between adjacent photosensors in each row of the CCD array, due to the way the content of each CCD row is read in order to be sent to the frame buffer. This can be verified by computing the autocovariance $C_{EE}(i, j)$ of the image of a spatially uniform pattern parallel to the image plane and illuminated by diffuse light.

Algorithm AUTO_COVARIANCE

Let $c = 1/N^2$, $N_{i'} = N - i' - 1$, and $N_{j'} = N - j' - 1$. Given an image E, for each $i', j' = 0, \dots, N - 1$ compute

$$C_{EE}(i', j') = c \sum_{i=0}^{N_{i'}} \sum_{j=0}^{N_{j'}} (E(i, j) - \overline{E(i, j)})(E(i + i', j + j') - \overline{E(i + i', j + j')}) \qquad (2.18)$$

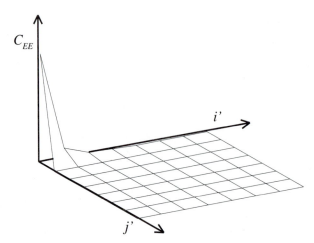

Figure 2.12   Autocovariance of the image of a uniform pattern for a typical image acquisition system, showing cross-talking between adjacent pixels along i'.

☞   The autocovariance should actually be estimated as the average of the autocovariance computed on many images of the same pattern. To minimize the effect of radiometric nonlinearities (see (2.13)), C_{EE} should be computed on a patch in the central portion of the image.

Figure 2.12 displays the graph of the average of the autocovariance computed on many images acquired by the same acquisition system used to generate Figure 2.11. The autocovariance was computed by means of (2.18) on a patch of 16×16 pixels centered in the image center. Notice the small but visible covariance along the horizontal direction: consistently with the physical properties of many CCD cameras, this indicates that the grey value of each pixel is not completely independent of that of its neighbors.

2.4   Camera Parameters

We now come back to discuss the geometry of a vision system in greater detail. In particular, we want to characterize the parameters underlying camera models.

2.4.1   Definitions

Computer vision algorithms reconstructing the 3-D structure of a scene or computing the position of objects in space need equations linking the coordinates of points in 3-D space with the coordinates of their corresponding image points. These equations are written in the camera reference frame (see (2.14) and section 2.2.4), but it is often assumed that

- the camera reference frame can be located with respect to some other, known, reference frame (the *world reference frame*), and

- the coordinates of the image points in the camera reference frame can be obtained from *pixel coordinates*, the only ones directly available from the image.

This is equivalent to assume knowledge of some camera's characteristics, known in vision as the camera's *extrinsic* and *intrinsic* parameters. Our next task is to understand the exact nature of the intrinsic and extrinsic parameters and why the equivalence holds.

Definition: Camera Parameters

The *extrinsic parameters* are the parameters that define the location and orientation of the camera reference frame with respect to a known world reference frame.

The *intrinsic parameters* are the parameters necessary to link the pixel coordinates of an image point with the corresponding coordinates in the camera reference frame.

In the next two sections, we write the basic equations that allow us to define the extrinsic and intrinsic parameters in practical terms. The problem of estimating the value of these parameters is called *camera calibration*. We shall solve this problem in Chapter 6, since calibration methods need algorithms which we discuss in Chapters 4 and 5.

2.4.2 Extrinsic Parameters

The camera reference frame has been introduced for the purpose of writing the fundamental equations of the perspective projection (2.14) in a simple form. However, *the camera reference frame is often unknown*, and a common problem is determining the location and orientation of the camera frame with respect to some known reference frame, *using only image information*. The extrinsic parameters are defined as *any set of geometric parameters that identify uniquely the transformation between the unknown camera reference frame and a known reference frame*, named the *world reference frame*.

A typical choice for describing the transformation between camera and world frame is to use

- a 3-D translation vector, $\mathbf{T}$, describing the relative positions of the origins of the two reference frames, and
- a 3×3 rotation matrix, R, an orthogonal matrix ($R^\top R = R R^\top = I$) that brings the corresponding axes of the two frames onto each other.

The orthogonality relations reduce the number of degrees of freedom of R to three (see section A.9 in the Appendix).

In an obvious notation (see Figure 2.13), the relation between the coordinates of a point $\mathbf{P}$ in world and camera frame, $\mathbf{P}_w$ and $\mathbf{P}_c$ respectively, is

$$\mathbf{P}_c = R(\mathbf{P}_w - \mathbf{T}), \tag{2.19}$$

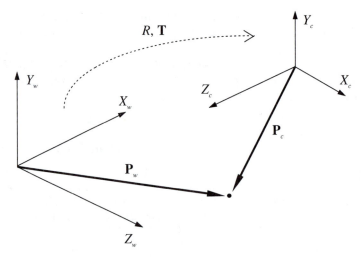

Figure 2.13 The relation between camera and world coordinate frames.

with

$$R = \begin{pmatrix} r_{11} & r_{12} & r_{13} \\ r_{21} & r_{22} & r_{23} \\ r_{31} & r_{32} & r_{33} \end{pmatrix}.$$

Definition: Extrinsic Parameters

The camera extrinsic parameters are the translation vector, **T**, and the rotation matrix, R (or, better, its free parameters), which specify the transformation between the camera and the world reference frame.

2.4.3 Intrinsic Parameters

The intrinsic parameters can be defined as the set of parameters needed to characterize the optical, geometric, and digital characteristics of the viewing camera. For a pinhole camera, we need three sets of intrinsic parameters, specifying respectively

- the perspective projection, for which the only parameter is the focal length, f;
- the transformation between camera frame coordinates and pixel coordinates;
- the geometric distortion introduced by the optics.

From Camera to Pixel Coordinates. To find the second set of intrinsic parameters, we must link the coordinates (x_{im}, y_{im}) of an image point in pixel units with the coordinates (x, y) of the same point in the camera reference frame. The coordinates

(x_{im}, y_{im}) can be thought of as coordinates of a new reference frame, sometimes called *image reference frame*.

The Transformation between Camera and Image Frame Coordinates

Neglecting any geometric distorsions possibly introduced by the optics and in the assumption that the CCD array is made of a rectangular grid of photosensitive elements, we have

$$x = -(x_{im} - o_x)s_x$$
$$y = -(y_{im} - o_y)s_y \tag{2.20}$$

with (o_x, o_y) the coordinates in pixel of the image center (the principal point), and (s_x, s_y) the effective size of the pixel (in millimeters) in the horizontal and vertical direction respectively.

Therefore, the current set of intrinsic parameters is f, o_x, o_y, s_x, s_y.

☞   The sign change in (2.20) is due to the fact that the horizontal and vertical axes of the image and camera reference frames have opposite orientation.

In several cases, the optics introduces image distortions that become evident at the periphery of the image, or even elsewhere using optics with large fields of view. Fortunately, these distortions can be modelled rather accurately as simple *radial distortions*, according to the relations

$$x = x_d(1 + k_1 r^2 + k_2 r^4)$$
$$y = y_d(1 + k_1 r^2 + k_2 r^4)$$

with (x_d, y_d) the coordinates of the distorted points, and $r^2 = x_d^2 + y_d^2$. As shown by the equations above, this distortion is a radial displacement of the image points. The displacement is null at the image center, and increases with the distance of the point from the image center. k_1 and k_2 are further intrinsic parameters. Since they are usually very small, radial distortion is ignored whenever high accuracy is not required in all regions of the image, or when the peripheral pixels can be discarded. If not, as $k_2 << k_1$, k_2 is often set equal to 0, and k_1 is the only intrinsic parameter to be estimated in the radial distortion model.

☞   The magnitude of geometric distortion depends on the quality of the lens used. As a rule of thumb, with optics of average quality and CCD size around 500×500, expect distortions of several pixels (say around 5) in the outer cornice of the image. Under these circumstances, a model with $k_2 = 0$ is still accurate.

It is now time for a summary.

Intrinsic Parameters

The camera intrinsic parameters are defined as the focal length, f, the location of the image center in pixel coordinates, (o_x, o_y), the effective pixel size in the horizontal and vertical direction (s_x, s_y), and, if required, the radial distortion coefficient, k_1.

2.4.4   Camera Models Revisited

We are now fully equipped to write relations linking directly the pixel coordinates of an image point with the world coordinates of the corresponding 3-D point, *without explicit reference to the camera reference frame* needed by (2.14).

Linear Version of the Perspective Projection Equations.   Plugging (2.19) and (2.20) into (2.14) we obtain

$$-(x_{im} - o_x)s_x = f\frac{\mathbf{R}_1^\top (\mathbf{P}_w - \mathbf{T})}{\mathbf{R}_3^\top (\mathbf{P}_w - \mathbf{T})}$$

$$-(y_{im} - o_y)s_y = f\frac{\mathbf{R}_2^\top (\mathbf{P}_w - \mathbf{T})}{\mathbf{R}_3^\top (\mathbf{P}_w - \mathbf{T})} \qquad (2.21)$$

where $\mathbf{R}_i$, $i = 1, 2, 3$, is a 3-D vector formed by the i-th row of the matrix R. Indeed, (2.21) relates the 3-D coordinates of a point in the world frame to the image coordinates of the corresponding image point, via the camera extrinsic and intrinsic parameters.

☞   Notice that, due to the particular form of (2.21), not all the intrinsic parameters are independent. In particular, the focal length could be absorbed into the effective sizes of the CCD elements.

Neglecting radial distortion, we can rewrite (2.21) as a simple matrix product. To this purpose, we define two matrices, M_{int} and M_{ext}, as

$$M_{int} = \begin{pmatrix} -f/s_x & 0 & o_x \\ 0 & -f/s_y & o_y \\ 0 & 0 & 1 \end{pmatrix}$$

and

$$M_{ext} = \begin{pmatrix} r_{11} & r_{12} & r_{13} & -\mathbf{R}_1^\top \mathbf{T} \\ r_{21} & r_{22} & r_{23} & -\mathbf{R}_2^\top \mathbf{T} \\ r_{31} & r_{32} & r_{33} & -\mathbf{R}_3^\top \mathbf{T} \end{pmatrix},$$

so that *the 3 × 3 matrix M_{int} depends only on the intrinsic parameters, while the 3 × 4 matrix M_{ext} only on the extrinsic parameters.* If we now add a "1" as a fourth coordinate of $\mathbf{P}_w$ (that is, express $\mathbf{P}_w$ in homogeneous coordinates), and form the product $M_{int}M_{ext}\mathbf{P}_w$, we obtain a linear matrix equation describing perspective projections.

The Linear Matrix Equation of Perspective Projections

$$\begin{pmatrix} x_1 \\ x_2 \\ x_3 \end{pmatrix} = M_{int} M_{ext} \begin{pmatrix} X_w \\ Y_w \\ Z_w \\ 1 \end{pmatrix}.$$

What is interesting about vector $[x_1, x_2, x_3]^\top$ is that the ratios (x_1/x_3) and (x_2/x_3) are nothing but the image coordinates:

$$x_1/x_3 = x_{im}$$

$$x_2/x_3 = y_{im}.$$

Moreover, we have separated nicely the two steps of the world-image projection:

- M_{ext} performs the transformation between the world and the camera reference frame;

- M_{int} performs the transformation between the camera reference frame and the image reference frame.

☞   In more formal terms, the relation between a 3-D point and its perspective projection on the image plane can be seen as a linear transformation from the *projective space*, the space of vectors $[X_w, Y_w, Z_w, 1]^\top$, to the *projective plane*, the space of vectors $[x_1, x_2, x_3]^\top$. This transformation is defined *up to an arbitrary scale factor* and so that the matrix M has only 11 independent entries (see review questions). This fact will be discussed in Chapter 6.

The Perspective Camera Model.   Various camera models, including the perspective and weak-perspective ones, can be derived by setting appropriate constraints on the matrix $M = M_{int} M_{ext}$. Assuming, for simplicity, $o_x = o_y = 0$ and $s_x = s_y = 1$, M can then be rewritten as

$$M = \begin{pmatrix} -fr_{11} & -fr_{12} & -fr_{13} & f\mathbf{R}_1^\top \mathbf{T} \\ -fr_{21} & -fr_{22} & -fr_{23} & f\mathbf{R}_2^\top \mathbf{T} \\ r_{31} & r_{32} & r_{33} & -\mathbf{R}_3^\top \mathbf{T} \end{pmatrix}.$$

When unconstrained, M describes the full-perspective camera model and is called *projection matrix*.

The Weak-Perspective Camera Model.   To derive the form of M for the weak-perspective camera model, we observe that the image $\mathbf{p}$ of a point $\mathbf{P}$ is given by

$$\mathbf{p} = M \begin{pmatrix} X_w \\ Y_w \\ Z_w \\ 1 \end{pmatrix} = \begin{pmatrix} f\mathbf{R}_1^\top(\mathbf{T} - \mathbf{P}) \\ f\mathbf{R}_2^\top(\mathbf{T} - \mathbf{P}) \\ \mathbf{R}_3^\top(\mathbf{P} - \mathbf{T}) \end{pmatrix}. \tag{2.22}$$

But $\|\mathbf{R}_3^\top(\mathbf{P} - \mathbf{T})\|$ is simply the distance of $\mathbf{P}$ from the projection center along the optical axis; therefore, the basic constraint for the weak-perspective approximation can be written as

$$\left| \frac{\mathbf{R}_3^\top(\mathbf{P}_i - \bar{\mathbf{P}})}{\mathbf{R}_3^\top(\bar{\mathbf{P}} - \mathbf{T})} \right| << 1, \tag{2.23}$$

where $\mathbf{P}_1, \mathbf{P}_2$ are two points in 3-D space, and $\bar{\mathbf{P}}$ the centroid of $\mathbf{P}_1$ and $\mathbf{P}_2$. Using (2.23), (2.22) can be written for $\mathbf{P} = \mathbf{P}_i$, $i = 1, 2$, as

$$\mathbf{p}_i \approx \begin{pmatrix} f\mathbf{R}_1^\top(\mathbf{T} - \mathbf{P}_i) \\ f\mathbf{R}_2^\top(\mathbf{T} - \mathbf{P}_i) \\ \mathbf{R}_3^\top(\bar{\mathbf{P}} - \mathbf{T}) \end{pmatrix}.$$

Therefore, the projection matrix M becomes

$$M_{wp} = \begin{pmatrix} -fr_{11} & -fr_{12} & -fr_{13} & f\mathbf{R}_1^\top\mathbf{T} \\ -fr_{21} & -fr_{22} & -fr_{23} & f\mathbf{R}_2^\top\mathbf{T} \\ 0 & 0 & 0 & \mathbf{R}_3^\top(\bar{\mathbf{P}} - \mathbf{T}) \end{pmatrix}.$$

The Affine Camera Model. Another interesting camera model, widely used in the literature for its simplicity, is the so-called *affine model*, a mathematical generalization of the weak-perspective model. In the affine model, the first three entries in the last row of the matrix M are equal to zero. All other entries are unconstrained. The affine model does not appear to correspond to any physical camera, but leads to simple equations and has appealing geometric properties. The affine projection does not preserve angles but does preserve parallelism.

The main difference with the weak-perspective model is the fact that, in the affine model, only the ratio of distances measured along parallel directions is preserved. We now move on to consider range images.

2.5 Range Data and Range Sensors

In many applications, one wants to use vision to measure distances; for example, to steer vehicles away from obstacles, estimate the shape of surfaces, or inspect manufactured objects. A single intensity image proves of limited use, as pixel values are related to surface geometry only indirectly; that is, through the optical and geometrical properties of the surfaces as well as the illumination conditions. All these are usually complex to model and often unknown. As we shall see in Chapter 9, reconstructing 3-D shape from a single intensity image is difficult and often inaccurate. Can we acquire images encoding shape *directly*? Yes: this is exactly what range sensors do.

Range Images

Range images are a special class of digital images. Each pixel of a range image expresses the *distance between a known reference frame and a visible point in the scene.* Therefore, a range image reproduces the 3-D *structure* of a scene, and is best thought of as a *sampled surface.*

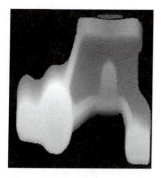  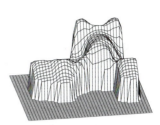

Figure 2.14   Range views of a mechanical component displayed as intensity image (left, the lighter the closer), cosine shaded (middle), and 3-D surface (right). Courtesy of R. B. Fisher, Department of Artificial Intelligence, University of Edinburgh.

2.5.1   Representing Range Images

Range images can be represented in two basic forms. One is a list of 3-D coordinates in a given reference frame, called *xyz form* or *cloud of points*, for which no specific order is required. The other is a matrix of depth values of points along the directions of the *x, y* image axes, called the r_{ij} *form*, which makes spatial information explicit. Notice that *xyz* data can be more difficult to process than r_{ij} data, as no spatial order is assumed. Range images are also referred to as *depth images, depth maps, xyz maps, surface profiles*, and *2.5-D images*.

☞   Obviously, r_{ij} data can always be visualized as a normal intensity image; the term "range image" refers indeed to the r_{ij} form. We will assume this in the following, unless otherwise specified. One can also display a range image as *cosine shaded*, whereby the grey level of each pixel is proportional to the norm of the gradient to the range surface. Figure 2.14 illustrates the three main methods of displaying range images.

☞   Dense range images prove useful for estimating the differential properties (local shape) of object surfaces.

2.5.2   Range Sensors

An *optical range sensor* is a device using optical phenomena to acquire range images. We concentrate on optical sensors as we are concerned with vision. Range sensors may measure depth at one point only, or the distance and shape of surface profiles, or of full surfaces. It is useful to distinguish between *active* and *passive* range sensors.

Definition: Active and Passive Range Sensors

Active range sensors project energy (e.g., a pattern of light, sonar pulses) on the scene and detect its position to perform the measure; or exploit the effect of controlled changes of some sensor parameters (e.g., focus).

Passive range sensors rely only on intensity images to reconstruct depth (e.g., stereopsis, discussed in Chapter 7).

Passive range sensors are the subject of Chapters 7, 8, and 9, and are not discussed further here. Active range sensors exploit a variety of physical principles; examples are radars and sonars, Moiré interferometry, focusing, and triangulation. Here, we sketch the first three, and concentrate on the latter in greater detail.

Radars and Sonars. The basic principle of these sensors is to emit a short electromagnetic or acoustic wave, or *pulse*, and detect the return (echo) reflected from surrounding surfaces. Distance is obtained as a function of the time taken by the wave to hit a surface and come back, called *time of flight*, which is measured directly. By sweeping such a sensor across the target scene, a full range image can be acquired. Different principles are used in imaging laser radars; for instance, such sensors can emit an amplitude-modulated laser beam and measure the phase difference between the transmitted and received signals.

Moiré Interferometry. A Moiré interference pattern is created when two gratings with regularly spaced patterns (e.g., lines) are superimposed on each other. Moiré sensors project such gratings onto surfaces, and measure the phase differences of the observed interference pattern. Distance is a function of such phase difference. Notice that such sensors can recover absolute distance only if the distance of one reference point is known; otherwise, only relative distances between scene points are obtained (which is desirable for inspection).

Active Focusing/Defocusing. These methods infer range from two or more images of the same scene, acquired under varying focus settings. For instance, *shape-from-focus* sensors vary the focus of a motorized lens continuously, and measure the amount of blur for each focus value. Once determined the best focused image, a model linking focus values and distance yields the distance. In *shape-from-defocus*, the blur-focus model is fitted to two images only to estimate distance.

In the following section, we concentrate on *triangulation-based* range sensors. The main reason for this choice is that they are based on intensity cameras, so we can exploit everything we know on intensity imaging. Moreover, such sensors can give accurate and dense 3-D coordinate maps, are easy to understand and build (as long as limited accuracy is acceptable), and are commonly found in applications.

2.5.3 Active Triangulation

We start by discussing the basic principle of active triangulation. Then, we discuss a simple sensor, and how to evaluate its performance. As we do not know yet how to calibrate intensity cameras, nor how to detect image features, you will be able to implement the algorithms in this section only after reading Chapters 4 and 5.

The basic geometry for an active triangulation system is shown in Figure 2.15. A light projector is placed at a distance b (called *baseline*) from the center of projection of

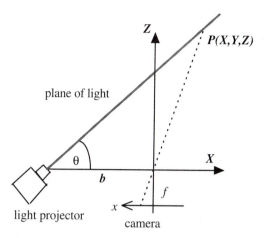

Figure 2.15 The basic geometry of active, optical triangulation (planar XZ view). The Y and y axes are perpendicular to the plane of the figure.

a pinhole camera.[9] The center of projection is the origin of the reference frame XYZ, in which all the sensor's measurements are expressed. The Z axis and the camera's optical axis coincide. The y and Y, and x and X axes are respectively parallel but point in opposite directions. Let f be the focal length. The projector emits a plane of light perpendicular to the plane XZ and forming a controlled angle, θ, with the XY plane. The Y axis is parallel to the plane of light and perpendicular to the page, so that only the profile of the plane of light is shown. The intersection of the plane of light with the scene surfaces is a planar curve called the *stripe*, which is observed by the camera. In this setup, the coordinates of a stripe point $\mathbf{P} = [X, Y, Z]^\top$ are given by

$$\begin{pmatrix} X \\ Y \\ Z \end{pmatrix} = \frac{b}{f \cot \theta - x} \begin{pmatrix} x \\ y \\ f \end{pmatrix} \tag{2.24}$$

☞   The focal length and the other intrinsic parameters of the intensity camera can be calibrated with the same procedures to be used for intensity cameras (Chapter 6).

Applying this equation to all the visible stripe points, we obtain the 3-D profile of the surface points under the stripe (a cross-section of the surface). We can acquire multiple, adjacent profiles by advancing the object under the stripe, or sweeping the stripe across the object, and repeat the computation for each relative position of stripe and object. The sequence of all profiles is a full range image of the scene.

---

[9]Notice that, in Figure 2.15, the center of projection is *in front* of the screen, not behind, and is not the origin of the camera frame. This does not alter the geometry of image formation. Why?.

In order to measure (x, y), we must identify the stripe points in the image. To facilitate this task, we try to make the stripe stand out in the image. We can do this by projecting laser light, which makes the stripe brighter than the rest of the image; or we can project a black line onto a matte white or light grey object, so that the only really dark image points are the stripe's. Both solutions are popular but have drawbacks. In the former case, concavities on shiny surfaces may create reflections that confuse the stripe detection; in the latter, stripe location may be confused by shadows, marks and dark patches. In both cases, no range data can be obtained where the stripe is invisible to the camera because of occlusions. Sensors based on laser light are called *3-D laser scanners*, and are found very frequently in applications. A real sensor, modelled closely after the basic geometry in Figure 2.15, is shown in Figure 2.16.

☞   To limit occlusions one often uses two or more cameras, so that the stripe is nearly always visible from at least one camera.

2.5.4 A Simple Sensor

In order to use (2.24) we must calibrate f, b and θ. Although it is not difficult to devise a complete calibration procedure based on the projection equations and the geometry of Figure 2.17, we present here a simple and efficient method, called *direct calibration*, which does not require any equations at all. Altogether we shall describe a small but complete range sensor, how to calibrate it, and how to use it for measuring range profiles of 3-D objects. The algorithms require knowledge of some simple image processing operations, that you will be able to implement after going through the next three chapters.

The direct calibration procedure builds a lookup table (LUT) linking image and 3-D coordinates. Notice that this is possible because a one-to-one correspondence exists between image and 3-D coordinates, thanks to the fact that the stripe points are constrained to lie in the plane of light. The LUT is built by measuring the image coordinates of a grid of known 3-D points, and recording both image and world coordinates for each point; the depth values of all other visible points are obtained by interpolation.

The procedure uses a few rectangular blocks of known heights δ (Figure 2.17). One block (call it G) must have a number (say n) of parallel, rectangular grooves. We assume the image size (in pixels) is $x_{max} \times y_{max}$.

Algorithm RANGE_CAL

Set up the system and reference frame as in Figure 2.17. With no object in the scene, the vertical stripe falls on $Z = 0$ (background plane) and should be imaged near $y = y_{max} - 1$.

1. Place block G under the stripe, with grooves perpendicular to the stripe plane. Ensure the stripe appears parallel to x (constant y).

2. Acquire an image of the stripe falling on G. Find the y coordinates of the stripe points falling on G's higher surface (i.e., not in the groove) by scanning the image columns.

3. Compute the coordinates $[x_i, y_Z]^\top$, $i = 1, \ldots, n$, of the centers of the stripe segments on G's top surface, by taking the centers of the segments in the scanline $y = y_Z$. Enter each image point $[x_i, y_Z]^\top$ and its corresponding 3-D points $[X, Z]^\top$ (known) into a table T.

4. Put another block under G, raising G's top surface by δ. Ensure that the conditions of step 1 still apply. Be careful not to move the XYZ reference frame.

5. Repeat steps 2, 3, 4 until G's top surface is imaged near $y = 0$.

6. Convert T into a 2-D lookup table L, indexed by image coordinates $[x, y]^T$, with x between 0 and $x_{max} - 1$, and y between 0 and $y_{max} - 1$, and returning $[X, Z]^T$. To associate values to the pixels not measured directly, interpolate linearly using the four nearest neighbors.

The output is a LUT linking coordinates of image points and coordinates of scene points.

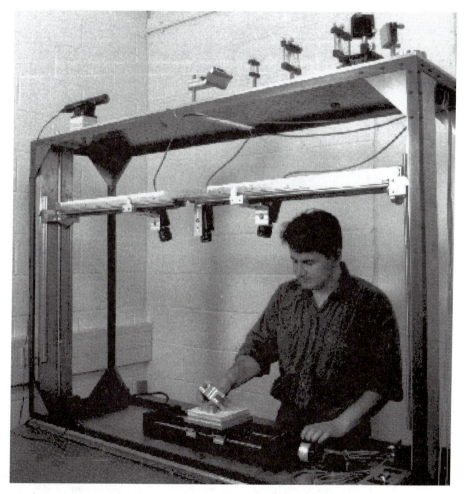

Figure 2.16 A real 3-D triangulation system, developed at Heriot-Watt University by A. M. Wallace and coworkers. Notice the laser source (top left), which generates a laser beam; the optical components forming the plane of laser light (top middle and left); the cameras; and the motorized platform (bottom middle) supporting the object and sweeping it through the stationary plane of light.

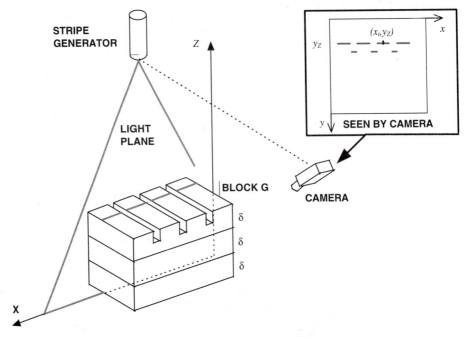

Figure 2.17   Setup for direct calibration of a simple profile range sensor.

And here is how to use L to acquire a range profile.

Algorithm RANGE_ACQ

The input is the LUT, L, built by RANGE_CAL.

1. Put an object under the stripe and acquire an image of the stripe falling on G.
2. Compute the image coordinates $[x, y]^\top$ of the stripe points by scanning each image column.
3. Index L using the image coordinates (x, y) of the stripe point, to obtain range points $[X, Z]^\top$.

The output is the set of 3-D coordinates corresponding to the stripe points imaged.

Notice that the numbers computed by such a sensor grow from the background plane ($Z = 0$), not from the camera.

☞ When a new block is added to the calibration scene, the stripe should move up by at least one or two pixels; if not, the calibration will not discriminate between Z levels. Be sure to use the *same* code for peak location in RANGE_ACQ *and* RANGE_CAL! The more sparse the calibration grid, the less accurate the range values obtained by interpolation in L.

When is a range sensor better than another for a given application? The following list of parameters is a basis for characterizing and comparing range sensors. Most parameters apply for non-triangulation sensors too.

Basic Parameters of Range Sensors

Workspace: the volume of space in which range data can be collected.

Stand-off distance: the approximate distance between the sensor and the workspace.

Depth of field: the depth of the workspace (along Z).

Accuracy: statistical variations of repeated measurements of a known true value (ground truth). Accuracy specifications should include at least the mean absolute error, the RMS error, and the maximum absolute error over N measures of a same object, with $N \gg 1$.

Resolution or precision: the smallest change in range that the sensor can measure or represent.

Speed: the number of range points measured per second.

Size and weight: important in some applications (e.g., only small sensors can be fitted on a robot arm).

☞   It is often difficult to know the *actual* accuracy of a sensor without carrying out your own measurements. Accuracy figures are sometimes reported without specifying to which error they refer to (e.g., RMS, absolute mean, maximum), and often omitting the experimental conditions and the optical properties of the surfaces used.

2.6 Summary

After working through this chapter you should be able to:

- ❑ explain how digital images are formed, represented and acquired
- ❑ estimate experimentally the noise introduced in an image by an acquisition system
- ❑ explain the concept of intrinsic and extrinsic parameters, the most common models of intensity cameras, and their applicability
- ❑ design (but not yet implement) an algorithm for calibrating and using a complete range sensor based on direct calibration

2.7 Further Readings

It is hard to find more on the content of this chapter on just one book. As a result if you want to know more you must be willing to do some bibliographic search. A readable account of basic optics can be found in the Feynman's Lecture on Physics [4]. A classic on the subject and beyond is the Born and Wolf [3] . The Born and Wolf also covers topics like image formation and spatial frequency filtering (though it is not always simple to go through). Our derivation of (2.13) is based on Horn and Sjoberg [6]. Horn [5] gives an extensive treatment of surface reflectance models. Of the many, very good textbooks on signal theory, our favorite is the Oppenheim, Willsky, and Young [11]. The discussion on camera models via the projection matrix is based on the appendix of Mundy and Zisserman's book *Geometric Invariants in Computer Vision* [9].

Our discussion of range sensors is largely based on Besl [1], which is a very good introduction to the principles, types and evaluation of range sensors. A recent, detailed review of commercial laser scanners can be found in [14]. Two laser-based, active triangulation range sensors are described in [12, 13]; the latter is based on direct calibration, the former uses a geometric camera model. References [8] and [2] are examples of triangulation sensors projecting patterns of lines generated using incoherent light (as opposed to laser light) onto the scene. Krotkow [7] and Nayar and Nakagawa [10] make good introductions to focus- based ranging.

2.8   Review

Questions

❏   *2.1*   How does an image change if the focal length is varied?

❏   *2.2*   Give an intuitive explanation of the reason why a pinhole camera has an infinite depth of field.

❏   *2.3*   Use the definition of F-number to explain geometrically why this quantity measures the fraction of the light entering the camera which reaches the image plane.

❏   *2.4*   Explain why the beat frequency of fluorescent room light (e.g., 60 Hz) can skew the results of EST_NOISE.

❏   *2.5*   *Intensity thresholding* is probably the simplest way to locate interesting objects in an image (a problem called *image segmentation*). The idea is that only the pixels whose value is above a threshold belong to interesting objects. Comment on the shortcomings of this technique, particularly in terms of the relation between scene radiance and image irradiance. Assuming that scene and illumination can be controlled, what would you do to guarantee successful segmentation by thresholding?

❏   *2.6*   The projection matrix M is a 3×4 matrix defined up to an arbitrary scale factor. This leaves only 11 of the 12 entries of T independent. On the other hand, we have seen that the matrix T can be written in terms of 10 parameters (4 intrinsic and 6 extrinsic independent parameters). Can you guess the independent intrinsic parameter that has been left out? If you cannot guess now, you have to wait for Chapter 6.

❏   *2.7*   Explain the problem of camera calibration, and why calibration is necessary at all.

❏   *2.8*   Explain why the length *in millimeters* of an image line of endpoints $[x_1, y_1]$ and $[x_2, y_2]$ is not simply $\sqrt{(x_2 - x_1)^2 - (y_2 - y_1)^2}$. What does this formula miss?

❏   *2.9*   Explain the difference between a range and an intensity image. Could range images be acquired using intensity cameras only (i.e., no laser light or the like)?

❏   *2.10*   Explain the reason for the word "shaded" in "cosine-shaded rendering of a range image". What assumptions on the illumination does a cosine-shaded image imply? How is the surface gradient linked to shading?

❏   *2.11*   What is the reason for step 1 in RANGE_CAL?

❏ *2.12* Consider a triangulation sensor which scans a whole surface profile by translating an object through a plane of laser light. Now imagine the surface is scanned by making the laser light sweep the object. In both cases the camera is stationary. What parts of the triangulation algorithm change? Why?

❏ *2.13* The performance of a range sensor based on (2.24) depend on the values of f, b, θ. How would you define and determine "optimal" values of f, b, θ for such a sensor?

Exercises

○ *2.1* Show that (2.1) and (2.2) are equivalent.

○ *2.2* Devise an experiment that checks the prediction of (2.13) on your own system. *Hint:* Use a spatially uniform object (like a flat sheet of matte gray paper) illuminated by perfectly diffuse light. Use optics with a wide field of view. Repeat the experiment by averaging the acquired image over time. What difference does this averaging step make?

○ *2.3* Show that, in the pinhole camera model, three collinear points in 3-D space are imaged into three collinear points on the image plane.

○ *2.4* Use the perspective projection equations to explain why, in a picture of a face taken frontally and from a very small distance, the nose appears much larger than the rest of the face. Can this effect be reduced by acting on the focal length?

○ *2.5* Estimate the noise of your acquisition system using procedures EST_NOISE and AUTO_COVARIANCE.

○ *2.6* Use the equations of section 2.3.2 to estimate the spatial aliasing of your acquisition system, and devise a procedure to estimate, roughly, the number of CCD elements of your camera.

○ *2.7* Write a program which displays a range image as a normal image (grey levels encode distance) or as a cosine shaded image.

○ *2.8* Derive (2.24) from the geometry shown in Figure 2.15. *Hint:* Use the law of sines and the pinhole projection equation. Why have we chosen to position the reference frame as in Figure 2.15?

○ *2.9* We can predict the sensitivity of measurements obtained through (2.24) by taking partial derivatives with respect to the formula's parameters. Compare such predictions with respect to b and f.

Projects

● *2.1* You can build your own pinhole camera, and join the adepts of *pinhole photography*. Pierce a hole about 5 mm in diameter on one side of an old tin box, 10 to 30 cm in depth. Spray the inside of box and lid with black paint. Pierce a pinhole in a piece of thick aluminium foil (e.g., the one used for milk tops), and fix the foil to the hole in the box with black tape. In a dark room, fix a piece of black and white photographic film on the hole in the box, and seal the box with black tape. The nearer the pinhole to the film, the wider the field of view. Cover the pinhole with a piece of black paper to be used as shutter. Your camera is ready. Indicatively, a 125-ASA film may require an exposure of about 5 seconds. Make sure that the camera does not move as you open and close the shutter. Some experimentation will be necessary, but results can be striking!

● *2.2* Although you will learn how to locate image features and extract straight lines automatically in the next chapter, you can get ready for an implementation of the profile scanner described in section 2.5.4, and set up the equipment necessary. All you need (in addition to camera, frame buffer and computer) is a projector creating a black stripe (easily done with a slide projector and an appropriate slide, or even with a flashlight) and a few, accurately cut blocks. You must also work out the best arrangement for projector, stripe and camera.

References

[1] P.J. Besl, Active, Optical Imaging Sensors, *Machine Vision and Applications*, Vol. 1, pp. 127-152 (1988).

[2] A. Blake, H.R. Lo, D. McCowen and P. Lindsey, Trinocular Active Range Sensing, *IEEE Transactions on Pattern Analysis and Machine Intelligence*, Vol. 15, pp. 477-483 (1993).

[3] M. Born and E. Wolf, *Principles of Optics*, Pergamon Press, New York (1959).

[4] R.P. Feynman, R.B. Leighton, and M. Sands, *The Feynman Lectures on Physics*, Addison-Wesley, Reading, Mass (1965).

[5] B.K P. Horn, *Robot Vision*, MIT Press, Cambridge, MA (1986).

[6] B.K.P. Horn and R.W. Sjoberg, Calculating the Reflectance Map, *Applied Optics*, Vol. 18, pp 1770-1779 (1979).

[7] E. Krotkow, Focusing, *International Journal of Computer Vision*, Vol. 1, pp. 223–237 (1987).

[8] M. Maruyama and S. Abe, Range Sensing by Projecting Multiple Slits with Random Cuts, *IEEE Transactions on Pattern Analysis and Machine Intelligence*, vol. 15, no. 6, pp. 647-651 (1988).

[9] J.L. Mundy and A. Zisserman, Appendix - Projective Geometry for Machine Vision. In *Geometric Invariants in Computer Vision*, Mundy, J.L. and Zisserman, A., eds., MIT Press, Cambridge, MA (1992).

[10] S.K. Nayar and Y. Nakagawa, Shape from Focus, *IEEE Transactions on Pattern Analysis and Machine Intelligence*, Vol. 16, pp. 824-831 (1994).

[11] A.V. Oppenheim, A.S. Willsky and I.T. Young, *Signals and Systems*, Prentice-Hall International Editions (1983).

[12] P. Saint-Marc, J.-C. Jezouin and G. Medioni, A Versatile PC-Based Range Finding System, *IEEE Transactions on Robotics and Automation*, Vol. RA-7, no. 2, pp. 250–256 (1991).

[13] E. Trucco and R.B. Fisher, Acquisition of Consistent Range Data Using Direct Calibration, *Proc. IEEE Int. Conf. on Robotics and Automation*, San Diego, pp. 3410–3415 (1994).

[14] T. Wohlers, 3-D Digitizers, *Computer Graphics World*, July, pp. 73-77 (1992).

3

Dealing with Image Noise

The mariachis would serenade
And they would not shut up till they were paid.

Tom Lehrer, *In Old Mexico*

Attenuating or, ideally, suppressing image noise is important because any computer vision system begins by processing intensity values. This chapter introduces a few, basic noise models and filtering methods, which constitute an initial but useful toolkit for many practical situations.

Chapter Overview

Section 3.1 discusses the concept of noise and how to quantify it. It also introduces *Gaussian* and *impulsive noise*, and their effects on images.

Section 3.2 discusses some essential linear and a nonlinear filtering methods, aimed to attenuate random and impulsive noise.

What You Need to Know to Understand this Chapter

• The basics of signal theory: sampling theorem (Appendix, section A.3), Fourier transforms, and linear filtering.

3.1 Image Noise

Chapter 2 introduced the concept of acquisition noise, and suggested a method to estimate it. But, in general, the term *noise* covers much more.

Noise

In computer vision, *noise* may refer to any entity, in images, data or intermediate results, that is not interesting for the purposes of the main computation.

For example, one can speak of noise in different cases:

- For image processing algorithms, like edge or line detection, noise might be the spurious fluctuations of pixel values introduced by the image acquisition system.
- For algorithms taking as input the results of some numerical computation, noise can be the errors introduced in the latter by random fluctuations or inaccuracies of input data, the computer's limited precision, round-off errors, and the like.
- For algorithms trying to group lines into meaningful objects, noise is the contours which do not belong to any meaningful object.

☞   In computer vision, what is considered noise for a task is often the interesting signal for a different task.[1]

Different types of noise are countered by different techniques, depending on the noise's nature and characteristics. This chapter concentrates on *image noise*. It must be clear that *noise filtering* is a classic topic of both signal and image processing, and the literature on the subject is *vast* (see section 3.4, Further Readings). This chapter is just meant to provide a few starting tools which prove useful in many practical situations.

It is now time to formalize better our *dramatis persona*.

Image Noise

We shall assume that the main image noise is *additive* and *random*; that is a spurious, random signal, $n(i, j)$, added to the true pixel values $I(i, j)$:

$$\hat{I}(i, j) = I(i, j) + n(i, j) \tag{3.1}$$

Noise Amount

The *amount of noise* in an image can be estimated by means of σ_n, the standard deviation of the random signal $n(i, j)$. It is important to know how strong is the noise with respect to the interesting signal. This is specified by the *signal-to-noise ratio*, or SNR:

$$SNR = \frac{\sigma_s}{\sigma_n} \tag{3.2}$$

where σ_s is the standard deviation of the signal (the pixel values $I(i, j)$). The SNR is often expressed in decibel:

$$SNR_{dB} = 10 \log_{10} \frac{\sigma_s}{\sigma_n} \tag{3.3}$$

---

[1] This observation was made by Jitendra Malik.

☞ Additive noise is an adequate assumption for the image acquisition systems introduced in Chapter 2, but in some cases the noise might not be additive. For instance, *multiplicative noise*, whereby $\hat{I} = nI$, models image degradation in television lines and photographs owing to grain size.

Notice that we assume that the resolution of the quantized grey levels is sufficient to sample the image appropriately; that is, to represent *all* significant variations of the image irradiance.[2] Coarse quantization can introduce spurious contours and is thus called *quantization noise*. Byte images (256 grey levels per pixel), introduced in Chapter 2, appear to be adequate for most practical purposes and are extremely popular.

3.1.1 Gaussian Noise

In the absence of information, one often assumes $n(i, j)$ to be modelled by a *white, Gaussian, zero-mean stochastic process*. For each location (i, j), this amounts to thinking of $n(i, j)$ as a random variable, distributed according to a zero mean Gaussian distribution function of fixed standard deviation, which is added to $I(i, j)$ and whose values are completely independent of each other and of the image in both space and time.

This simple model predicts that noise values are distributed symmetrically around zero and, consequently, pixel values $\hat{I}(i, j)$ around their true values $I(i, j)$; this is what you expect from good acquisition systems, which, in addition, should guarantee low noise levels. Moreover, it is easier to deal formally with Gaussian distributions than with many other statistical models. To illustrate the effect of Gaussian noise on images, Figure 3.1 (a) shows a synthetic grey-level "checkerboard" pattern and the profile of the grey levels along a horizontal scanline. Figure 3.1 (b) shows the same image corrupted by additive Gaussian noise, and the profile of the grey levels along the same scanline.

☞ The Gaussian noise model is often a convenient approximation dictated by ignorance: if we do not know and *cannot estimate* the noise characteristics, we take it to be Gaussian. Be aware, however, that white Gaussian noise is just an approximation of additive real noise! You should always try and discover as much as possible about the origin of the noise; e.g., investigating which sensor acquired the data, and design suppression methods optimally tailored to its characteristics. This is known as *image restoration*, another vast chapter of image processing.

3.1.2 Impulsive Noise

Impulsive noise, also known as *spot* or *peak noise*, occurs usually in addition to the one normally introduced by acquisition. Impulsive noise alters random pixels, making their values very different from the true values and very often from those of neighboring pixels too. Impulsive noise appears in the image as a sprinkle of dark and light spots. It can be caused by transmission errors, faulty elements in the CCD array, or external noise corrupting the analog-to-digital conversion.

---

[2] Of course, what a significant variation is depends on what you are after. This is discussed further in Chapter 4.

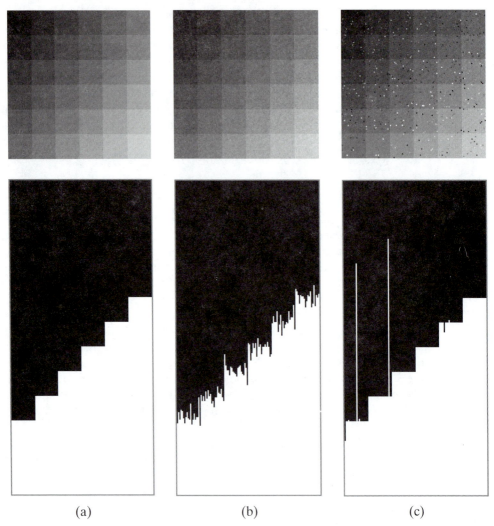

(a)   (b)   (c)

Figure 3.1   (a) Synthetic image of a 120×120 grey-level "checkerboard" and grey-level profile along a row. (b) After adding zero-mean Gaussian noise ($\sigma = 5$). (c) After adding salt and pepper noise (see text for parameters).

Salt-and-pepper noise is a model adopted frequently to simulate impulsive noise in synthetic images. The noisy image values $I_{sp}(h, k)$ are given by

$$I_{sp}(h, k) = \begin{cases} I(h, k) & x < l \\ i_{min} + y(i_{max} - i_{min}) & x \geq l \end{cases} \tag{3.4}$$

where I is the true image, $x, y \in [0, 1]$ are two uniformly distributed random variables, l is a parameter controlling how much of the image is corrupted, and i_{min}, i_{max} how severe is the noise. You can obtain *saturated salt-and-pepper noise* turning y into a two-valued variable ($y = 0$ or $y = 1$), and setting $i_{min} = 0$ and $i_{max} = 255$.

To illustrate the effects of salt and pepper noise on images, Figure 3.1 (right) shows the "checkerboard" pattern and the same scanline of Figure 3.1 (left) corrupted by salt and pepper noise with $i_{min} = 0$, $i_{max} = 255$, and $l = .99$.

3.2 Noise Filtering

Problem Statement: Noise Suppression, Smoothing, and Filtering

Given an image I corrupted by noise n, attenuate n as much as possible (ideally, eliminate it altogether) without altering I significantly.

Attenuating or, if possible, suppressing image noise is important as the result of most computations on pixel values might be distorted by noise. An important example is computing image derivatives, which is the basis of many algorithms: any noise in the signal can result in serious errors in the derivatives (see Exercise 3.3). A common technique for noise smoothing is *linear filtering*, which consists in convolving the image with a constant matrix, called *mask* or *kernel*.[3] As a reminder, here is the basic linear filtering algorithm.

Algorithm LINEAR_FILTER

Let I be a $N \times M$ image, m an odd number smaller than both N and M, and A the kernel of a linear filter, that is a $m \times m$ mask. The filtered version I_A of I at each pixel (i, j) is given by the discrete convolution

$$I_A(i, j) = I * A = \sum_{h=-\frac{m}{2}}^{\frac{m}{2}} \sum_{k=-\frac{m}{2}}^{\frac{m}{2}} A(h, k) I(i - h, j - k) \tag{3.5}$$

where $*$ indicates discrete convolution, and $m/2$ integer division (e.g., $3/2 = 1$).

A linear filter replaces the value $I(i, j)$ with a weighted sum of I values in a neighborhood of (i, j); the weights are the entries of the kernel. The effects of a linear filter on a signal can be better appreciated in the frequency domain. Through the *convolution theorem*, the Fourier transform of the convolution of I and A is simply the product of their Fourier transforms $\mathcal{F}(I)$ and $\mathcal{F}(A)$. Therefore, the result of convolving a signal with A is to attenuate (or suppress) the signal frequencies corresponding to low (or zero) values of $|\mathcal{F}(A)|$, spectrum of the filter A.

---

[3] The name *linear* is due to the fact that the convolution with a constant kernel models space- and time-invariant linear systems. From this point of view, the kernel is the *impulse response* of the filter.

3.2.1 Smoothing by Averaging

If all entries of A in (3.5) are non-negative, the filters performs *average smoothing*. The simplest smoothing kernel is the *mean filter*, which replaces a pixel value with the mean of its neighborhood; for instance, with $m = 3$

$$A_{avg} = \frac{1}{9} \begin{bmatrix} 1 & 1 & 1 \\ 1 & 1 & 1 \\ 1 & 1 & 1 \end{bmatrix} \tag{3.6}$$

☞   If the sum of all kernel entries is not one, as it happens for averaging kernels, $I_A(i, j)$ must be divided by the sum of the entries, to avoid that the filtered image becomes brighter than the original.

Why does such a filter attenuate noise? Intuitively, averaging takes out small variations: Averaging m^2 noisy values around pixel (i, j) divides the standard deviation of the noise by $\sqrt{m^2} = m$.

Frequency Behavior of the Mean Filter.   In the frequency domain we have that the Fourier transform of a 1-D mean filter kernel of width $2W$ is the "sinc" function

$$sinc(\omega) = \frac{2\sin(\omega W)}{\omega},$$

(Figure 3.3 shows an example in 2-D). Since the signal frequencies falling inside the main lobe are weighted more than the frequencies falling in the secondary lobes, the mean filter can be regarded as an approximate "low-pass" filter.

Limitations of Averaging.   Averaging is simple but has problems, including at least the following.

1. Signal frequencies shared with noise are lost; this implies that sharp signal variations are filtered out by averaging, and the image is blurred. As we shall see in Chapter 4, blurring affects the accuracy of feature localization.
2. Impulsive noise is only attenuated and diffused, not removed.
3. The secondary lobes in the Fourier transform of the mean filter's let noise into the filtered image.

3.2.2 Gaussian Smoothing

Gaussian smoothing is a particular case of averaging, in which the kernel is a 2-D Gaussian. Its effect is illustrated by Figure 3.2, which shows the results of Gaussian smoothing applied to the noisy "checkerboards" of Figure 3.1(center and right), corrupted by Gaussian and impulsive noise respectively. Notice that impulsive noise has only been attenuated; in fact, each spike has also been spread in space.

Frequency Behavior of the Gaussian Kernel.   The Fourier transform of a Gaussian is still a Gaussian and, hence, has no secondary lobes. This makes the Gaussian kernel

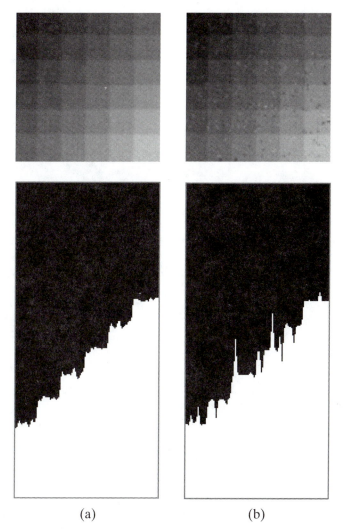

(a)   (b)

Figure 3.2 (a) Results of applying Gaussian filtering (kernel width 5 pixel, $\sigma = 1$) to the "checkerboard" image corrupted by Gaussian noise, and grey-level profile along a row. (b) Same for the "checkerboard" image corrupted by salt and pepper noise.

a better low-pass filter than the mean filter. A comparison of the mean and Gaussian filters in both the spatial and frequency domain in 2-D is shown in Figure 3.3.

Separability of the Gaussian Kernel. Gaussian smoothing can be implemented efficiently thanks to the fact that the kernel is *separable*:

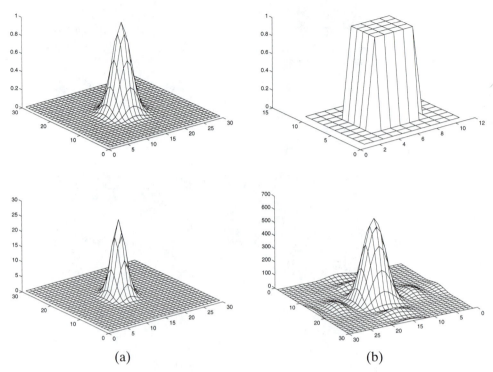

(a)   (b)

Figure 3.3 (a) The plot of a 5×5 Gaussian kernel of width 5 (top) and its Fourier transform (bottom). (b) The same for a mean-filter kernel.

$$I_G = I * G =$$

$$= \sum_{h=-\frac{m}{2}}^{\frac{m}{2}} \sum_{k=-\frac{m}{2}}^{\frac{m}{2}} G(h, k) I(i - h, j - k) =$$

$$= \sum_{h=-\frac{m}{2}}^{\frac{m}{2}} \sum_{k=-\frac{m}{2}}^{\frac{m}{2}} e^{-\frac{h^2 + k^2}{2\sigma^2}} I(i - h, j - k) =$$

$$= \sum_{h=-\frac{m}{2}}^{\frac{m}{2}} e^{-\frac{h^2}{2\sigma^2}} \sum_{k=-\frac{m}{2}}^{\frac{m}{2}} e^{-\frac{k^2}{2\sigma^2}} I(i - h, j - k) \tag{3.7}$$

This means that convolving an image I with a 2-D Gaussian kernel G is the same as convolving first all rows, then all columns with a 1-D Gaussian having the same σ. The advantage is that time complexity increases linearly with mask size, instead of quadratically (see Exercise 3.4). The next box gives the obvious algorithm for a separable kernel implemented in the special case of the Gaussian kernel.

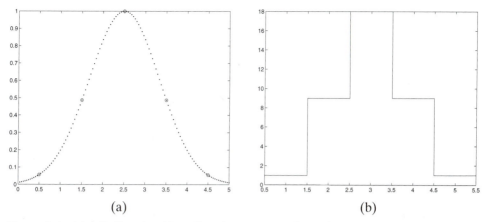

Figure 3.4   (a) 1-D Gaussian (dotted) and real samples (circles) for 5×5 kernel. (b) Plot of corresponding integer kernel.

Algorithm SEPAR_FILTER

To convolve an image I with a $m \times m$ 2-D Gaussian kernel G with $\sigma = \sigma_G$

1. Build a 1-D Gaussian mask g, of width m, with $\sigma_g = \sigma_G$;
2. Convolve each row of I with g, yielding a new image I_r;
3. Convolve each column of I_r with g.

Building Gaussian Kernels.   Thanks to the separability of the Gaussian kernel, we can consider only 1-D masks. To build a discrete Gaussian mask, one has to sample a continuous Gaussian. To do so, we must determine the mask width given the Gaussian kernel we intend to use, or, conversely, the σ of the continuous Gaussian given the desired mask width. A relation between σ and the mask width w (typically an odd number) can be obtained imposing that w subtends most of the area under the Gaussian. An adequate choice is $w = 5\sigma$, which subtends 98.76% of the area. Fitting this portion of the Gaussian between the endpoints of the mask, we find that a 3-pixel mask corresponds to $\sigma_3 = 3/5 = 0.6$ pixel; a 5-pixel mask to $\sigma_5 = 5/5 = 1$ pixel; in general,

$$\sigma_w = \frac{w}{5} \tag{3.8}$$

Sampling a continuous Gaussian yields real kernel entries. Filtering times can be greatly reduced by *approximated integer kernels* so that image values being integers too, no floating point operations are necessary at all. To build an integer kernel, you simply normalize the real kernel to make its smallest entry 1, round off the results, and divide by the sum of the entries. Figure 3.4 shows the plot of a 1-D Gaussian profile, the real samples taken, and the 5×5 integer kernel ([1, 9, 18, 9, 1]).

Algorithm INT_GAUSS_KER

To build an approximate, integer kernel G_i:

1. Compute a floating point kernel $G(h, k)$ the same size as G_i; let $g_{min} = G(0, 0)$ be the minimum value of G.

2. Determine the normalization factor $f = 1/g_{min}$.

3. Compute the entries of the non-normalized filter as $G_i(h, k) = int[fG(h, k)]$, where int indicates closest integer.

3.2.3 Are our Samples Really Gaussian?

You must be aware that problems lurk behind the straightforward recipe we gave for building discrete Gaussian kernels. It is instructive to take a closer look at least at one, sampling. The pixelization imposes a fixed sampling interval of 1 pixel. If the pixel width is taken as unit, by virtue of the sampling theorem (see Appendix, section A.3), we cannot reconstruct completely from its samples a signal containing frequencies higher than 0.5 pixel^{-1}, as any significant component at $|\omega| > \omega_c = 2\pi(0.5) = \pi$ is lost. Notice that ω_c is fixed *only* by the pixelization step, not by the signal.

What are the consequences for building discrete Gaussian kernels? In the *continuum*, the Fourier transform of the Gaussian

$$g(x, \sigma) = e^{-\frac{x^2}{2\sigma^2}}$$

is the Gaussian $g(\omega, \sigma')$ with $\sigma' = 1/\sigma$. As $g(\omega, \sigma')$ is not bandlimited, sampling $g(x, \sigma)$ on the pixel grid implies *necessarily* the loss of all components with $|\omega| > \omega_c$. For relatively small σ this means that the Fourier transform of $g(x, \sigma)$, $g(\omega, \sigma')$, is substantially different from zero well outside the interval $[-\pi, \pi]$, as shown in Figure 3.5. To avoid aliasing, the best we can do is to try keeping most of the energy of $g(\omega, \sigma')$ within the interval $[-\pi, \pi]$. Applying the "98.86% of the area" criterion in the frequency domain, we find

$$5\sigma' = \frac{5}{\sigma} \leq 2\pi$$

or

$$\sigma \geq \frac{5}{2\pi} = 0.796.$$

The preceding inequality tells you that you cannot sample appropriately a Gaussian kernel whose σ is less than 0.8 (in pixel units) no matter how many spatial samples you keep!

We can also interpret this result in terms of the minimum size for a Gaussian kernel. Since $\sigma = w/5$, for $w = 3$ we have $\sigma = 0.6$. Therefore, you cannot build a faithful Gaussian kernel with just 3 samples. For $w = 5$, instead, we have $\sigma = 1$ which means that 5 samples are enough. What happens if you ignore all this? Figure 3.6 shows that the inverse FFT of the FFT of the original Gaussian $g(x, \sigma)$ is significantly different from $g(x, \sigma)$ for $\sigma = 0.6$ ($w = 3$). In accordance with our prediction, a much smaller difference is found for $\sigma = 1$ ($w = 5$).

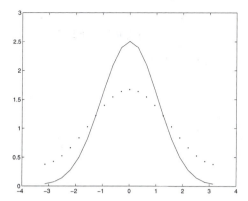

Figure 3.5   The Fourier transforms of two sampled Gaussians, for $w = 3$ ($\sigma = 0.6$, dotted line) and $w = 5$ ($\sigma = 1$, solid line). Notice that a smaller portion of the transform corresponding to $\sigma = 1$ is lost between $-\pi$ and π.

Gaussian Smoothing by Repeated Averaging.   *Repeated averaging* (RA) is a simple and efficient way to approximate Gaussian smoothing. It is based on the fact that, by virtue of the central limit theorem, convolving a 3×3 averaging mask n times with an image I approximates the convolution of I with a Gaussian mask of $\sigma = \sqrt{n/3}$ and size $3(n + 1) - n = 2n + 3$.

☞   Notice that RA leads to a different relation between σ and n from the one we obtained from the area criterion (Exercise 3.6).

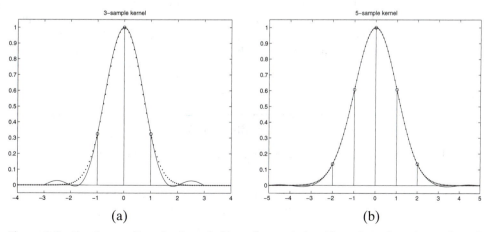

(a)   (b)

Figure 3.6   Continuous Gaussian kernels (dotted), sampled real kernels, and continuous kernels reconstructed from samples (solid), for $\sigma = 0.6$ ($w = 3$) (a) and $\sigma = 1$ ($w = 5$) (b) respectively.

Algorithm REP_AVG

Let $A * B$ indicate the convolution of matrices A and B. Let I be the input image. Define the 3×3 RA mask

$$R = \frac{1}{24} \begin{bmatrix} 1 & 2 & 1 \\ 2 & 12 & 2 \\ 1 & 2 & 1 \end{bmatrix} \tag{3.9}$$

To convolve I with an approximated Gaussian kernel of $\sigma = \sqrt{n/3}$:

1. $I_{res} = I$
2. For $i = 1, \ldots, n$, $I_{res} = I_{res} * R$

You might be tempted to combine separability and repeated averaging, as this would yield a very efficient algorithm indeed. But are you sure that the kernel defined in REP_AVG is separable? Using separability with a nonseparable kernel means that the result of REP_AVG is different from the application of the 2-D mask, which may result in errors in further processing; image differentiation is once again an apt example.

A safe way to combine separability and repeated averaging is *cascading*. The idea is that smoothing with Gaussian kernels of increasingly large standard deviations can also be achieved by convolving an image repeatedly with the same Gaussian kernel. In this way, each filtering pass of REP_AVG is surely separable (see Exercise 3.7).

3.2.4 Nonlinear Filtering

In section 3.2.1, we listed the main problems of the averaging filter: blur, poor feature localization, secondary lobes in the frequency domain, and incomplete suppression of peak noise. Gaussian filters solve only the third one, as the Fourier transform of a Gaussian has no secondary lobes. The remaining problems are tackled efficiently by *nonlinear filtering*; that is, filtering methods that cannot be modelled by convolution.

The *median filter* is a useful representative of this class. A median filter just replaces each pixel value $I(i, j)$ with the median of the values found in a local neighborhood of (i, j). As with averaging, the larger the neighborhood, the smoother the result.

Algorithm MED_FILTER

Let I be the input image, I_m the filtered image, and n an odd number.
    For each pixel (i, j):

1. Compute the median $m(i, j)$ of the values in a $n \times n$ neighborhood of (i, j), $\{I(i + h, j + k), h, k \in [-n/2, n/2]\}$, where $n/2$ indicates integer division.
2. Assign $I_m(i, j) = m(i, j)$

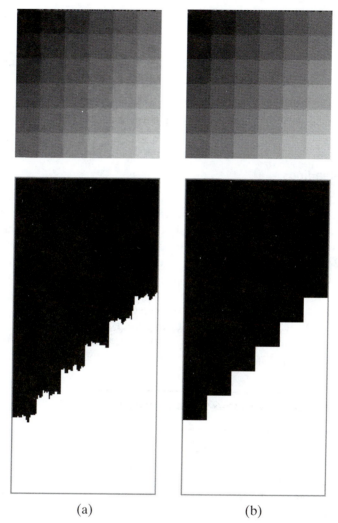

(a)   (b)

Figure 3.7 (a) Results of applying median filtering (3-pixel wide) to the "checkerboard" image corrupted by Gaussian noise, and grey-level profile along the same row of Figure 3.2. (b) Same for the "checkerboard" image corrupted by impulsive noise.

Figure 3.7 shows the effects of median filtering on the "checkerboard" image corrupted by Gaussian and impulsive noise (Figure 3.1 center and right, respectively). Compare these results with those obtained by Gaussian smoothing (Figure 3.2): Median filtering has suppressed impulsive noise completely. Contours are also blurred less by the median than by the Gaussian filter; therefore, a median filter preserves discontinuities better than linear, averaging filters.

3.3 Summary

After working through this chapter you should be able to:

❑ explain the concept of noise, image noise, and why noise smoothing is important for computer vision

❑ design noise-smoothing algorithms using Gaussian and median filtering

❑ decide whether it is appropriate to use linear or median smoothing filters in specific situations

3.4 Further Readings

Noise filtering and image restoration are classic topics of noise and image processing. Detailed discussions of image processing methods are found in several books; for instance, [4, 3, 10, 8]. Papoulis [7] is a good reference text for Fourier transforms.

Repeated averaging for computer vision was first reported by Brady *et al.* [1]. Cai [2] discusses several linear filtering methods in the context of diffusion smoothing. Witkin [11] and Lindeberg [5] provide a good introduction to *scale-space represen-tations*, the study of image properties when smoothing with Gaussians of increasing standard deviation (the *scale parameter*). One reason for keeping multiple scales is that some image features may be lost after filtering with large kernels, but small kernels could keep in too much noise. Alternative methods for representing signals at multiple scales include pyramids [9] and wavelets [6] (see also references therein).

3.5 Review

Questions

❑ *3.1* Explain the concept of image noise, how it can be quantified, and how it can affect computer vision computations.

❑ *3.2* How would you estimate the quantization noise in a range image in terms of mean and standard deviation? Notice that this allows you to compare directly quantization and acquisition noise.

❑ *3.3* Explain why a non-negative kernel works as a low-pass filters, and in which assumptions it can suppress noise.

❑ *3.4* What is a separable kernel? What are the advantages of separability?

❑ *3.5* What are the problems of the mean filter for noise smoothing? Why and in what sense is Gaussian smoothing better?

❑ *3.6* Explain why the sampling accuracy of a 1-D Gaussian filter with $\sigma = 0.6$ cannot be improved using more than three spatial samples.

❑ *3.7* What is repeated averaging? What are its effects and benefits?

❑ *3.8* What is the difference between cascading and repeated averaging?

❑ *3.9* Can you think of any disadvantage of cascading? (*Hint:* Which standard deviations do you achieve?)

Exercises

○ *3.1* Design an algorithm for Gaussian smoothing using separability and cascading.

○ *3.2* Use EST_NOISE (Chapter 2) to determine the standard deviation of a Gaussian smoothing filter for the attenuation of acquisition noise.

○ *3.3* Show that the derivative of a 1-D signal, $I = I(x)$, amplifies the higher frequency components of the signal. Then, prove that even a slight amount of noise, $n = n(x)$, can be responsible for a large difference between the derivative of I and $I + n$. (*Hint:* Assume that n can be written as $\epsilon \sin(\omega x)$ for some *very small* ϵ and *very large* ω.)

○ *3.4* Compare the time complexity of convolution with a $n \times n$ kernel when using (a) direct convolution with the 2-D mask, (b) a separable kernel, and (c) cascading with a separable kernel.

○ *3.5* Design and implement a way of verifying experimentally that repeated averaging approximates Gaussian smoothing (*Hint:* With $n = 2$, $(I * m) * m = I * (m * m)$, which must approximate $I * G$, G a Gaussian mask.)

○ *3.6* Use the "98.76%-area" criterion to determine the standard deviation σ_{RA} of a sampled Gaussian kernel the same size as the one obtained by applying repeated averaging n times with a 3×3 mask. Compare σ_{RA} with the RA standard deviation $\sqrt{n/3}$. Which method smooths the image more? Which values of n make the difference noticeable? Verify your answers experimentally.

○ *3.7* Prove that convolving a 1-D signal twice with a Gaussian kernel of standard deviation σ is equivalent to convolving the signal with a Gaussian kernel of $\sigma_c = \sqrt{2}\sigma$, scaled by the area of the Gaussian filter. (*Hint:* Make use of the identity

$$\int_{-\infty}^{+\infty} e^{-\frac{1}{2}Ax^2 + Zx}\, dx = \sqrt{\frac{2\pi}{A}}\, e^{\frac{z^2}{2A}},$$

with $A > 0$.)

○ *3.8* Consider the 1-D step profile

$$f(i) = \begin{cases} 4 & i \in [0, 3] \\ 8 & i \in [4, 7] \end{cases}.$$

Work out the result of median filtering with $n = 3$, and compare the result with the output of filtering with the averaging mask $1/4\,[1\ 2\ 1]$.

○ *3.9* Median filtering can degrade thin lines. This can be partially avoided by using nonsquare neighborhoods. What neighborhood shape would you use to preserve horizontal or vertical lines, 1-pixel wide?

Project

- *3.1*   Write programs implementing Gaussian and median noise filtering. The code should allow you to specify the filter's width. The Gaussian implementation should be made as efficient as possible.

References

[1]   M. Brady, J. Ponce, A. Yuille and M. Asada, Describing Surfaces, *Computer Vision, Graphics and Image Processing*, Vol. 32, no. 1, pp. 1-28 (1985).

[2]   L.D. Cai, *Scale-Based Surface Understanding Using Diffusion Smoothing*, PhD Thesis, Department of Artificial Intelligence, University of Edinburgh (1990).

[3]   R.C. Gonzalez and R.E. Woods, *Digital Image Processing*, Addison-Wesley, Reading (MA) (1992).

[4]   R.M. Haralick and L.G. Shapiro, *Computer and Robot Vision*, Vol. I, Addison-Wesley, Reading (MA) (1992).

[5]   T. Lindeberg, Scale-Space for Discrete Signals, *IEEE Transactions on Pattern Analysis and Machine Intelligence*, Vol. PAMI-12, no. 3, pp. 234-254 (1990).

[6]   S.G. Mallat, A Theory of Multiresolution Image Processing: the Wavelet Representation, *IEEE Transactions on Pattern Analysis and Machine Intelligence*, Vol. PAMI-11, no. 6, pp. 674–693 (1989).

[7]   A. Papoulis, *The Fourier Integral and Its Applications*, McGraw-Hill, New York (1962).

[8]   W.K. Pratt, *Digital Image Processing*, Wiley, New York (1991).

[9]   A. Rosenfeld, *Multiresolution Image Processing*, Springer-Verlag, New York (1984).

[10]   A. Rosenfeld and A.C. Kak, *Digital Picture Processing*, Academic Press, London (1976).

[11]   A.P. Witkin, Scale-Space Filtering, *Proc. 8th Int. Conf. on Artificial Intelligence IJCAI-83*, Karlsruhe, pp. 1019–1022 (1983).

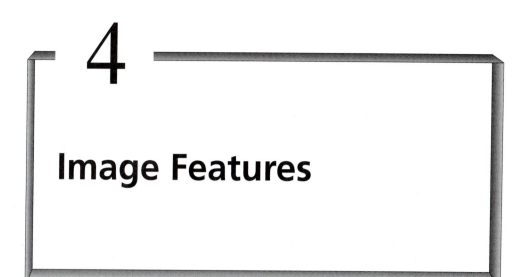

Image Features

Quel naso dritto come una salita
Quegli occhi allegri da italiano in gita.[1]

Paolo Conte, *Bartali*

This and the following chapter consider the detection, location and representation of special parts of the image, called *image features*, usually corresponding to interesting elements of the scene.

Chapter Overview

Section 4.1 introduces the concept of image feature, and sketches the fundamental issues of *feature detection*, on which many computer vision algorithms are based.

Section 4.2 deals with *edges*, or contour fragments, and how to detect them. Edge detectors are the basis of the line and curve detectors presented in the next chapter.

Section 4.3 presents features which do not correspond necessarily to geometric elements of the scene, but are nevertheless useful.

Section 4.4 discusses surface features and *surface segmentation* for range images.

What You Need to Know to Understand this Chapter

- Working knowledge of Chapter 2 and 3.
- Basic concepts of signal theory.
- Eigenvalues and eigenvectors of a matrix.
- Elementary differential geometry, mainly surface curvatures (Appendix, section A.5).

---

[1] The nose as straight as an uphill road; The merry eyes of an Italian on holidays.

4.1 What Are Image Features?

In computer vision, the term *image feature* refers to two possible entities:

1. *a global property of an image* or part thereof, for instance the average grey level, the area in pixel *(global feature)*; or

2. *a part of the image with some special properties*, for instance a circle, a line, or a textured region in an intensity image, a planar surface in a range image *(local feature)*.

The sequence of operations of most computer vision systems begins by detecting and locating some features in the input images. In this and the following chapter, we concentrate on the second definition above, and illustrate how to detect special parts of intensity and range images like points, curves, particular structures of grey levels, or surface patches. The reason for this choice is that most algorithms in the following chapters assume that specific, local features have already been located. Here, we provide ways of doing that. Global features are indeed used in computer vision, but are less useful to solve the problems tackled by Chapters 7, 8, 9, 10 and 11. We assume therefore the following definition.

Definition: Image Features

Image features are local, meaningful, detectable parts of the image.

Meaningful means that the features are associated to interesting scene elements via the image formation process. Typical examples of meaningful features are sharp intensity variations created by the contours of the objects in the scene, or image regions with uniform grey levels, for instance images of planar surfaces. Sometimes the image features we look for are not associated obviously to any part or property of the scene, but reflect particular arrangements of image values with desirable properties, like *invariance* or ease of detectability. For instance, section 4.3 discusses an example of features which prove adequate for tracking across several images (Chapter 8). On the other hand, the number of pixels of grey level 134 makes a rather unuseful feature, as, in general, it cannot be associated to any interesting properties of the scene, as individual grey levels change with illumination and viewpoint.

Detectable means that location algorithms must exist, otherwise a particular feature is of no use! Different features are, of course, associated to different detection algorithms; these algorithms output collections of *feature descriptors*, which specify the position and other essential properties of the features found in the image. For instance, a descriptor for line features could specify the coordinates of the segment's central point, the segment's length, and its orientation. Feature descriptors are used by higher-level programs; for instance, in this book, chains of edge points (section 4.2) are used by line detectors (Chapter 5); lines, in turn, are used by calibration (Chapter 6) and recognition algorithms (Chapter 10).

☞ In 3-D computer vision, feature extraction is an intermediate step, not the goal of the system. We do not extract lines, say, just to obtain line maps; we extract lines to navigate robots in corridors, to decide whether an image contains a certain object, to calibrate the intrinsic parameters of a camera, and so on. The important corollary is that *it does not make much sense to pursue "perfect" feature extraction per se*, as the adequacy of a feature detection algorithm should be *ultimately* assessed in the context of the *complete* system.[2] Of course reasonably general performance criteria can and should be applied to test feature extraction modules independently (see section 4.2.4).

4.2 Edge Detection

4.2.1 Basics

Definition: Edges

Edge points, or simply *edges*, are pixels at or around which the image values undergo a sharp variation.

Problem Statement: Edge Detection

Given an image corrupted by acquisition noise, locate the edges most likely to be generated by scene elements, not by noise.

Figure 4.1 illustrates our definition. It shows an intensity image and the intensity profile along the scanline shown: notice how the main sharp variations correspond to significant contours.[3] Notice that image noise too causes intensity variations, which results in spurious edges; a good edge detection algorithm, or *edge detector*, should suppress most of them.

☞ The term "edge" is also used to refer to *connected chains* of edge points, that is, contour fragments. Edge points are sometimes called *edgels* (for "edge elements").

There are various reasons for our interest in edges. The contours of potentially interesting scene elements like solid objects, marks on surfaces, and shadows, all generate intensity edges. Moreover, image lines, curves and contours, which are often the basic elements for stereopsis, calibration, motion analysis and recognition, are detected from chains of edge points. Finally, line drawings are common and suggestive images for humans. Throughout, we refer to intensity images, but it makes perfect sense to apply edge detection to range images as well (see review questions).

Our next task is to make the problem more precise. Edge detection in computer vision is typically a three-step process.

---

[2] Incidentally, this is true of *any* intermediate module of a vision system.

[3] Image edges are commonly presented as "discontinuities in the underlying irradiance function," but it seems more accurate to speak of "sharp image variations" than "discontinuities". The reason is that the scene radiance is low-pass filtered by the optics (Chapter 2) and the resulting image brightness cannot have *real* 0-order discontinuities.

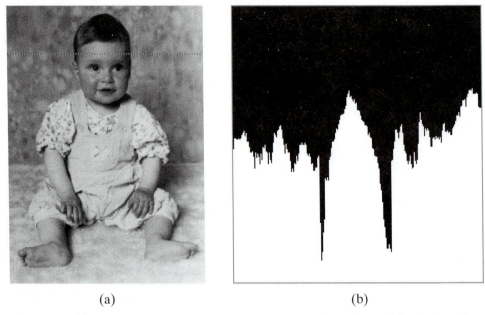

(a)   (b)

Figure 4.1   (a) A 325 × 237-pixel image, with scanline $i = 56$ highlighted. (b) The intensity profile along the highlighted scanline. Notice how the main intensity variations indicate the borders of the hair region along the scanline.

The Three Steps of Edge Detection

Noise Smoothing. Suppress as much of the image noise as possible, without destroying the true edges. In the absence of specific information, assume the noise white and Gaussian.

Edge Enhancement. Design a filter responding to edges; that is, the filter's output is large at edge pixels and low elsewhere, so that edges can be located as the local maxima in the filter's output.

Edge Localization. Decide which local maxima in the filter's output are edges and which are just caused by noise. This involves:

- thinning wide edges to 1-pixel width (*nonmaximum suppression*);
- establishing the minimum value to declare a local maxima an edge (*thresholding*).

    Edge detection algorithms are found in their tens in the literature of computer vision and image processing. Many produce similar results. Instead of taking you through a plethora of algorithms, we introduce directly the *Canny edge detector*, probably the most used edge detector in today's machine vision community. Canny's detector is *optimal* in a precise, mathematical sense; going through the main ideas behind its derivation is an instructive example of good practice in the design of low-level vision algorithms. We shall also sketch two other edge detection algorithms.

4.2.2 The Canny Edge Detector

To arrive at Canny's edge detector, we need to:

1. formulate a mathematical model of edges and noise;
2. formulate quantitative performance criteria, formalizing desirable properties of the detector (e.g., good immunity to noise);
3. synthesize the best filter given models and performance criteria. We shall be looking for a linear filter (see Chapter 3), as it is easy to manipulate and implement.

Here is the skeleton of the algorithm we are going to derive. We shall build the missing algorithms in the next sections.

Algorithm CANNY_EDGE_DETECTOR

Given an image I:

1. apply CANNY_ENHANCER to I;
2. apply NONMAX_SUPPRESSION to the output of CANNY_ENHANCER;
3. apply HYSTERESIS_THRESH to the output of NONMAX_SUPPRESSION.

Modelling Edges and Noise.   Edges of intensity images can be modelled according to their *intensity profiles*. For most practical purposes, a few models are sufficient to cover all interesting edges, and these are illustrated in Figure 4.2 for the 1-D case.

☞   You should think of the 1-D signals shown in Figure 4.2 as cross-sections of 2-D images along a line of arbitrary orientation (not necessarily a row or column).

Step edges (Figure 4.2 (a)) are probably the most common type in intensity images. They occur typically at contours of image regions of different intensities. In most cases, the transition occurs over several pixels, not just one; one speaks then of *ramp edges*. *Ridge edges* (Figure 4.2 (b)) are generated by thin lines. Clearly, wide ridges can be modelled by two step edges. *Roof edges* (Figure 4.2 (c)) are relatively rare, but may appear along the intersection of surfaces. Notice that steps and ridges correspond to sharp variations of the intensity values, while roofs correspond to variations of their first derivatives.

A good step-edge detector will actually find all edges necessary for most purposes, and for this reason we concentrate on *step edge detectors*. The output we want is a list of *step edge descriptors*, each of which should include the essential properties shown in the box that follows:

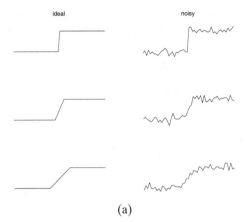

(a)

Left: ideal step (top, transition occurs over one pixel) and ramp edges. Right: corresponding noisy version, obtained by adding Gaussian noise (standard deviation 5% of the step height).

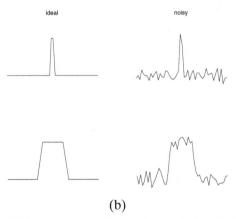

(b)

Left: ideal ridge edges. Right: corresponding noisy version, obtained as for step edges.

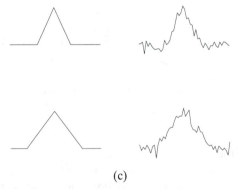

(c)

Left: ideal roof edges. Right: corresponding noisy version, obtained as for step edges.

Figure 4.2 Three types of 1-D edge profiles.

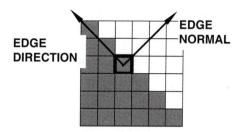

Figure 4.3 Illustration of edge normal and edge direction. The edge position considered is (2,2), with the origin in the upper left corner.

The Essential Edge Descriptor

Edge normal: for edges in 2-D images, the direction (unit vector) of the maximum intensity variation at the edge point. This identifies the direction perpendicular to the edge (Figure 4.3).

Edge direction: the direction perpendicular to the edge normal, and therefore tangent to the contour of which the edge is part (Figure 4.3). This identifies the direction tangent to the edge.

Edge position or center: the image position at which the edge is located, along the perpendicular to the edge. This is usually saved in a binary image (1 for edge, 0 for no edge).

Edge strength: a measure of the local image contrast; i.e., how marked the intensity variation is across the edge (along the normal).

We model the *ideal, 1-D step edge* as

$$G(x) = \begin{cases} 0 & x < 0 \\ A & x \geq 0 \end{cases} \qquad (4.1)$$

And here is a summary of the assumptions we are making.

Assumptions

- The edge enhancement filter is linear.
- The filter must be optimal for noisy step edges.
- The image noise is additive, white and Gaussian.

Criteria for Optimal Edge Detection.   We must now express the characteristics we expect of an optimal edge detector, that is, formalize optimality criteria.

Criterion 1: Good Detection. *The optimal detector must minimise the probability of false positives (detecting spurious edges caused by noise), as well as that of missing real edges.*

This is achieved by maximising the signal-to-noise ratio (SNR), defined here as the ratio of the root-mean-squared (RMS) responses of the filter to the ideal step edge (4.1) and the noise, respectively. Let f be the impulse response of the 1-D filter, and assume the filter's width is $2W$. Consider the 1-D signal in (4.1) centered in $x = 0$ (the edge's center). The response of a linear filter f to this edge is

$$\int_{-W}^{W} G(-t)f(t)dt = A \int_{-W}^{0} f(t)dt, \tag{4.2}$$

and the good-detection criterion becomes

$$SNR = \frac{A\| \int_{-W}^{0} f(t)dt \|}{n_0 \sqrt{\int_{-W}^{W} f^2(t)dt}} \tag{4.3}$$

where n_0^2 is the RMS noise amplitude per unit length.[4]

Criterion 2: Good Localization. *The edges detected must be as close as possible to the true edges.*

This can be formalized by the reciprocal of the RMS distance of the detected edge from the center of the true edge. It can be proven that, in our assumptions, this results in

$$LOC = \frac{A\| f'(0) \|}{n_0 \sqrt{\int_{-W}^{W} f'^2(t)dt}}. \tag{4.4}$$

We omit the rather complex derivation (see Further Readings). We now notice that (4.3) and (4.4) identify two performance measures for step edge detectors (call them Σ and Λ), *which depend only on the filter*, and not on either noise or step magnitude. In particular, Σ and Λ are defined by

$$SNR = \frac{A}{n_0}\Sigma(f) \qquad LOC = \frac{A}{n_0}\Lambda(f'). \tag{4.5}$$

The product $\Sigma\Lambda$ is a measure of how well the filter f satisfies both criteria simultaneously, so we must look for the f maximing $\Sigma\Lambda$. One can prove that this product is maximized for $f(x) = G(-x)$ in $[-W, W]$. Therefore, *the optimal, 1-D step edge detector is a simple difference operator*, or *box filter*.

Unfortunately, something is still missing. The response of a difference operator to a noisy edge contains many local maxima (Figure 4.4), not just the one we are after. We need, therefore, a third criterion, which is really a *constraint*.

Single Response Constraint. *The detector must return one point only for each true edge point; that is, minimize the number of local maxima around the true edge created by noise.* Without going into details, this is formalized by imposing that the mean distance between local maxima caused by the noise (which can be estimated from the statistics of the noise) only be some fraction of the filter's half-width W.

---

[4] Incidentally, you have just learned (or been reminded of) the expression of the RMS response of a linear filter to white, Gaussian noise; that is, the denominator of (4.3).

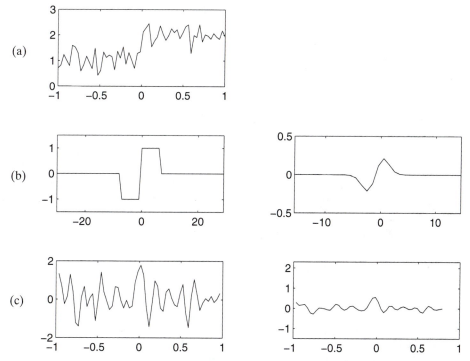

Figure 4.4 (a) Noisy, step edge, corrupted by additive Gaussian noise (standard deviation is 15% of the uncorrupted step's height). (b) The box filter (left) and the first derivative of a Gaussian (right). (c) Response to the noisy edge of the box filter (left) and of the first derivative of a Gaussian. The latter contains fewer local maxima thanks to the smoothing effect of the Gaussian. Notice the different amplitudes of the two responses due to the different amplitudes of the filters.

One consequence of our formalization is definitely worth noting. Suppose we take a spatially scaled version of the filter, that is, $f_w(x) = f(x/w)$. One can prove that

$$\Sigma(f_w) = \sqrt{w}\,\Sigma(f) \qquad \Lambda(f'_w) = \frac{1}{\sqrt{w}}\Lambda(f') \tag{4.6}$$

that is, *a larger filter improves detection, but worsens localization by the same amount*, and *vice versa*. We can summarize this important result as follows.

The Localization-Detection Tradeoff

We can reach an optimal compromise between the location and detection criteria by adjusting the filter's spatial scale, but we cannot improve both criteria simultaneously.

Optimal Step Edge Enhancement.   Given our edge and noise models, and the design criteria, we can obtain the optimal enhancing filter f as the function maximizing the product $\Lambda(f)\Sigma(f)$ (criteria 1 and 2), under the constraint of single response. *We have turned edge detection into a constrained optimization problem*, that is, a well-formalized mathematical problem with clear assumptions.

The bad news is that it can be proven that the solution is unique, but it is very difficult to find closed-form solutions. We can, however, evaluate numerically criteria (4.3) and (4.4) for candidate operators f, and pick the best performer.[5] In this way, one realises that *a very good approximation of the ideal step edge detector is the first derivative of a Gaussian*, which proves only 20% worse than the ideal operator with respect to (4.3) and (4.4).[6]

So far we have worked with 1-D signals, but how about 2-D images? The key to generalization is that *the 1-D treatment still applies in the direction of the edge normal*. Since we do not know *a priori* where the normal is, we should compute, at each image point, the result of applying 1-D directional filters in all possible directions. Therefore, the optimal, 2-D edge detector would involve a set of 1-D filters spanning a large number of orientations. Such filters would be implemented by narrow masks, elongated along the edge direction; the mask's shorter cross-section is the ideal 1-D detector, and the longer (in the perpendicular direction) is Gaussian, which contributes to noise smoothing.

This laborious method can be simplified significantly. One can use a *circular Gaussian filter*, which amounts to applying Gaussian smoothing to the image followed by gradient estimation, and use the gradient to estimate all the directional derivatives required. This is less accurate than applying directional filters but adequate for most practical cases. In this way, we estimate the edge strength as

$$s(i, j) = \|\nabla(G * I)\| \tag{4.7}$$

and the edge normal as

$$\mathbf{n} = \frac{\nabla(G * I)}{\|\nabla(G * I)\|} \tag{4.8}$$

Here is an algorithm implementing a practical approximation of the optimal step edge enhancer.[7]

Algorithm CANNY_ENHANCER

The input is I, an intensity image corrupted by noise. Let G be a Gaussian with zero mean and standard deviation σ.

---

[5] This can be done for any edge class, not only step edges. All integrals evaluated using the step edge model change; see Canny's original article (Further Readings) for details.

[6] And 90% of the single-response criterion, although this figure is rather complicated to achieve.

[7] Notice that this algorithm *implements edge descriptors through a set of images*. An alternative would be to define a data structure gathering all the properties of an edge.

1. Apply Gaussian smoothing to I (algorithm LINEAR_FILTER of Chapter 3 with a Gaussian kernel discretising G), obtaining $J = I * G$.

2. For each pixel (i, j):

   (a) compute the gradient components, J_x and J_y (Appendix, section A.2);
   (b) estimate the edge strength

$$e_s(i, j) = \sqrt{J_x^2(i, j) + J_y^2(i, j)}$$

   (c) estimate the orientation of the edge normal

$$e_o(i, j) = \arctan \frac{J_y}{J_x}$$

The output is a *strength image*, E_s, formed by the values $e_s(i, j)$, and an *orientation image*, E_o, formed by the values $e_o(i, j)$.

☞ To make the implementation as efficient as possible, you can use the fact that $\nabla(G * I) = \nabla G * I$ (that is, convolve the image with the 1-D first derivative of a Gaussian) and Gaussian separability. You can also use lookup tables to store the square root values.

☞ The value of σ to be used depends on the length of interesting connected contours, the noise level, and the localization-detection trade off.

We must now find the position of edge centers and discard as many noisy edges as possible in the output of the enhancing filter. This is done in two steps: *nonmaximum suppression* and *thresholding*.

Nonmaximum Suppression. The strength image, E_s, output by CANNY_ENHANCER may contain wide ridges around the local maxima. Nonmaximum suppression thins such ridges to produce 1-pixel wide edges. Here is an essential algorithm, using a rather coarse quantisation of edge normal directions (4-levels).

Algorithm NONMAX_SUPPRESSION

The input is the output of CANNY_ENHANCER, that is, the edge strength and orientation images, E_s and E_o. Consider the four directions $d_1 \ldots d_4$, identified by the 0°, 45°, 90° and 135° orientations (with respect to the horizontal axis image reference frame).
  For each pixel (i, j):

1. find the direction, $\hat{d}_k$, which best approximates the direction $E_o(i, j)$ (the normal to the edge);

2. if $E_s(i, j)$ is smaller than at least one of its two neighbors along $\hat{d}_k$, assign $I_N(i, j) = 0$ (suppression); otherwise assign $I_N(i, j) = E_s(i, j)$.

The output is an image, $I_N(i, j)$, of the thinned edge points (that is, $E_s(i, j)$ after suppressing nonmaxima edge points).

Figure 4.5 Strength images output by CANNY_ENHANCER run on Figure 4.1, after nonmaximum suppression, showing the effect of varying the filter's size, that is, the standard deviation, σ_f, of the Gaussian. Left to right: $\sigma_f = 1, 2, 3$ pixel.

Figure 4.5 shows the output of our implementation of CANNY_ENHANCER, after nonmaximum suppression, when run on the image in Figure 4.1 (left)[8] with three values of standard deviation of the filtering Gaussian. Notice how smaller filters capture shorter edges, but several of these do not belong to any interesting contour in the image (e.g., background edges).

Thresholding. The image output by NONMAX_SUPPRESSION, I_N, still contains the local maxima created by noise. How do we get rid of these? We can try to discard all pixels of value less than a threshold, but this has two problems:

- if we set a low threshold in the attempt of capturing true but weak edges, some noisy maxima will be accepted too (*false contours*);
- the values of true maxima along a connected contours may fluctuate above and below the threshold, fragmenting the resulting edge (*streaking*).

A solution is *hysteresis thresholding*.

Algorithm HYSTERESIS_THRESH

The input is I_N, the output of NONMAX_SUPPRESSION, E_o, the edge orientation image, and τ_l, τ_h, two thresholds such that $\tau_l < \tau_h$.

For all the edge points in I_N, and scanning I_N in a fixed order:

1. Locate the next unvisited edge pixel, $I_N(i, j)$, such that $I_N(i, j) > \tau_h$;

---

[8] Without the highlighted scanline, obviously!

2. Starting from $I_N(i, j)$, follow the chains of connected local maxima, in both directions perpendicular to the edge normal, as long as $I_N > \tau_l$. Mark all visited points, and save a list of the locations of all points in the connected contour found.

The output is a set of lists, each describing the position of a connected contour in the image, as well as the strength and the orientation images, describing the properties of the edge points.

Hysteresis thresholding reduces the probability of false contours, as they must produce a response higher than τ_h to occur, as well as the probability of streaking, which requires now much larger fluctuations to occur than in the single-threshold case.

☞   If τ_h is large, τ_l can also be set to 0.

Notice that HYSTERESIS_THRESH performs *edge tracking*: it finds chains of connected edge maxima, or connected contours. The descriptors for such chains, saved by HYSTERESIS_THRESH in addition to edge point descriptors, can be useful for curve detection.

☞   Notice that Y-junctions will be split by NONMAX_SUPPRESSION. How serious this is depends on what the edges are computed for. A possible solution is to modify HYSTERESIS_THRESH so that it recognizes Y-junctions, and interrupt all edges.

Figure 4.6 shows the output of our implementation of NONMAX_SUPPRESSION and HYSTERESIS_THRESH when run on the images in Figure 4.5. All contours are one pixel wide, as desired.

Figure 4.6 Output of HYSTERESIS_THRESH run on Figure 4.5, showing the effect of varying the filter's size. Left to right: $\sigma_f = 1, 2, 3$ pixel. The grey levels has been inverted (black on white) for clarity.

4.2.3 Other Edge Detectors

Early edge detection algorithms were less formalized mathematically than Canny's. We sketch two examples, the *Roberts* and the *Sobel* edge detectors, which are easily implemented in their essential form.

Algorithm ROBERTS_EDGE_DET

The input is formed by an image, I, and a threshold, τ.

1. apply noise smoothing as appropriate (for instance, Gaussian smoothing in the absence of information on noise: see Chapter 3), obtaining a new image I_s;

2. filter I_s (algorithm LINEAR_FILTER, Chapter 3) with the masks

$$\begin{bmatrix} 1 & -1 \\ -1 & 1 \end{bmatrix} \quad \begin{bmatrix} -1 & 1 \\ 1 & -1 \end{bmatrix}$$

obtaining two images I_1 and I_2;

3. estimate the gradient magnitude at each pixel (i, j) as

$$G(i, j) = \sqrt{I_1^2(i, j) + I_2^2(i, j)},$$

obtaining an image of magnitude gradients, G;

4. mark as edges all pixels (i, j) such that $G(i, j) > \tau$.

The output is the location of edge points obtained in the last step.

Algorithm SOBEL_EDGE_DET

Same as for ROBERTS_EDGE_DET, but replace step 2. with the following.

2. filter I_s (algorithm LINEAR_FILTER, Chapter 3) with the masks

$$\begin{bmatrix} -1 & -2 & -1 \\ 0 & 0 & 0 \\ 1 & 2 & 1 \end{bmatrix} \quad \begin{bmatrix} -1 & 0 & 1 \\ -2 & 0 & 2 \\ -1 & 0 & 1 \end{bmatrix}$$

obtaining two images I_1 and I_2.

Notice that the element special to these two detectors is the edge-enhancing filter (see Review Questions). Figure 4.7 shows an example of Sobel edge detection.

4.2.4 Concluding Remarks on Edge Detection

Evaluating Edge Detectors. The ultimate evaluation for edge detectors which are part of larger vision systems is whether or not a particular detector improves the performance of the global system, other conditions being equal. For instance, within an

Figure 4.7 Left: output of Sobel edge enhancer run on Figure 4.1. Middle: edges detected by thresholding the enhanced image at 35. Right: same, thresholding at 50. Notice that some contours are thicker than one pixel (compare with Figure 4.5).

inspection system, the detector leading to the best accuracy in the target measurements, and acceptably fast, is to be preferred.

However, it is useful to evaluate edge detectors *per se* as well. We run edge detectors in the hope of finding the contours of interesting scene elements; therefore, one could think that an edge detector is good "if it finds object contours", and in a sense this is true. But it is also imprecise and unfair, because edge detectors do not know about "objects contours" at all; they look for intensity variations, and we must evaluate algorithms for what they actually know and do. Specific edge detectors can be evaluated

- *theoretically* (e.g., for edge enhancement filters, using Canny's criteria in section 4.2.2);

- *experimentally*, estimating various *performance indices* in a large number of experiments with synthetic images (Appendix, section A.1)), in which all edge and noise parameters are perfectly known and vary in realistic ranges. Performance indices include

  - the number of spurious edges (which is an estimate *a posteriori* of the probability of false detection),
  - the number of true edges missed (which is an estimate *a posteriori* of the probability of misses),
  - the average and RMS errors of estimates of edge position and orientation.

Subpixel-Precision Edge Detection.   All the edge detectors in this chapter identify the pixel *which contains* the center of the true edge center. In fact, the center could be anywhere within the pixel, so that the average accuracy is 0.5 pixel. In precision applications, half a pixel may correspond to unacceptably large errors in millimiters, and it is important to locate the edge position with *subpixel precision*. For instance, in commercial laser scanners (Chapter 2) an accuracy of about 0.25mm is a common target, and half a pixel may correspond to less than 0.5mm. The easiest way to achieve subpixel resolution is to locate the peak of a parabola interpolating three values in the output of CANNY_ENHANCER, namely at the edge pixel and at its two neighbours along the edge normal. Of course, any information available on edge profiles and noise should be exploited to improve on the parabola method.

4.3   Point Features: Corners

Although the mathematics of edge detection may seem involved, edges can be characterized intuitively in geometric terms: they are the projection of object boundaries, surface marks, and other interesting elements of a scene. We now give an example of image features that can be characterized more easily than edges in mathematical terms, but do not correspond necessarily to any geometric entities of the observed scene. These features can be interpreted as *corners*, but not only in the sense of intersections of image lines; they capture corner structures in patterns of intensities. Such features prove stable across sequences of images, and are therefore interesting to track objects across sequences (Chapter 8).

How do we detect corner features? Consider the spatial image gradient, $[E_x, E_y]^\top$ (the subscripts indicate partial differentiation, e.g., $E_x = \frac{\partial E}{\partial x}$). Consider a generic image point, p, a neighbourhood Q of p, and a matrix, C, defined as

$$C = \begin{bmatrix} \sum E_x^2 & \sum E_x E_y \\ \sum E_x E_y & \sum E_y^2 \end{bmatrix}, \tag{4.9}$$

where the sums are taken over the neighbourhood Q. This matrix characterizes the *structure* of the grey levels. How?

The key to the answer is in the eigenvalues of C and their geometric interpretation. Notice that C is symmetric, and can therefore be diagonalized by a rotation of the coordinate axes; thus, with no loss of generality, we can think of C as a diagonal matrix:

$$C = \begin{bmatrix} \lambda_1 & 0 \\ 0 & \lambda_2 \end{bmatrix}.$$

The two eigenvalues, λ_1 and λ_2, are both nonnegative (why?); let us assume $\lambda_1 \geq \lambda_2$. The geometric interpretation of λ_1 and λ_2 can be understood through a few particular cases. First, consider a perfectly uniform Q: the image gradient vanishes everywhere, C becomes the null matrix, and we have $\lambda_1 = \lambda_2 = 0$. Second, assume that Q contains an ideal black and white step edge: we have $\lambda_2 = 0, \lambda_1 > 0$, and the eigenvector associated with λ_1 is parallel to the image gradient. Note that C is rank deficient in both cases, with rank 0 and 1 respectively. Third, assume that Q contains the corner of a black square against a white background: as there are two principal directions in Q, we expect $\lambda_1 \geq \lambda_2 > 0$, and the larger the eigenvalues, the stronger (higher contrast) their corresponding image lines. At this point, you have caught on with the fact that *the eigenvectors encode edge directions, the eigenvalues edge strength*. A corner is identified

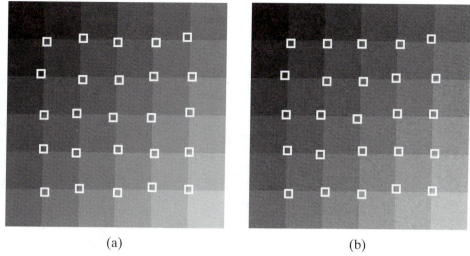

(a)   (b)

Figure 4.8 Corners found in a 8-bit, synthetic checkerboard image, corrupted by two realizations of synthetic Gaussian noise of standard deviation 2. The corner is the bottom right point of each 15×15 neighbourhood (highlighted).

by two strong edges; therefore, as $\lambda_1 \geq \lambda_2$, *a corner is a location where the smaller eigenvalue, λ_2, is large enough.*

Time for examples. Figure 4.8 shows the corners found in a synthetic image of a checkerboard, with and without additive noise. Figure 4.9 shows the corners found in the image of a building, and the histogram of the λ_2 values. The shape of this histogram is rather typical for most natural images. If the image contains uniform regions, or many almost ideal step edges, the histogram has a second peak at $\lambda_2 = 0$. The tail (right) of the histogram is formed by the points for which λ_2 is large, which are precisely the points (or, equivalently, the neighbourhoods) we are interested in. Figure 4.10 shows another example with a road scene.

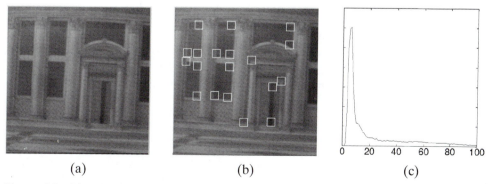

(a)   (b)   (c)

Figure 4.9 (a): original image of a building. (b): the 15×15 pixel neighbourhoods of some of the image points for which $\lambda_2 > 20$. (c): histogram of λ_2 values across the image.

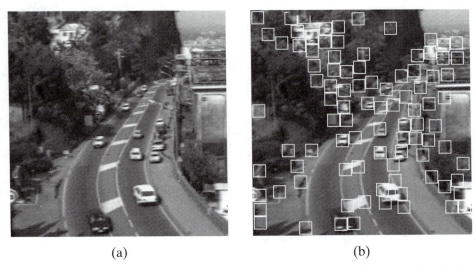

(a)   (b)

Figure 4.10 (a): image of an outdoor scene. The corner is the bottom right point of each 15×15 neighbourhood (highlighted). (b): corners found using a 15×15 neighbourhood.

We reiterate that our feature points include high-contrast image corners and T-junctions generated by the intersection of object contours (as the corners in Figure 4.8), but also corners of the local intensity pattern not corresponding to obvious scene features (as some of the corners in Figure 4.10). In general terms, at corners points *the intensity surface has two well-pronounced, distinctive directions, associated to eigenvalues of C both significantly larger than zero.*

We now summarize the procedure for locating this new type of image features.

Algorithm CORNERS

The input is formed by an image, I, and two parameters: the threshold on λ_2, τ, and the linear size of a square window (neighbourhood), say $2N + 1$ pixels.

1. Compute the image gradient over the entire image I;

2. For each image point p:

    (a) form the matrix C of (4.9) over a $(2N + 1) \times (2N + 1)$ neighbourhood Q of p;
    (b) compute λ_2, the smaller eigenvalue of C;
    (c) if $\lambda_2 > \tau$, save the coordinates of p into a list, L.

3. Sort L in decreasing order of λ_2.

4. Scanning the sorted list top to bottom: for each current point, p, delete all points appearing further on in the list which belong to the neighbourhood of p.

The output is a list of feature points for which $\lambda_2 > \tau$ and whose neighbourhoods do not overlap.

Algorithm CORNERS has two main parameters: the threshold, τ, and the size of the neighbourhood, $(2N + 1)$. The threshold, τ, can be estimated from the histogram of λ_2 (Exercise 4.6), as the latter has often an obvious valley near zero (Figure 4.9).

☞   Notice that such valley is not *always* present (Exercise 4.7).

Unfortunately, there is no simple criterion for the estimation of the optimal size of the neighbourhood. Experience indicates that choices of N between 2 and 10 are adequate in most practical cases.

☞   In the case of corner points, the value of N is linked to the location of the corner within the neighbourhood. As you can see from Figure 4.9, for relatively large values of N the corner tends to move away from the neighbourhood center (see Exercise 4.8 for a quantitative analysis of this effect).

4.4 Surface Extraction from Range Images

Many 3-D objects, especially man-made, can be conveniently described in terms of the shape and position of the surfaces they are made of. For instance, you can describe a cone as an object formed by two surface patches, one conical and one planar, the latter perpendicular to the axis of the former. Surface-based descriptions are used for object classification, pose estimation, and reverse engineering, and are ubiquitous in computer graphics.

As we have seen in Chapter 2, range images are basically a sampled version of the visible surfaces in the scene. Therefore, ignoring the distortions introduced by sensor imperfections, *the shape of the image surface[9] and the shape of the visible scene surfaces are the same*, and any geometric property holding for one holds for the other too. This section presents a well-known method to find patches of various shapes composing the visible surface of an object. The method, called *HK segmentation*, partitions a range image into regions of homogeneous shape, called *homogeneous surface patches*, or just *surface patches* for short.[10] The method is based on differential geometry; Appendix, section A.5 gives a short summary of the basic concepts necessary.

☞   The solution to several computer vision problems involving 3-D object models are simpler when using 3-D features than 2-D features, as image formation must be taken into account for the latter.

---

[9] That is, the image values regarded as a surface defined on the image plane.

[10] Notice that surface patches are the basic ingredients for building a surface-based CAD model of an object automatically.

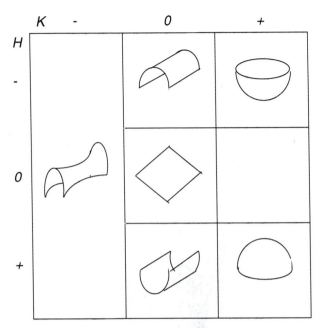

Figure 4.11   Illustration of the local shapes resulting from the HK classification.

Problem Statement: HK Segmentation of Range Images

Given a range image I in r_{ij} form, compute a new image registered with I and the same size, in which each pixel is associated with a local shape class selected from a given dictionary.

To solve this problem, we need two tools: a dictionary of shape classes, and an algorithm determining which shape class approximates best the surface at each pixel.

4.4.1 Defining Shape Classes

Since we want to estimate surface shape at each point (pixel), we need a *local* definition of shape. Differential geometry provides a convenient one: using the sign of the *mean curvature H* and of the *Gaussian curvature K*, we can classify the local surface shape as shown in Table 4.1, and illustrated by Figure 4.11.

In the table, *concave* and *convex* are defined with respect to the viewing direction: a hole in the range surface is concave, and its principal curvatures (Appendix, section A.5) negative. At cylindrical points, one of the two principal curvatures vanishes, as for instance at any point of a simple cylinder or cone (not the vertex). At elliptic points, both principal curvatures have the same sign, and the surface looks locally like either

K	H	Local shape class
0	0	plane
0	+	concave cylindrical
0	−	convex cylindrical
+	+	concave elliptic
+	−	convex elliptic
−	any	hyperbolic

Table 4.1 Surface patches classification scheme.

the inside of a bowl (if concave) or the tip of a nose (if convex). At hyperbolic points, the principal curvatures are nonzero and have different signs; the surface looks like a saddle.

Notice that this classification is *qualitative*, in the sense that only the sign of the curvatures, not their magnitude,[11] influences the result. This offers some robustness, as sign can often be estimated correctly even when magnitude estimates become noisy.

4.4.2 Estimating Local Shape

Given Table 4.1, all we have to do is to recall the appropriate expressions of H and K, evaluate them at each image point, and use the signs of H and K to index Table 4.1. Here is how to compute H, K from a range image, h, in r_{ij} form (subscripts indicate again partial differentiation).

$$K = \frac{h_{xx}h_{yy} - h_{xy}^2}{(1 + h_x^2 + h_y^2)^2} \tag{4.10}$$

$$2H = \frac{(1 + h_x^2)h_{yy} - 2h_x h_y h_{xy} + (1 + h_y^2)h_{xx}}{(1 + h_x^2 + h_y^2)^{3/2}} \tag{4.11}$$

Unfortunately, we cannot expect good results without sorting out a few details. First, the input image contains noise, and this distorts the numerical estimates of derivatives and curvatures; noise smoothing is therefore required. Notice that the worst noise may be due to quantisation (if the 8-bit image does not capture all significant depth variations of the scene), or to limited sensor accuracy.

---

[11] With the exception of zero, of course.

☞ The low acquisition noise of state-of-the-art laser scanners should not jeopardise seriously the quality of the HK segmentation. Segmentation can fail, however, because image quantisation and resolution are not sufficient given the objects and the stand-off distance.

☞ Gaussian smoothing, as any averaging filter, tends to underestimate high curvatures, and to introduce spurious curvatures around contours (Exercise 4.9).

Second, the result may still contain small, noisy patches, even when smoothing is applied to the data. Small patches can be eliminated by additional filtering (Exercise 4.10). Third, planar patches should yield $H = K = 0$, but numerical estimates of H and K will never be exactly zero. To decide which small numbers can be safely considered zero, we can establish *zero-thresholds* for H and K. In this case, the accuracy with which planar patches are extracted depends (among others) on the noise level, the orientation of the plane and the values of the thresholds. Fourth, estimating derivatives and curvatures does not make sense at surface discontinuities. To skip discontinuities, one could therefore run an edge detector on R (e.g., CANNY_EDGE_DETECTOR for step edges), keep a map of edge points, and skip them when estimating H and K.

To summarize, here is a basic HK segmentation algorithm.

Algorithm RANGE_SURF_PATCHES

The input is a range image, I, in r_{ij} form, and a set of six shape labels, $\{s_1, \ldots s_6\}$, associated to the classes of Table 4.1.

1. Apply Gaussian smoothing to I, obtaining I_s.
2. Compute the images of the derivatives, $I_x, I_y, I_{xy}, I_{xx}, I_{yy}$ (Appendix, section A.2).
3. Compute the H, K images using (4.11) and (4.10).
4. Compute the shape image, S, by assigning a shape label s_i to each pixel, according to the rules in Table 4.1.

The output is the shape image S.

Figure 4.12 and 4.13 show two examples of HK segmentation. The range data were acquired and processed using the range image acquisition and processing systems developed by the Computer Vision Group of Heriot-Watt University. The objects are mechanical components formed by planar and curved surfaces. The figures show the input range data, the data after smoothing (both as grey-level image and as 3-D plot), and the patches detected.

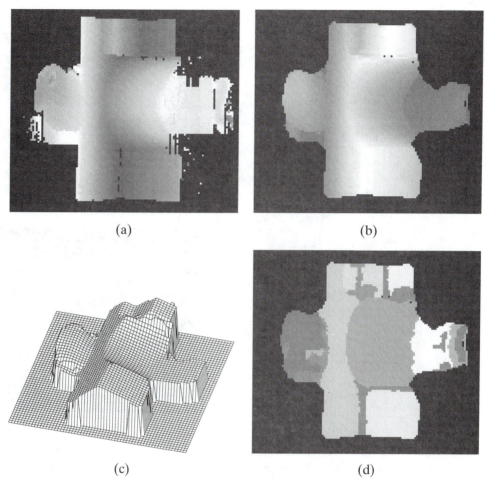

(a)   (b)

(c)   (d)

Figure 4.12   (a): Input range image, grey coded (the darker the closer to the sensor). (b): After smoothing, grey coded. (c): Same as top right, as 3-D isometric plot. (d): The patches detected by HK segmentation. Courtesy of M. Umasuthan, Heriot-Watt University.

In order to be used by subsequent tasks like classification or pose estimation, the output of RANGE_SURF_PATCHES is often converted into a list of *symbolic patch descriptors*. In each descriptor, a surface patch is associated with a number of attributes, which may include a unique identifier, position of patch center, patch area, information on normals and curvatures, contour representations, and pointers to neighbouring patches. Closed-form surface models (e.g., quadrics) are often fitted to the surface patches extracted by the HK segmentation, and only the model's coefficients and type (e.g., plane, cylinder, cone) stored in the symbolic descriptors.

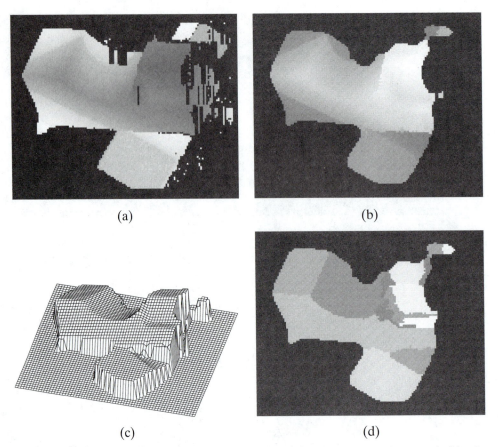

(a)   (b)

(c)   (d)

Figure 4.13   (a): input range image, grey coded (the darker the closer to the sensor). (b): after smoothing, grey coded. (c): same as top right, as 3-D isometric plot. (d): the patches detected by HK segmentation. Courtesy of M. Umasuthan, Heriot-Watt University.

4.5 Summary

After working through this chapter you should be able to:

❏ explain what image features are and how they relate to the 3-D world;

❏ design detectors for edges and point features, and performance tests for the related algorithms;

❏ design a simple HK segmentation program for range images.

4.6 Further Readings

Several books review the theory and algorithms of large collections of edge detectors, for instance [8, 9, 14]. John Canny's description of his edge detector is found in [4].

Spacek [15] derives a slightly different, optimal edge detector. De Micheli *et al.* [6] and Pratt [14] give examples of discussions on performance evaluation of edge detectors. A very good electronic textbook on image processing, including material on feature detection, is HIPR (Hypermedia Image Processing Reference), published by Wiley (on-line information at `http://www.wiley.co.uk/electronic/hipr`).

Second-order derivative filters for edge detection have been very popular in the eighties; a classic reference is Marr's book [13]. These filters look for the zero-crossings of the second derivative of a Gaussian-filtered image. Their disadvantages in comparison with Canny's detector include worse directional properties (being isotropic, their output contains contributions from the direction perpendicular to the edge normal; this does increase noise without contribute to detection); moreover, they always produce closed contours, which do not always correspond to interesting edges. For a theoretical analysis of the main properties of first-order and second order derivative filters see Torre and Poggio [18].

The point feature (corner) detector CORNERS is based on Tomasi and Kanade's one [16]; an application to motion-based reconstruction is described in [17], and discussed in Chapter 8. Further corner detectors are reported in [10, 3, 9].

Besl and Jain [1], Hoffman and Jain [11], and Fan [7] are some variations of the HK segmentation method. Hoover *et al.* [12] and Trucco and Fisher [19] report useful experimental assessments of H, K segmentation algorithms from range images. For examples of reverse engineering from range images, see [5, 2].

4.7 Review

Questions

❑ *4.1* Describe the difference between local and global features, and give examples of image features from both classes.

❑ *4.2* Given our definition and classification of edges, discuss the differences between edges in intensity and range images. Do our edge types make sense for range images?

❑ *4.3* Would you apply intensity edge detection algorithms to range images? Would the algorithms require modifications? Why?

❑ *4.4* Consider the equation giving the orientation of the edge normal in algorithm CANNY_ENHANCER. Why can you ignore the aspect ratio of the pixel here, but not in the situation proposed by Question 2.8?

❑ *4.5* In section 4.2.2, we used the image gradient to estimate the edge normal. Discuss the practical approximations implied by this choice.

❑ *4.6* Explain why Y-junctions are split by HYSTHERESIS_THRESH.

❑ *4.7* Discuss the differences between Sobel's edge enhancement scheme preceded by a Gaussian smoothing pass, and Canny's edge enhancement scheme. What are the main *design* differences? Do you expect different results? Why?

❑ *4.8* You can suppress short edge chains in the output of CANNY_EDGE_ DETECTOR by filtering the input image with wider Gaussians. How would you achieve the same result with a Sobel edge detector?

❑ *4.9* How would you design an experimental comparison of Canny's and Sobel's edge detectors?

❑ *4.10* Why is the case $K > 0$, $H = 0$ not featured in Table 4.1?

❑ *4.11* Explain why HK segmentation cannot be applied to intensity images in the hope to find homogeneous scene surfaces.

Exercises

○ *4.1* Consider the SNR measure formalizing the good detection criterion (4.3). Show that, if the filter has any symmetric components, the SNR measure will worsen (decrease). This shows that the best detection is achieved by purely anti-symmetric filters.

○ *4.2* Prove the detection-localization uncertainty equations (4.6) for step edge detection. (*Hint:* Substitute $x_1 = A/x_0$ in the localization and detection criteria.)

○ *4.3* Modify ROBERTS_EDGE_DET and SOBEL_EDGE_DET to produce full edge descriptors.

○ *4.4* Use the result of the previous exercise to explain why the Canny's enhancement is better than Sobel's.

○ *4.5* Suppose you are told that an edge detector has an accuracy of 0.3 pixel. Convert this figure to millimeters (scene measurements), listing the assumptions and information required to do so.

○ *4.6* Write an algorithm which determines a suitable threshold for which the valley near zero values is present, and one for which it is not, in a unimodal histogram shaped as the one in Figure 4.9 (c).

○ *4.7* What does the shape of the histogram used in CORNERS depend on? How does it vary with the image? Give an example of image in which the valley near zero values is present, and one in which it is not.

○ *4.8* Construct a synthetic image of a white square against a black background. Verify that the four vertices lie within the neighborhood of the image points with largest λ_2 returned by the procedure CORNERS. Study the location of the vertices within the neighborhood as a function of (*a*) the size of the neighborhood, and (*b*) the σ of the spatial filter employed in the computation of the spatial image gradient.

○ *4.9* Consider the symmetric, 1-D profile $f(0) = 0$, $f(1) = 0$, $f(2) = 3$, $f(3) = 5$, $f(4) = 6$ and $f(4 + i) = f(4 - i)$, $i = 1 \ldots 4$. Work out the result of filtering with the averaging mask [1 1 1]. Does the (highest) curvature at the midpoint change? How? What happens at the borders of the profile? What happens at the borders of a rectangular profile?

○ *4.10* Section 4.4.2 stated that small, noisy regions in the H and K images can be eliminated by filtering. One method is *image morphology*. You shrink each image region by *d* pixels, then expand it again by *d* pixels. We expect that, for appropriate values of *d*, small regions disappear completely, large regions grow back to their original shape. Write an algorithm performing shrinking and expansion which does not find connected regions to begin with (this problem is a classic of image processing: consult [8, 9, 14] but only after trying!).

○ *4.11* Write an algorithm producing full image descriptors from the output image, *S*, of RANGE_SURF_PATCHES. This involves finding all connected regions of *S*, a region being formed by pixel with the same label s_i (another classic of image processing: consult [8, 9, 14] but only after trying!).

Projects

● *4.1* Implement CANNY_EDGE_DETECTOR, and test its performance as suggested in section 4.2.4 (see also Appendix, section A.1). Time permitting, implement Roberts' and Sobel's detectors, apply the same performance evaluation tests, and compare the results with those of the Canny's detector.

● *4.2* Implement algorithm CORNERS and an interface displaying corners superimposed on the original images. For each fixed pair of values of the algorithm parameters, measure the robustness of your implementation with synthetic images of squares corrupted by increasing amounts of Gaussian noise, as follows. Record the RMS distance of the estimated corners from the true positions, and the number of spurious and missed corners. Plot these values in three graphs, against the standard deviation of the noise. Do the graphs agree with your expectations?

● *4.3* Implement algorithm RANGE_SURF_PATCHES and test its performance as follows. Build a synthetic set of test range images, containing a few simple surfaces (e.g., spheres, cylinders, planes, polyhedra). Observe *H*, *K* errors and size of patches classified correctly, varying patch sizes, orientations, zero thresholds, and smoothing parameters in reasonable ranges. Which parameters are the most critical (i.e., small changes produce dramatic changes in output measures)? Does this vary with surface type?

References

[1]   P.J. Besl and R.C. Jain, Segmentation through Variable-Order Surface Fitting, *IEEE Transactions on Pattern Analysis and Machine Intelligence*, Vol. PAMI-10, pp. 167-192 (1988).

[2]   P. Besl and N. McKay, A Method for Registration of 3-D Shapes, *IEEE Transactions on Pattern Analysis and Machine Intelligence*, Vol. PAMI-14, no. 2, pp. 239-256 (1992).

[3]   M.J. Brady and H. Wang, Vision for Mobile Robots, *Philosophical Transactions of the Royal Society of London B*, Vol 337, pp. 341–350 (1992).

[4]   J. Canny, A Computational Approach to Edge Detection, *IEEE Transactions on Pattern Analysis and Machine Intelligence*, Vol. PAMI-8, pp. 679-698 (1986).

[5]   Y. Chen and G. Medioni, Object Modeling by Registration of Multiple Range Images, *Image and Vision Computing*, Vol. 10, no. 3, pp. 145-155 (1992).

[6]   E. De Micheli, B. Caprile, P. Ottonello and V. Torre, Localization and Noise in Edge Detection, *IEEE Transactions on Pattern Analysis and Machine Intelligence*, Vol. PAMI-11, no. 10, pp. 1106-1117 (1989).

[7]   T.-J. Fan, *Describing and Recognising 3-D Objects Using Surface Properties*, Springer-Verlag (1990).

[8]   R.C. Gonzalez and R.E. Woods, *Digital Image Processing*, Addison-Wesley, Reading (MA) (1992).

[9]   R.M. Haralick and L.G. Shapiro, *Computer and Robot Vision*, Volume I, Addison-Wesley (1992).

[10] C. Harris, Geometry from Visual Motion. In *Active Vision*, A. Blake and A. Yuille eds., MIT Press, Cambridge (MA), pp. 263–284 (1992).

[11] R. Hoffman and A.K. Jain, Segmentation and Classification of Range Images, *IEEE Transactions on Pattern Analysis and Machine Intelligence*, Vol. PAMI-9, no. 5, pp. 608-620 (1987).

[12] A. Hoover, G. Jean-Baptiste, X. Xiang, P.J. Flynn, H. Bunke, D.B. Goldgof, K. Bowyer, D.W. Eggert, A.W. Fitzgibbon and R.B. Fisher, An Experimental Comparison of Range Image Segmentation Algorithm, *IEEE Transactions on Pattern Analysis and Machine Intelligence*, Vol. PAMI-18, no. 7, pp. 673-689 (1996).

[13] D Marr, *Vision*, Freeman, San Francisco (1982).

[14] W.K. Pratt, *Digital Image Processing*, Wiley, New York (1991).

[15] L.A. Spacek: Edge Detection and Motion Detection, *Image and Vision Computing*, Vol. 4, pp. 43-51 (1986).

[16] C. Tomasi and T. Kanade, Shape and Motion from Image Streams: a Factorization Method – 3. Detection and Tracking of Point Features, Technical Report CMU-CS-91-132, Carnegie Mellon University, Pittsburgh (PA) (1991).

[17] C. Tomasi and T. Kanade, Shape and Motion from Image Streams under Orthography: a Factorization Method, *International Journal of Computer Vision* Vol 9, pp. 137–154 (1992).

[18] Torre, V. and Poggio, T. On edge detection, *IEEE Transactions on Pattern Analysis and Machine Intelligence*, Vol. PAMI-8, no. 2, pp. 147–163 (1986).

[19] E. Trucco and R.B. Fisher, Experiments in Curvature-Based Segmentation of Range Data, *IEEE Transactions on Pattern Analysis and Machine Intelligence*, Vol. PAMI-17, no.2, pp. 57-62 (1995).

5

More Image Features

There was an old man of West Dumpet
Who possessed a large nose like a trumpet

Edward Lear, *The Book of Nonsense*

This chapter develops further the discussion on image features, introducing more features and related detection algorithms.

Chapter Overview

Section 5.1 introduces *grouping* and *model fitting*, and their relation.

Section 5.2 presents the *Hough transform*, a class of algorithms for locating lines and curves in images.

Section 5.3 describes three algorithms for *fitting ellipses* to noisy image data.

Section 5.4 introduces *deformable contours*, a special class of algorithms for curve detection and description.

Section 5.5 tackles the problem of forming groups of line segments likely to belong to the same object.

What You Need to Know to Understand this Chapter

- Working knowledge of the previous chapters.
- Least-squares parameter fitting (Appendix, sections A.6 and A.7).

5.1 Introduction: Line and Curve Detection

Lines and curves are important features in computer vision because they define the contours of objects in the image. This chapter presents methods to detect lines, ellipses, and general closed contours. Here is a statement of our task.

Problem Statement: Line and Curve Detection

Given the output of an edge detector run on an image, I, find all the instances of a given curve (e.g., a line or an ellipse) or parts thereof (e.g., line segments or arcs of ellipse) in I.

Strictly speaking, the detection of image lines can be performed directly; that is, by *template matching*: One can look for the peaks of the convolution between the image and a set of masks matched to long, linear edges in all possible orientations. There are at least two disadvantages: First, accurate line location requires a large set of masks; second, we again run into all the problems connected to filter design that we discussed in Chapter 4. Trying to apply template matching to curves worsen the problem. Therefore, we follow an alternative plan: We start with CANNY_EDGE_DETECTOR, and we feed the resulting edge image to the line or ellipse detectors. We shall depart from this plan only when looking for general-shape, closed image contours (section 5.4), which will be fit to intensity images directly.

Under this assumption, line and curve detection splits logically into two subproblems:

Grouping: Which image points compose each instance of the target curve in the image?

Model fitting: Given a set of image points probably belonging to a single instance of the target curve, find the best curve interpolating the points.

In the previous chapter, we have seen that edge detection procedures like HYSTERESIS_THRESH, for example, solve the grouping problem by providing an ordered list of edge points (or *chain*). Let us now understand the problem of *model fitting*. Assume we know that certain image points lie, say, on a line. Because of pixelization and errors introduced by image acquisition and edge detection, there is no line going *exactly* through all the points; we must look for the best compromise line we can find. So we write the equation of a generic line (the *mathematical model*), $\mathbf{a}^\top \mathbf{x} = ax + by + c = 0$, and look for the parameter vector, $\mathbf{a}_o$, which results in a line going as near as possible to each image point. The vector $\mathbf{a}_o$ is computed by defining a distance function, D, between the line and the set of image points, and looking for the minimum of D over all possible $\mathbf{a}$. Very often, D is a squared distance, and finding $\mathbf{a}_o$ implies solving a least squares problem.

This chapter introduces algorithms for curve fitting and grouping, as well as algorithms that perform both grouping and model fitting at the same time, but are suitable only for lines and simple curves.

5.2   The Hough Transform

The Hough transform (henceforth called HT) was introduced to detect complex patterns of points in binary images and became quickly a very popular algorithm to detect lines and simple curves. The key idea is to *map a difficult pattern detection problem* (finding instances of a given curve) *into a simple peak detection problem* in the space of the parameters of the curve. We start with an image of contour points, such as the output of an edge detector. The contour points needs not to be grouped in edge chains.

5.2.1   The Hough Transform for Lines

We begin by introducing the two basic steps of HT work using the case of lines:

1. *Transform line detection into a line intersection problem.* Any line, $y = mx + n$, is identified by a unique parameter pair, (m, n). Therefore, the line is represented by a point in the m, n plane (the *parameter space*). Conversely, any point $\mathbf{p} = [x, y]^{\top}$ in the image corresponds to a line $n = x(-m) + y$ in parameter space, which, as m and n vary, represents all possible image lines through $\mathbf{p}$. Therefore, a line defined by N collinear image points, $\mathbf{p}_1 \ldots \mathbf{p}_N$, is identified in parameter space by the intersection of the lines associated with $\mathbf{p}_1 \ldots \mathbf{p}_N$, as illustrated in Figure 5.1 for $N = 2$.

2. *Transform line intersection in a simple peak detection problem*, or search for a maximum. Imagine to divide the m, n plane into a finite grid of cells, the resolution of which depends on the accuracy we need, and to associate a counter, $c(m, n)$, initially set to zero, to each cell. Assume for simplicity that the image contains only one line, (m', n'), formed by points $\mathbf{p}_1 \ldots \mathbf{p}_N$. For each image point, $\mathbf{p}_i$, increment

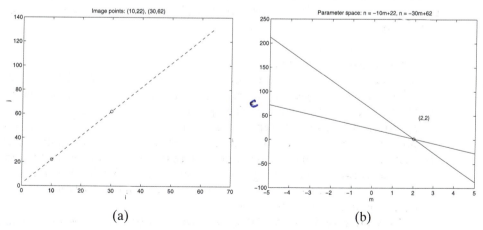

(a)   (b)

Figure 5.1   Illustration of the basic idea of the Hough transform for lines. The two image points (a) are mapped onto two lines in parameter space (b). The coordinates of the intersection of these lines are the parameters (m, n) of the image line through the two points.

all counters on the corresponding line in parameter space. All the parameter-space lines, $l_1 \ldots l_N$, associated to $\mathbf{p}_1 \ldots \mathbf{p}_N$, go through (m', n'), so that $c(m', n') = N$. Any other counter on $l_1 \ldots l_N$ is 1. Therefore, the image line is identified simply by the peak of $c(m, n)$ in parameter space.

To make things work, several (if straightforward) extensions are necessary.

Keeping the Parameter Space Finite. Both m and n can take on values in $[-\infty, \infty]$, which implies that we cannot sample the whole parameter space. To minimize the risk of missing interesting parameter ranges, we can sample wide intervals for both m and n, but at the price of reducing resolution, as there is a limit to the size of the discrete parameter space that we can search in acceptable time. Moreover, the (m, n) parametrization does not capture the bundle $x = k$, with k a constant. The *polar representation* $\rho = x \cos \theta + y \sin \theta$ where ρ represents the distance between the image origin and the line, and θ the line orientation, solves both problems: The intervals of variation of ρ and θ are finite, and any line can be represented.

☞   Notice that an image point is now represented by a *sinusoid*, not a line, in parameter space.

Simultaneous Detection of Multiple Lines. Edge images may contain many lines. To find them all, you simply look for all local maxima of $c(m, n)$.

Effect of Nonlinear Contours. Edge images usually contain points not belonging to any line, for instance curved contours, or just noise introduced by the edge detector. These points result in a spread of low, random counter values throughout the parameter space. Consequently, a multitude of local, noisy peaks appears, and we must divide the peaks created by noise from those identifying lines. The simplest way to achieve this is to threshold $c(m, n)$.

Noise. Because of pixelization and the limited accuracy of edge detection, not all the parameter-space lines corresponding to the points of an image line intersect at the same point. Consequently, the image points do not contribute to one counter only, but to *several* counters within a small neighborhood of the correct one, so that the peak identifying the line is spread over that neighborhood. This phenomenon is reinforced by the presence of noisy points. Depending on the resolution of the parameter space and the accuracy required, one can estimate the true parameters just as the local maximum (m', n'), or as a weighted average of the values (m, n) in a neighborhood of (m', n') such that $c(m, n) > \tau c(m', n')$, where τ is a fixed fraction (e.g., 0.9) and the weights are proportional to the counters' values.

As an example, Figure 5.2 (a) shows a synthetic 64×64 image of two lines. Only a subset of the lines' points are present, and spurious points appear at random locations. Figure 5.2 (b) shows the counters in the associated (m, n) parameter space.

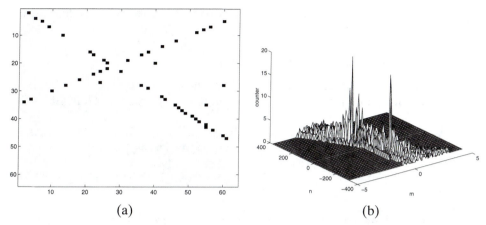

Figure 5.2 (a) An image containing two lines, sampled irregularly, and several random points. (b) Plot of the counters in the corresponding parameter space (how many points contribute to each cell (m, n)). Notice that the main peaks are obvious, but there are many secondary peaks.

We are now ready for the following algorithm:

Algorithm HOUGH_LINES

The input is E, a $M \times N$ binary image in which each pixel $E(i, j)$ is 1 if an edge pixel, 0 otherwise. Let ρ_d, θ_d be the arrays containing the discretized intervals of the ρ, θ parameter spaces ($\rho \in [0, \sqrt{M^2 + N^2}], \theta \in [0, \pi]$), and R, T, respectively, their number of elements.

1. Discretize the parameter spaces of ρ and θ using sampling steps $\delta\rho$, $\delta\theta$, which must yield acceptable resolution and manageable size for ρ_d and θ_d.

2. Let $A(R, T)$ be an array of integer counters (accumulators); initialize all elements of A to zero.

3. For each pixel, $E(i, j)$, such that $E(i, j) = 1$ and for $h = 1 \dots T$:

    (a) let $\rho = i \cos \theta_d(h) + j \sin \theta_d(h)$;
    (b) find the index, k, of the element of ρ_d closest to ρ;
    (c) increment $A(k, h)$ by one.

4. Find all local maxima (k_p, h_p) such that $A(k_p, h_p) > \tau$, where τ is a user-defined threshold.

The output is a set of pairs $(\rho_d(k_p), \theta_d(h_p))$, describing the lines detected in E in polar form.

☞   If an estimate $m_g(\mathbf{p})$ of the edge direction at image point $\mathbf{p}$ is available, and we assume that $m_g(\mathbf{p})$ is also the direction of the line through $\mathbf{p}$, a *unique* cell ($m_g, n_g = y - m_g x$) can be identified. In this case, instead of the whole line, we increment only the counter at (m_g, n_g); to allow for the uncertainty associated with edge direction estimates, we increment all the cells on a small segment centered in (m_g, n_g), the length of which depends inversely on the reliability of the direction estimates. This can speed up considerably the construction of the parameter space.

5.2.2 The Hough Transform for Curves

The HT is easily generalized to detect curves $y = f(x, \mathbf{a})$, where $\mathbf{a} = [a_1, \ldots, a_P]^\top$ is a vector of P parameters. The basic algorithm is very similar to HOUGH_LINES.

Algorithm HOUGH_CURVES

The input is as in HOUGH_LINES. Let $y = f(x, \mathbf{a})$ be the chosen parametrization of a target curve.

1. Discretize the intervals of variation of $a_1, \ldots, a_P$ with sampling steps yielding acceptable resolution and manageable size for the parameter space. Let $s_1, \ldots, s_p$ be the sizes of the discretized intervals.

2. Let $A(s_1, s_2 \ldots s_P)$ be an array of integer counters (accumulators), and initialize all its elements to zero.

3. For each pixel $E(i, j)$ such that $E(i, j) = 1$, increment all counters on the curve defined by $y = f(x, \mathbf{a})$ in A.

4. Find all local maxima $\mathbf{a}_m$ such that $A(\mathbf{a}_m) > \tau$, where τ is a user-defined threshold.

The output is a set of vectors, $\mathbf{a}_1 \ldots \mathbf{a}_P$, describing the curve instances detected in E.

☞   The size of the parameter space increases exponentially with the number of model parameters, and the time needed to find all maxima becomes rapidly unacceptable. This is a serious limitation. In particular, assuming for simplicity that the discretized intervals of all parameters have the same size N, the cost of an exhaustive search for a curve with p parameters is proportional to N^p. This problem can be tackled by variable-resolution parameter spaces (see Question 5.6).

5.2.3 Concluding Remarks on Hough Transforms

The HT algorithm is a *voting algorithm*: Each point "votes" for all combinations of parameters which may have produced it if it were part of the target curve. From this point of view, the array of counters in parameter space can be regarded as a *histogram*. The final total of votes, $c(\mathbf{m})$, in a counter of coordinates $\mathbf{m}$ indicates the relative likelihood of the hypothesis "a curve with parameter set $\mathbf{m}$ exists in the image."

The HT can also be regarded as *pattern matching*: the class of curves identified by the parameter space is the class of patterns. Notice that the HT is more efficient than direct template matching (comparing all possible appearances of the pattern with the image).

The HT has several attractive features. First, as all points are processed independently, it copes well with occlusion (if the noise does not result in peaks as high as those created by the shortest true lines). Second, it is relatively robust to noise, as spurious points are unlikely to contribute consistently to any single bin, and just generate background noise. Third, it detects multiple instances of a model in a single pass.

The major limitation of the HT is probably the rapid increase of the search time with the number of parameters in the curve's representation. Another limitation is that non-target shapes can produce spurious peaks in parameter space: For instance, line detection can be disturbed by low-curvature circles.

5.3 Fitting Ellipses to Image Data

Many objects contain circular shapes, which almost always appear as *ellipses* in intensity images (but see Exercise 5.5); for this reason, ellipse detectors are useful tools for computer vision. The ellipse detectors we consider take an image of edge points in input, and find the best ellipse fitting the points. Therefore this section concentrates on *ellipse fitting*, and *assumes that we have identified a set of image points plausibly belonging to a single arc of ellipse.*

Problem Statement: Ellipse Fitting

Let $\mathbf{p}_1 \ldots \mathbf{p}_N$ be a set of N image points, $\mathbf{p}_i = [x_i, y_i]^\top$. Let $\mathbf{x} = \left[x^2, xy, y^2, x, y, 1 \right]^\top$, $\mathbf{p} = [x, y]^\top$, and

$$f(\mathbf{p}, \mathbf{a}) = \mathbf{x}^\top \mathbf{a} = ax^2 + bxy + cy^2 + dx + ey + f = 0$$

the implicit equation of the generic ellipse, characterized by the parameter vector $\mathbf{a} = [a, b, c, d, e, f]^\top$.

Find the parameter vector, $\mathbf{a}_o$, associated to the ellipse which fits $\mathbf{p}_1 \ldots \mathbf{p}_N$ best in the least squares sense, as the solution of

$$\min_{\mathbf{a}} \sum_{i=1}^{N} [D(\mathbf{p}_i, \mathbf{a})]^2 \tag{5.1}$$

where $D(\mathbf{p}_i, \mathbf{a})$ is a suitable distance.

☞   Notice that the equation we wrote for $f(\mathbf{p}, \mathbf{a})$ is really a *generic conic*. We shall have more to say about this point later.

What is a suitable distance? There are two main answers for ellipse fitting, the *Euclidean distance* and the *algebraic distance*.

5.3.1 Euclidean Distance Fit

The first idea is to try and minimize the Euclidean distance between the ellipse and the measured points. In this case, problem (5.1) becomes

$$\min_{\mathbf{a}} \sum_{i=1}^{N} \| \mathbf{p} - \mathbf{p}_i \|^2 \tag{5.2}$$

under the constraint that $\mathbf{p}$ belongs to the ellipse:

$$f(\mathbf{p}, \mathbf{a}) = 0.$$

Geometrically, the Euclidean distance seems the most appropriate. Unfortunately, it leads only to an approximate, numerical algorithm. How does this happen? Let us try Lagrange multipliers to solve problem (5.2). We define an objective function

$$L = \sum_{i=1}^{N} \|\mathbf{p} - \mathbf{p}_i\|^2 - 2\lambda f(\mathbf{p}, \mathbf{a}),$$

and set $\frac{\partial L}{\partial x} = \frac{\partial L}{\partial y} = 0$ which yield

$$\mathbf{p} - \mathbf{p}_i = \lambda \nabla f(\mathbf{p}, \mathbf{a}). \tag{5.3}$$

Since we do not know $\mathbf{p}$, we try to express it as a function of computable quantities. To do this, we introduce two approximations:

1. We consider a first-order approximation of the curve

$$0 = f(\mathbf{p}, \mathbf{a}) \approx f(\mathbf{p}_i, \mathbf{a}) + [\mathbf{p} - \mathbf{p}_i]^{\top} \nabla f(\mathbf{p}_i, \mathbf{a}). \tag{5.4}$$

2. We assume that the $\mathbf{p}_i$ are close enough to the curve, so that $\nabla f(\mathbf{p}) \approx \nabla f(\mathbf{p}_i)$.

Approximation 2 allows us to rewrite (5.3) as

$$\mathbf{p} - \mathbf{p}_i = \lambda \nabla f(\mathbf{p}_i, \mathbf{a}),$$

which, plugged into (5.4), gives

$$\lambda = \frac{-f(\mathbf{p}_i, \mathbf{a})}{\|\nabla f(\mathbf{p}_i, \mathbf{a})\|^2}.$$

Substituting in (5.3), we finally find

$$\|\mathbf{p} - \mathbf{p}_i\| = \frac{|f(\mathbf{p}_i, \mathbf{a})|}{\|\nabla f(\mathbf{p}_i, \mathbf{a})\|}.$$

This is the equation we were after: It allows us to replace, in problem (5.2), the unknown quantity $\|\mathbf{p} - \mathbf{p}_i\|$ with a function we can compute. The resulting algorithm is as follows:

Algorithm EUCL_ELLIPSE_FIT

The input is a set of N image points, $\mathbf{p}_1, \ldots, \mathbf{p}_N$. We assume the notation introduced in the problem statement box for ellipse fitting.

1. Set $\mathbf{a}$ to an initial value $\mathbf{a}_0$.

2. Using $\mathbf{a}_0$ as initial point, run a numerical minimization to find the solution of

$$\min_{\mathbf{a}} \sum_{i=1}^{N} \frac{f(\mathbf{p}_i, \mathbf{a})^2}{\|\nabla f(\mathbf{p}_i, \mathbf{a})\|^2}.$$

The output is the solution vector $\mathbf{a}_m$, defining the best-fit ellipse.

☞   A reasonable initial value is the solution of the closed-form algorithm discussed next (section 5.3.2).

How satisfactory is EUCL_ELLIPSE_FIT? Only partially. We started with the *true* (Euclidean) distance, the best possible, but were forced to introduce approximations, and arrived at a nonlinear minimization that can be solved only numerically. We are not even guaranteed that the best-fit solution is an ellipse: It could be *any* conic, as we imposed no constraints on $\mathbf{a}$. Moreover, we have all the usual problems of numerical optimization, including how to find a good initial estimate for $\mathbf{a}$ and how to avoid getting stuck in local minima.

☞   The good news, however, is that EUCL_ELLIPSE_FIT can be used for *general conic fitting*. Of course, there is a risk that the result is not the conic we expect (see Further Readings).

A logical question at this point is: If using the *true* distance implies anyway approximations and a numerical solution, can we perhaps find an *approximate* distance leading to a closed-form solution without further approximations? The answer is yes, and the next section explains how to do it.

5.3.2 Algebraic Distance Fit

Definition: Algebraic Distance

The *algebraic distance* of a point $\mathbf{p}$ from a curve $f(\mathbf{p}, \mathbf{a}) = 0$ is simply $|f(\mathbf{p}, \mathbf{a})|$.

The algebraic distance is different from the true geometric distance between a curve and a point; in this sense, we start off with an approximation. However, this is the *only* approximation we introduce, since the algebraic distance turns problem (5.1) into a linear problem that we can solve in closed form and with no further approximations.

Problem (5.1) becomes

$$\min_{\mathbf{a}} \sum_{i=1}^{N} |\mathbf{x}_i^\top \mathbf{a}|^2 \tag{5.5}$$

To avoid the trivial solution $\mathbf{a} = 0$, we must enforce a constraint on $\mathbf{a}$. Of the several constraints possible (see Further Readings), we choose one which forces the solution to be an ellipse:

$$b^2 - 4ac = \mathbf{a}^\top \begin{bmatrix} 0 & 0 & -2 & 0 & 0 & 0 \\ 0 & 1 & 0 & 0 & 0 & 0 \\ -2 & 0 & 0 & 0 & 0 & 0 \\ 0 & 0 & 0 & 0 & 0 & 0 \\ 0 & 0 & 0 & 0 & 0 & 0 \\ 0 & 0 & 0 & 0 & 0 & 0 \end{bmatrix} \mathbf{a} = \mathbf{a}^\top C \mathbf{a} = -1. \tag{5.6}$$

☞ Notice that this can be regarded as a "normalized" version of the elliptical constraint $b^2 - 4ac < 0$, as $\mathbf{a}$ is only defined up to a scale factor.

We can find a solution to this problem *with no approximations*. First, we rewrite problem (5.5) as

$$\min_{\mathbf{a}} \|\mathbf{a}^\top X^\top X \mathbf{a}\| = \min_{\mathbf{a}} \|\mathbf{a}^\top S \mathbf{a}\|, \tag{5.7}$$

where

$$X = \begin{bmatrix} x_1^2 & x_1 y_1 & y_1^2 & x_1 & y_1 & 1 \\ x_2^2 & x_2 y_2 & y_2^2 & x_2 & y_2 & 1 \\ & & \cdots & & & \\ x_N^2 & x_N y_N & y_N^2 & x_N & y_N & 1 \end{bmatrix} \tag{5.8}$$

In the terminology of constrained least squares, X is called the *design matrix*, $S = X^\top X$ the *scatter matrix*, and C the *constraint matrix*. Again using Lagrange multipliers, we obtain that problem (5.7) is solved by

$$S \mathbf{a} = \lambda C \mathbf{a}. \tag{5.9}$$

This is a so-called *generalized eigenvalue problem*, which can be solved in closed form. It can be proven that the solution, $\mathbf{a}_o$, is the eigenvector corresponding to the *only* negative eigenvalue. Most numerical packages will find the solution of problem (5.9) for you, taking care of the fact that C is rank-deficient. The resulting algorithm is very simple.

Algorithm ALG_ELLIPSE_FIT

The input is a set of N image points, $\mathbf{p}_1, \ldots, \mathbf{p}_N$. We assume the notation introduced in the problem statement box for ellipse fitting.

1. Build the design matrix, X, as per (5.8).
2. Build the scatter matrix, $S = X^\top X$.
3. Build the constraint matrix, C, as per (5.6).
4. Use a numerical package to compute the eigenvalues of the generalized eigenvalue problem, and call λ_n the only negative eigenvalue.

The output is the best-fit parameter vector, $\mathbf{a}_o$, given by the eigenvector associated to λ_n.

Figure 5.3 shows the result of ALG_ELLIPSE_FIT run on an elliptical arc corrupted by increasing quantities of Gaussian noise.

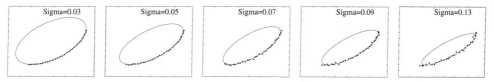

Figure 5.3 Example of best-fit ellipses found by ALG_ELLIPSE_FIT for the same arc of ellipse, corrupted by increasingly strong Gaussian noise. From left to right, the noise varies from 3% to 20% of the data spread (figure courtesy of Maurizio Pilu, University of Edinburgh).

ALG_ELLIPSE_FIT tends to be *biased towards low-eccentricity solutions*, indeed a characteristic of *all* methods based on the algebraic distance. Informally, this means that the algorithm prefers fat ellipses to thin ellipses, as shown in Figure 5.4. The reason is best understood through the geometric interpretation of the algebraic distance.

Geometric Interpretation of the Algebraic Distance (Ellipses)

Consider a point $\mathbf{p}_i$ *not* lying on the ellipse $f(\mathbf{p}, \mathbf{a}) = 0$. The algebraic distance, $f(\mathbf{p}_i, \mathbf{a})$, is proportional to

$$Q = 1 - \left[\frac{(r+d)^2}{r^2} \right],$$

where r is the distance of the ellipse from its center along a line which goes through $\mathbf{p}_i$, and d is the distance of $\mathbf{p}_i$ from the ellipse along the same line (Figure 5.5).

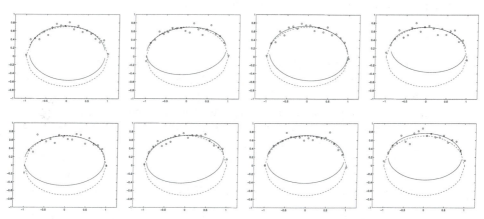

Figure 5.4 Illustration of the low-eccentricity bias introduced by the algebraic distance. ALG_ELLIPSE_FIT was run on 20 samples covering half an ellipse, spaced uniformly along *x*, and corrupted by different realizations of rather strong, Gaussian noise with constant standard deviation ($\sigma = 0.08$, about 10% of the smaller semiaxis). The best-fit ellipse (solid) is systematically biased to be "fatter" than the true one (dashed).

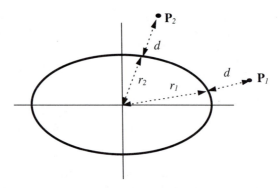

Figure 5.5 Illustration of the distances d and r in the geometric interpretation of the algebraic distance, Q. At a parity of d, Q is larger at $\mathbf{P}_2$ than at $\mathbf{P}_1$.

☞ Notice that this interpretation is valid for *any* conic. For hyperbolae, the center is the intersection of the asymptotes; for parabolae, the center is at infinity.

For any fixed d, Q is maximum at the intersection of the ellipse with its smaller axis (e.g., $\mathbf{P}_2$ in Figure 5.5) and minimum at the intersection of the ellipse with its larger axis (e.g., $\mathbf{P}_1$ in Figure 5.5). Therefore, the algebraic distance is maximum (high weight) for observed points around the flat parts of the ellipse, and minimum (low weight) for observed points around the pointed parts. As a consequence, a fitting algorithm based on Q tends to believe that most data points are concentrated in the flatter part of the ellipse, which results in "fatter" best-fit ellipses.

5.3.3 Robust Fitting

One question which might have occurred to you is: Where do the data points for ellipse fitting come from? In real applications, and without *a priori* information on the scene, finding the points most likely to belong to a specific ellipse is a difficult problem. In some cases, it is reasonable to expect that the data points can be selected by hand. Failing that, we rely on edge chaining as described in HYSTERESIS_THRESH. In any case, it is very likely that the data points contain *outliers*.

Outliers are data points which violate the statistical assumptions of the estimator. In our case, an outlier is an edge point *erroneously* assumed to belong to an ellipse arc. Both EUCL_ELLIPSE_FIT and ALG_ELLIPSE_FIT, as least-squares estimators, assume that all data points can be regarded as true points corrupted by additive, Gaussian noise; hence, even a small number of outliers can degrade their results badly. *Robust estimators* are a class of methods designed to tolerate outliers.[1] A robust distance that

---

[1] Section A.7 in the Appendix gives a succinct introduction to robust estimators.

often works well is the absolute value, which is adopted by the following algorithm for robust ellipse fitting.

Algorithm ROB_ELLIPSE_FIT

The input is a set of N image points, $\mathbf{p}_1, \ldots, \mathbf{p}_N$. We assume the notation introduced in the problem statement box for ellipse fitting.

1. run ALG_ELLIPSE_FIT, and call its solution $\mathbf{a}_0$.

2. Using $\mathbf{a}_0$ as initial value, run a numerical minimization to find a solution of

$$\min_{\mathbf{a}} \sum_{i=1}^{N} |\, \mathbf{x}_i^\top \mathbf{a} \,|$$

The output is the solution, $\mathbf{a}_m$, which identifies the best-fit ellipse.

Figure 5.6 illustrates the problems caused by outliers to ALG_ELLIPSE_FIT, and allows you to compare the result of ALG_ELLIPSE_FIT with those of ROB_ELLIPSE_FIT, started from the solution of ALG_ELLIPSE_FIT, in conditions of severe noise (that is, lots of outliers *and* Gaussian noise). Both algorithms were run on 40 points from half an ellipse, spaced uniformly along x, and corrupted by different realizations of Gaussian noise with constant standard deviation ($\sigma = 0.05$, about 7% of the smaller semiaxis, a). About 20% of the points were turned into outliers by adding a uniform deviate in $[-a, a]$ to their coordinates. Notice the serious errors caused to ALG_ELLIPSE_FIT by the outliers, which are well tolerated by ROB_ELLIPSE_FIT.

5.3.4 Concluding Remarks on Ellipse Fitting

With moderately noisy data, ALG_ELLIPSE_FIT should be your first choice. With seriously noisy data, the eccentricity of the best-fit ellipse can be severely underestimated (the more so, the smaller the arc of ellipse covered by the data). If this is a problem for your application, you can try EUCL_ELLIPSE_FIT, starting it from the solution of ALG_ELLIPSE_FIT. With data containing many outliers, the results of both EUCL_ELLIPSE_FIT and ALG_ELLIPSE_FIT will be skewed; in this case, ROB_ELLIPSE_FIT, started from the solution of ALG_ELLIPSE_FIT should do the trick (but you are advised to take a look at the references in section A.7 in the Appendix if robustness is a serious issue for your application). If speed matters, your best bet is ALG_ELLIPSE_FIT alone, assuming you use a reasonably efficient package to solve the eigenvalue problem, and the assumptions of the algorithms are plausibly satisfied.

What "moderately noisy" and "seriously noisy" mean quantitatively depends on your data (number and density of data points along the ellipse, statistical distribution, and standard deviation of noise). In our experience, ALG_ELLIPSE_FIT gives good fits with more than 10 points from half an ellipse, spaced uniformly along x, and

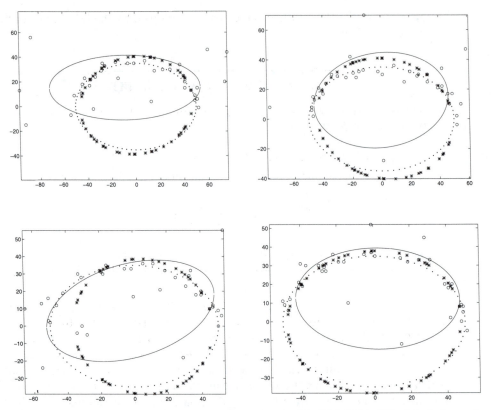

Figure 5.6 Comparison of ALG_ELLIPSE_FIT and ROB_ELLIPSE_FIT when fitting to data severely corrupted by outliers. The circles show the data points, the asterisks suggest the robust fit, the solid line show the algebraic fit, and the dots the true (uncorrupted) ellipse.

corrupted by Gaussian noise of standard deviation up to about 5% of the smaller semiaxis. Section 5.7 suggests Further Readings on the evaluation and comparison of ellipse-fitting algorithms.

5.4 Deformable Contours

Having discussed how to fit simple curves, we now move on to the general problem of fitting a curve of arbitrary shape to a set of image edge points. We shall deal with *closed contours* only.

A widely used computer vision model to represent and fit general, closed curves is the *snake*, or *active contour*, or again *deformable contour*. You can think of a snake as an elastic band of arbitrary shape, sensitive to the intensity gradient. The snake is located initially near the image contour of interest, and is attracted towards the target contour by forces depending on the intensity gradient.

☞   Notice that the snake is applied to the *intensity image*, not to an image of edge points as the line and ellipse detectors of the previous sections.

We start giving a description of the deformable contour model using the notion of *energy functional* and continuous image coordinates (no pixelization). We then discuss a simple, iterative algorithm fitting a deformable contour to a chain of egde points of a real, pixelized image.

5.4.1   The Energy Functional

The key idea of deformable contours is to associate an *energy functional* to each possible contour shape, in such a way that the image contour to be detected corresponds to a minimum of the functional. Typically, the energy functional used is a sum of several terms, each corresponding to some force acting on the contour.

Consider a contour, $c = c(s)$, parametrized by its arc length,[2] s. A suitable energy functional, $\mathcal{E}$, consists of the sum of three terms:

$$\mathcal{E} = \int \left(\alpha(s) E_{cont} + \beta(s) E_{curv} + \gamma(s) E_{image} \right) ds, \qquad (5.10)$$

where the integral is taken along the contour c and each of the energy terms, E_{cont}, E_{curv}, and E_{image}, is a function of c or of the derivatives of c with respect to s. The parameters α, β and γ control the relative influence of the corresponding energy term, and can vary along c. Let us now define more precisely the three energy terms in (5.10).

5.4.2   The Elements of the Energy Functional

Each energy term serves a different purpose. The terms E_{cont} and E_{curv} encourage *continuity* and *smoothness* of the deformable contour, respectively; they can be regarded as a form of internal energy. E_{image} accounts for *edge attraction*, dragging the contour toward the closest image edge; it can be regarded as a form of external energy. What functions can achieve these behaviors?

Continuity Term.   We can exploit simple analogies with physical systems to devise a rather natural form for the continuity term:

$$E_{cont} = \left\| \frac{dc}{ds} \right\|^2 .$$

---

[2] Given an arbitrary parametrization of a curve, $c = c(t)$, with t the parameter and $0 \leq t \leq T$, the *arc length* s is defined as

$$s = \int_0^t \left\| \frac{dc}{d\tau} \right\| d\tau.$$

In the arc length parametrization, the tangent vector $dc)ds$ is always a unit vector (Exercise 5.7).

In the discrete case, the contour $\mathbf{c}$ is replaced by a chain of N image points, $\mathbf{p}_1, \ldots, \mathbf{p}_N$, so that

$$E_{cont} = \|\mathbf{p}_i - \mathbf{p}_{i-1}\|^2.$$

A better form for E_{cont}, preventing the formation of clusters of snake points, is

$$E_{cont} = \left(\bar{d} - \|\mathbf{p}_i - \mathbf{p}_{i-1}\| \right)^2, \tag{5.11}$$

with $\bar{d}$ the average distance between the pairs $(\mathbf{p}_i, \mathbf{p}_{i-1})$. If $\|\mathbf{p}_i - \mathbf{p}_{i-1}\| >> \bar{d}$, we have

$$E_{cont} \simeq \|\mathbf{p}_i - \mathbf{p}_{i-1}\|^2,$$

while for smaller distances (5.11) promotes the formation of equally spaced chains of points and avoids the formation of point clusters.

Smoothness Term.   The aim of the smoothness term is to *avoid oscillations* of the deformable contour. This is achieved by introducing an energy term penalizing high contour curvatures. Since E_{cont} encourages equally spaced points on the contour, the curvature is well approximated by the second derivative of the contour (Exercise 5.8); hence, we can define E_{curv} as

$$E_{curv} = \|\mathbf{p}_{i-1} - 2\mathbf{p}_i + \mathbf{p}_{i+1}\|^2. \tag{5.12}$$

Edge Attraction Term.   The third term corresponds to the energy associated to the external force attracting the deformable contour towards the desired image contour. This can be achieved by a simple function:

$$E_{image} = -\|\nabla I\|, \tag{5.13}$$

where ∇I is the spatial gradient of the intensity image I, computed at each snake point. Clearly, E_{image} becomes very small (negative) wherever the norm of the spatial gradient is large, (that is, near images edges), making $\mathcal{E}$ small and attracting the snake towards image contours. Note that E_{image}, unlike E_{cont} and E_{curv}, depends only on the contour, not on its derivatives with respect to the arc length.

5.4.3   A Greedy Algorithm

We are now ready to describe a method for fitting a snake to an image contour. The method is based on the minimization of the energy functional (5.10). First of all, let us summarize the assumptions and state the problem.

Assumptions

Let I be an image and $\bar{p}_1, \ldots, \bar{p}_N$ the chain of image locations representing the initial position of the deformable contour, which we assume close to the image contour of interest.

Problem Statement

Starting from $\bar{p}_1, \ldots, \bar{p}_N$, find the deformable contour $p_1, \ldots, p_N$ which fits the target image contour best, by minimizing the energy functional

$$\sum_{i=1}^{N} \left(\alpha_i E_{cont} + \beta_i E_{curv} + \gamma_i E_{image} \right),$$

with $\alpha_i, \beta_i, \gamma_i \geq 0$, and E_{cont}, E_{curv}, and E_{image} as in (5.11), (5.12), and (5.13) respectively.

Of the many algorithms proposed to fit deformable contours, we have selected a *greedy algorithm*. A greedy algorithm makes *locally optimal choices*, in the hope that they lead to a *globally* optimal solution. Among the reasons for selecting the greedy algorithm instead of other methods, we emphasize its simplicity and low computational complexity.

The algorithm is *conceptually simple* because it does not require knowledge of the calculus of variations.[3] It has a *low computational complexity* because it converges in a number of iterations proportional to the number of contour points times the number of locations in which each point can move at each iteration, whereas other snake algorithms take much longer.

The core of a greedy algorithm for the computation of a deformable contour consists of two basic steps. First, at each iteration, each point of the contour is moved within a small neighborhood to the point which minimizes the energy functional. Second, before starting a new iteration, the algorithm looks for corners in the contour, and takes appropriate measures on the parameters $\beta_1, \ldots, \beta_N$ controlling E_{curv}. Let us discuss these two steps in more detail.

Step 1: Greedy Minimization. The neighborhood over which the energy functional is locally minimized is typically small (for instance, a 3×3 or 5×5 window centered at each contour point). Keeping the size of the neighborhood small lowers the computational load of the method (the complexity being linear in the size of the neighborhood). The local minimization is done by direct comparison of the energy functional values at each location.

Step 2: Corner Elimination. During the second step, the algorithm searches for corners as curvature maxima along the contour. If a curvature maximum is found at point $\mathbf{p}_j$, β_j is set to zero. Neglecting the contribution of E_{curv} at $\mathbf{p}_j$ makes it possible to keep the deformable contour piecewise smooth.

☞   For a correct implementation of the method, it is important to normalize the contribution of each energy term. For the terms E_{cont} and E_{curv}, it is sufficient to divide by the largest

---

[3] The calculus of variations is the mathematical technique for determining the minimum of a functional, in the same way as calculus provides the tools for determining the minimum of an ordinary function.

value in the neighborhood in which the point can move. For E_{image}, instead, it may be useful to normalize the norm of the spatial gradient, $\|\nabla I\|$, as

$$\frac{\|\nabla I\| - m}{M - m},$$

with M and m maximum and minimum of $\|\nabla I\|$ over the neighborhood, respectively.

The iterations stop when a predefined fraction of all the points reaches a local minimum; however, the algorithm's greed does not guarantee convergence to the global minimum. It usually works very well as far as the initialization is not too far from the desired solution. Let us enclose the algorithm details in the usual box.

Algorithm SNAKE

The input is formed by an intensity image, I, which contains a closed contour of interest, and by a chain of image locations, $\mathbf{p}_1, \ldots, \mathbf{p}_N$, defining the initial position and shape of the snake.

Let f be the minimum fraction of snake points that must move in each iteration before convergence, and $U(\mathbf{p})$ a small neighborhood of point $\mathbf{p}$. In the beginning, $\mathbf{p}_i = \bar{\mathbf{p}}_i$ and $d = \bar{d}$ (used in E_{cont}).

While a fraction greater than f of the snake points move in an iteration:

1. for each $i = 1, \ldots, N$, find the location of $U(\mathbf{p}_i)$ for which the functional $\mathcal{E}$ defined in (5.10) is minimum, and move the snake point $\mathbf{p}_i$ to that location;

2. for each $i = 1, \ldots, N$, estimate the curvature k of the snake at $\mathbf{p}_i$ as

$$k = |\mathbf{p}_{i-1} - 2\mathbf{p}_i + \mathbf{p}_{i+1}|,$$

and look for local maxima. Set $\beta_j = 0$ for all $\mathbf{p}_j$ at which the curvature has a local maximum and exceeds a user-defined minimum value;

3. update the value of the average distance, $\bar{d}$.

On output this algorithm returns a chain of points $\mathbf{p}_i$ that represent a deformable contour.

☞   We still have to assign values to α_i, β_i, and γ_i. One possible choice is to initialize all of them to 1; another possibility is $\alpha_i = \beta_i = 1$ and $\gamma_i = 1.2$, which gives edges attraction more relevance in the minimization stage.

☞   To prevent the formation of noisy corners, it may be useful to add an extra condition: point $\mathbf{p}_j$ is a corner if and only if the curvature is locally maximum at $\mathbf{p}_j$ *and* the norm of the intensity gradient at $\mathbf{p}_j$ is sufficiently large. This ignores corners formed too far away from image edges.

Examples of the application of SNAKE to synthetic and real images are shown in Figures 5.7 and 5.8.

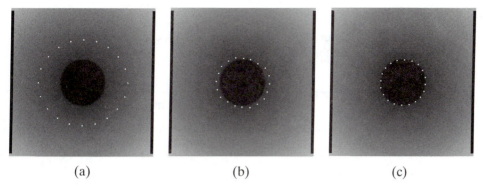

<table>
<tr><td>(a)</td><td>(b)</td><td>(c)</td></tr>
</table>

Figure 5.7   (a–c): Initial position of the snake, intermediate position (6th iteration), final result (20th iteration). Parameters were $\alpha_i = \beta_i = 1, \gamma_i = 1.2$. SNAKE used 7×7 local neighborhoods, and stopped when only 4 or fewer points changed in an iteration.

Figure 5.8   (a): Initial position of the snake. (b): Intermediate position (84th iteration). (c): Final result (130th iteration). Parameters were $\alpha_i = \beta_i = 1, \gamma_i = 1.2$. SNAKE used 7×7 local neighborhoods, and stopped when only 9 or fewer points changed in an iteration.

5.5 Line Grouping

We close this chapter by touching upon the difficult problem of *grouping*. For our purposes, we can state the problem as follows.

Problem Statement: Grouping

Given the set of features detected in an image, decide which groups of features are likely to be part of a same object, *without* knowing what objects we are looking at.

The features in question could be edge points, lines, or curve segments. We limit our discussion to the particular case of *straight line segments*, which are important features in many practical cases and allow us to introduce the main concepts of grouping. Moreover, line segments are easily represented through their endpoints; we adopt this convention here.

Notice that our problem statement generalizes the aims of grouping given at the beginning of this chapter, in two directions. First, we are now interested in grouping more complex features than edge points; second, we consider features not lying necessarily next to each other on the image plane.

As we cannot use object models, we must resort to properties of the image to form the desired groups of features. We consider three such properties: *proximity*, *parallelism*, and *collinearity*. Using these, we devise a method based on empirical estimates of the probability of a group arising by accident of viewpoint or object position. Grouping by means of properties like proximity, parallelism, and collinearity is often called *perceptual grouping*, or *perceptual organization*.[4]

Let us start by discussing how to exploit *proximity*. Clearly, two line segments close to each other in space project onto image segments also close to each other, irrespectively of viewpoint and positioning. The converse is not true: As illustrated in Figure 5.9, two 3-D line segments might be far apart, but appear close on the image plane, for example, due to accidental positioning in 3-D space. Notice that "accidental" indicates that you would expect the two 3-D segments in Figure 5.9 to appear, in general, far apart from each other. This observation leads us to a precise problem statement of *grouping by proximity*.

Problem Statement: Grouping by Proximity

Given a set of image line segments, projections of 3-D line segments, determine which pairs are *most likely* to be projections of 3-D line segments which are close to each other in the scene.

---

[4] Our brief discussion does not do justice to the vast body of psychophysical studies on perceptual organization. If you are interested to know more, look into the Further Readings.

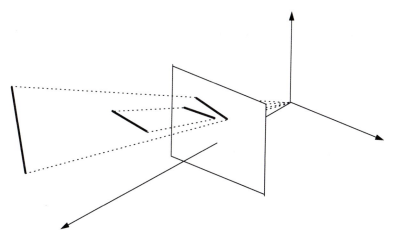

Figure 5.9 Due to accidental positioning, two 3-D lines far apart from each other project onto close image lines.

What we need is a way of quantifying "most likely"; in other words, we are looking for a *significance measure of proximity*, μ_{pro}, which takes on large values for pairs of image lines corresponding to pairs of 3-D lines which are close to each other in the scene. Intuition suggests that pairs of *long* image segments are better than short ones for our purposes. Therefore, if we denote with g the smaller distance between the endpoints of two image segments, and with l_s the length of the shorter segment (Figure 5.10(a)), a reasonable formula for μ_{pro} is

$$\mu_{pro} = \left(\frac{l_s}{g}\right)^2. \tag{5.14}$$

Under appropriate assumptions on the distribution of the segments in 3-D space, you can show that $1/\mu_{pro}$ *is proportional to the probability that the two endpoints are close by accident* (Exercise 5.9).

We can devise in a similar way significance measures for parallelism and collinearity, μ_{par} and μ_{col}:

$$\mu_{par} = \frac{l_s^2}{\theta s l_l} \tag{5.15}$$

$$\mu_{col} = \frac{l_s^2}{\theta s (l_s + g)} \tag{5.16}$$

with θ the angle between the two segments, s the separation (measured as the distance of the midpoint of the shorter segment from the longer one), and l_l the length of the longer segment (see Figure 5.10(b) and (c)).

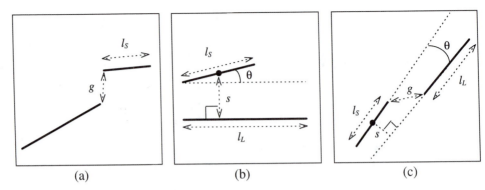

Figure 5.10 Illustration of the relevant quantities in grouping by proximity (a), parallelism (b), and collinearity (c).

☞ It goes almost without saying that in the case of parallelism the two segments are supposed to *overlap*,[5] while in the case of collinearity they are not.

As in the case of proximity, $1/\mu_{par}$ and $1/\mu_{col}$ can be interpreted as proportional to the *probability of accidental parallelism* and *collinearity*, respectively. Notice that the smaller θ and s, the larger μ_{par} and μ_{col}. In addition, μ_{par} gets larger for $l_l \simeq l_s$, while μ_{col} decreases with the distance g between the endpoints of the segment pair and is independent of the length of the longer segment.

We can now give a simple grouping algorithm based on μ_{pro}, μ_{par}, and μ_{col}.

Algorithm PERC_GROUPING

The input is S, a set of image line segments projections of line segments in space. Let t_{pro}, t_{par}, and t_{col} three positive real thresholds.

1. For all pairs of endpoints of image segments in S:
   - (a) compute μ_{pro} with (5.14);
   - (b) depending on whether the segments in the pair overlap, compute μ_{par} from (5.15) or μ_{col} with (5.16).

2. Order the significant measures obtained, separately and in decreasing order.

3. Group the segment pairs for which $\mu_{pro} > t_{pro}$, or $\mu_{par} > t_{par}$, or $\mu_{col} > t_{col}$.

The output is formed by groups of image segments most likely to be projections of 3-D segments part of a same object.

---

[5] Two segments a and b do not overlap if the projection of a over b does not overlap with b and vice versa. Otherwise, the two segments overlap.

☞   In case of conflict (the same image pair being assigned a significance measure above threshold for both proximity and either parallelism or collinearity), proximity is discarded.

☞   The values of t_{pro}, t_{par}, and t_{col} depend on the application and should be determined empirically. However, one is often interested only in finding, say, the ten best segment pair for each property; in this case, the problem of setting values for t_{pro}, t_{par}, and t_{col} is avoided.

PERC_GROUPING does a reasonable job of identifying close, parallel, and collinear segment pairs. If S becomes large, precomputing the pairs of segments which do not lie too far apart on the image plane leads to a more efficient implementation.

5.6 Summary

After working through this chapter, you should be able to:

❏   explain the relation between the image features introduced in this chapter and those introduced in Chapter 4

❏   detect lines, ellipses and more general curves in digital images

❏   design an algorithm for fitting deformable contours

❏   grouping image lines

5.7 Further Readings

Our treatment of the Hough transform is largely based on Illingworth and Kittler's review [8], which describes a number of applications and developments. Haralick and Shapiro's book [7] give algorithmic details, and Davies [4] reports a detailed analysis of Hough methods. The Hypermedia Image Processing Reference (see Further Readings, Chapter 4) covers both line and curve detection.

Our discussion of ellipse fitting using the Euclidean distance is based on Sampson [17] and Haralick and Shapiro [7]. The ellipse-specific algorithm, ALG_ELLIPSE_FIT, is due to Pilu, Fitzgibbon and Fisher [14]. Strang [19] (Chapter 6) and Press et al. [15] (Chapter 11) include good discussions of the generalized eigenvalue problem. The geometric interpretation of the algebraic distance for general conic fitting is discussed by Bookstein [3] and Sampson [17]. Rosin [16] studies the effects of various constraints for algebraic distance fitting. Kanatani [9] (Chapter 8 and 10) discusses *weighted least squares fitting*, a way of adapting least squares to the general case in which the standard deviation of the noise is different for each data point. Fitzgibbon and Fisher [5] report a useful, experimental comparison of several conic-fitting algorithms, while a review of applications of robust statistics in computer vision, due to Meer et al., can be found in [13].

Deformable contours were introduced by Kass, Witkin, and Terzopoulos in a now classical paper [10]. The algorithm discussed in this chapter is based on the work of Williams and Shah [20]. Further interesting work on snakes, addressing minimization and contour tracking issues, is reported in [1] and [2].

Much research has been devoted to perceptual grouping and perceptual organization [6, 21]; see [18] for a recent survey. The significance measures discussed in this chapter are essentially based on the work of Lowe [12], who discusses perceptual grouping at length in [11] from both the computational and psychophysical viewpoints.

5.8 Review

Questions

☐ *5.1* Line detection can be implemented as *template matching*, which consists in filtering images with masks responding to lines in different orientations, then applying a threshold to select line pixels. Compare template matching with the Hough transform as methods to detect image lines.

☐ *5.2* What are the disadvantages of template matching as a curve detection technique?

☐ *5.3* Discuss how the availability of edge direction information can reduce the search space of the Hough transform for the case of circles. Generalize for higher-order parameter spaces.

☐ *5.4* Explain the presence of the secondary peaks around the main two peaks in Figure 5.2.

☐ *5.5* Explain why the (ρ, θ) parametrization for lines (section 5.2.1) leads to a better discretization of the Hough parameter space than the (m, n) one. Compare the accuracies of the parameter estimates you can hope to achieve in both cases, given that the search for the maxima must take place in a reasonable time.

☐ *5.6* When using Hough transform algorithms, searching for maxima in parameter space becomes increasingly onerous as the number of model parameters increases. Time and space could be saved by sampling the parameter spaces more coarsely far from significant maxima. Discuss.

☐ *5.7* The finite size of the image implies that, on average, the length in pixel of the visible portions of lines close to the image center C is greater than that of lines distant from C. How does this bias the Hough transform? How could you counter this bias?

☐ *5.8* Explain why the result of EUCL_ELLIPSE_FIT could be any conic, not just an ellipse.

☐ *5.9* What is the purpose of the parameters α_i, β_i, and γ_i in the energy functional which defines a deformable contour?

☐ *5.10* Could you suggest an energy term that keeps a deformable contour far away from a certain image point?

☐ *5.11* Is the square in (5.14) really necessary?

Exercises

○ *5.1* Modify algorithm HOUGH_LINE to take advantage of information about the edge direction.

○ 5.2   Determine the equation of the curve identified in the (ρ, θ) plane by a point (x, y) when searching for lines.

○ 5.3   Write an algorithm, HOUGH_CIRCLE, using the Hough transform to detect circles.

○ 5.4   Write the equation of the weighted average suggested in section 5.2.1 to estimate the parameters of a line in the presence of noise.

○ 5.5   Prove that the image of a 3-D circle lying entirely in front of the viewer is an ellipse. What happens to the image if the 3-D circle lies on both sides of the image plane. (*Hint:* Think for instance of the profile of the Earth as seen from an airplane.)

○ 5.6   You are looking for lines in an image containing 100 edge points (notice that this is a small number for real applications). You decide to apply your fitting routines to all possible groups of points containing at least 5 points (brute force fitting). How many groups do you have to search? Assume that the time cost of fitting a line to n points is n^2 (in some relevant units), and that the time cost of forming m groups from 100 points is mn^2. How many points should a group contain to make grouping more expensive than brute force fitting?

○ 5.7   Show that the tangent vector to a curve parametrized by its arc length is a unit vector. (*Hint:* Use the chain rule of differentiation to find the relation between the tangent vector with respect to an arbitrary parametrization and the arc length. Now, looking at the definition of arc length . . .)

○ 5.8   Given a curve in Cartesian coordinates $\mathbf{c}(s) = [x(s), y(s)]^\top$, with s the arc length, the curvature k is given by the expression

$$k = \left| x'y'' - x''y' \right|.$$

Show that this definition of curvature reduces to the expression used in section 5.4 if $\mathbf{c} = [x, y(x)]^\top$ and the x are equally spaced.

○ 5.9   In the assumption of a background of line segments uniformly distributed in the image with respect to orientation, position, and scale, the density of lines varies with the inverse of the square of their length. Show that the expected number N of endpoints of lines of length l within a radius g of a given endpoint is thus given by

$$N = \frac{\pi g^2}{l^2}.$$

Projects

● 5.1   Implement algorithm HOUGH_LINES; then, write and implement an algorithm using the Hough transform to detect circles. Both algorithms should take advantage of information on edge directions when available, using your implementation of (or at last the output assumed from) relevant algorithms in Chapter 4. Using synthetic images, compare the time needed to detect a single line and a single circle formed by the same number of points. Corrupt the edge images with spurious edge points by adding saturated salt-and-pepper noise (Chapter 3), and estimate the robustness of both algorithms by plotting the errors in the parameter estimates as a function of the amounts of noise (quantified by the probability that a pixel is changed).

- 5.2   Implement algorithm SNAKE. Using synthetic images, study the difference between the deformable contours obtained for different choices of α_i, β_i, and γ_i.

References

[1]   A.A. Amini, S. Tehrani, and T.E. Weymouth, Usying Dynamic Programming for Minimizing the Energy of Active Contours in the Presence of Hard Constraints, *Proc. 2nd Intern. Conf. on Computer Vision*, Tampa Bay (FL), pp. 95–99 (1988).

[2]   A. Blake, R. Curwen, and A. Zisserman, A Framework for Spatio-Temporal Control in the Tracking of Visual Contours, *International Journal of Computer Vision*, Vol. 11, pp. 127–145 (1993).

[3]   F.L. Bookstein, Fitting Conic Sections to Scattered Data, *Computer Graphics and Image Processing*, Vol. 9, pp. 56-91 (1979).

[4]   E.R. Davies, *Machine Vision: Theory, Algorithms, Practicalities*, Academic Press, London (1990).

[5]   A.W. Fitzgibbon and R.B. Fisher, A Buyer's Guide to Conic Fitting, *Proc. 5th British Machine Vision Conference*, Birmingham, pp. 513-522 (1995).

[6]   B. Flinchbaugh and B. Chandrasekaran, A Theory of Spatio-Temporal Aggregation for Vision, *Artificial Intelligence*, Vol. 17 (1981).

[7]   R.M. Haralick and L.G. Shapiro, *Computer and Robot Vision*, Volume I, Addison-Wesley (1992).

[8]   J. Illingworth and J. Kittler, A Survey of the Hough Transform, *Computer Vision, Graphics, and Image Processing*, Vol. 44, pp. 87-116 (1988).

[9]   K. Kanatani: *Geometric Computation for Computer Vision*, Oxford University Press, Oxford (1993).

[10] M. Kass, A. Witkin, and D. Terzopoulos, Snakes: Active Contours Models, *Proc. First Inter. Conf. Comput. Vision*, London (UK), pp. 259–269 (1987).

[11] D.G. Lowe: *Perceptual Organization and Visual Recognition*, Kluwer, Boston (MA) (1985).

[12] D.G. Lowe, 3-D Object Recognition from Single 2-D Images, *Artificial Intelligence* Vol. 31, pp. 355–395 (1987).

[13] P. Meer, D. Mintz, A. Rosenfeld and D.Y. Kim: Robust Regression Methods for Computer Vision: a Review, *International Journal of Computer Vision*, Vol. 6, pp. 59–70 (1991).

[14] M. Pilu, A.W. Fitzgibbon and R.B. Fisher, Ellipse-Specific Least-Square Fitting, *IEEE Int. Conf. on Image Processing*, Lausanne (1996).

[15] W.H. Press, S.A. Teulosky, W.T. Vetterling and B.P. Flannery, *Numerical Recipes in C*, second edition, Cambridge University Press (1992).

[16] P.L. Rosin, A Note on the Least Squares Fitting of Ellipses, *Pattern Recognition Letters*, Vol. 14, pp. 799–808 (1993).

[17] P.D. Sampson, Fitting Conic Sections to "Very Scattered" Data: an Iterative Refinement of the Bookstein Algorithm, *Computer Graphics and Image Processing*, Vol. 18, pp. 97–108 (1982).

[18] S. Sarkar and K.L. Boyer, Perceptual Organization in Computer Vision: a Review and a Proposal for a Classificatory Structure, *IEEE Transactions on Systems, Man, and Cybernetics*, Vol. 23, pp. 382–399 (1993).

[19] G. Strang, *Linear Algebra and Its Applications*, Harcourt Brace, Orlando (FL) (1988).

[20] D.J. Williams and M. Shah, A Fast Algorithm for Active Contours and Curvature Estimation, *CVGIP: Image Understanding*, Vol. 55, pp. 14–26 (1992).

[21] A. Witkin and J. Tenenbaum, On the Role of Structure in Vision, in *Human and Machine Vision*, J. Beck, B. Hope and A. Rosenfeld eds., Academic Press (New York), pp. 481–543 (1983).

6

Camera Calibration

> For the ancient Egyptians, exactitude was symbolized by a feather
> that served as a weight on scales used for the weighing of souls.
>
> Italo Calvino, *Six Memos for the Next Millennium*

This chapter tackles the problem of camera calibration; that is, determining the value of the extrinsic and intrinsic parameters of the camera.

Chapter Overview

Section 6.1 defines and motivates the problem of camera calibration and its main issues.

Section 6.2 discusses a method based on simple geometric properties for estimating the camera parameters, given a number of correspondences between scene and image points.

Section 6.3 describes an alternative, simpler method, which recovers the projection matrix first, then computes the camera parameters as functions of the entries of the matrix.

What You Need to Know to Understand this Chapter

- Working knowledge of geometric camera models (Chapter 2).
- SVD and constrained least-squares (Appendix, section A.6).
- Familiarity with line extraction methods (Chapter 5).

6.1 Introduction

We learned in Chapters 4 and 5 how to identify and locate image features, and we are therefore fully equipped to deal with the important problem of *camera calibration*; that is, estimating the values of the intrinsic and extrinsic parameters of the camera model, which was introduced in Chapter 2.

The key idea behind calibration is to *write the projection equations linking the known coordinates of a set of 3-D points and their projections, and solve for the camera parameters*. In order to get to know the coordinates of some 3-D points, camera calibration methods rely on one or more images of a *calibration pattern*: that is, a 3-D object of known geometry, possibly located in a known position in space and generating image features which can be located accurately. Figure 6.1 shows a typical calibration pattern, consisting of two planar grids of black squares on a white background. It is easy to know the 3-D position of the vertices of each square once the position of the two planes has been measured, and locate the vertices on the images, for instance as intersection of image lines, thanks to the high contrast and simple geometry of the pattern.

Problem Statement

Given one or more images of a calibration pattern, estimate

1. the intrinsic parameters,
2. the extrinsic parameters, or
3. both.

☞   The accuracy of calibration depends on the accuracy of the measurements of the calibration pattern; that is, its construction tolerances. To be on the safe side, the calibration pattern should be built with tolerances one or two order of magnitudes smaller than the desired accuracy of calibration. For example, if the desired accuracy of calibration is 0.1mm, the calibration pattern should be built with tolerances smaller than 0.01mm.

Although there are techniques inferring 3-D information about the scene from *uncalibrated* cameras, some of which will be described in the next two chapters, effective camera calibration procedures open up the possibility of using a wide range of existing algorithms for 3-D reconstruction and recognition, all relying on the knowledge of the camera parameters.

This chapter discusses two algorithms for camera calibration. The first method recovers directly the intrinsic and extrinsic camera parameters; the second method estimates the projection matrix first, *without* solving explicitly for the various parameters,[1] which are then computed as closed-form functions of the entries of the projection matrix. The choice of which method to adopt depends largely on which algorithms are to be applied next (Section 6.4).

---

[1] Recall that the projection matrix links world and image coordinates, and its entries are functions of the intrinsic and extrinsic parameters.

Figure 6.1   The typical calibration pattern used in this chapter.

6.2 Direct Parameter Calibration

We start by identifying the parameters to be estimated, and cast the problem in geometric terms.

6.2.1 Basic Equations

Consider a 3-D point, P, defined by its coordinates $[X^w, Y^w, Z^w]^\top$ in the world reference frame. As usual in calibration, the world reference frame is known.

☞   This means to pick an accessible object defining three mutually orthogonal directions intersecting in a common point. In this chapter, this object is the calibration pattern; in vision systems for indoor robot navigation, for instance, it can be a corner of a room.

Let $[X^c, Y^c, Z^c]^\top$ be the coordinates of P in the *camera* reference frame (with $Z^c > 0$ if P is visible). As usual, the origin of the camera frame is the center of projection, and the Z axis is the optical axis. The position and orientation of the camera frame is unknown, since, unlike the image and world reference frames, the camera frame is inaccessible directly. This is equivalent to saying that the we do not know the extrinsic parameters; that is, the 3×3 rotation matrix R and 3-D translation vector $\mathbf{T}$ such that

$$\begin{bmatrix} X^c \\ Y^c \\ Z^c \end{bmatrix} = R \begin{bmatrix} X^w \\ Y^w \\ Z^w \end{bmatrix} + \mathbf{T}. \tag{6.1}$$

In components, (6.1) can be written as

$$X^c = r_{11}X^w + r_{12}Y^w + r_{13}Z^w + T_x$$
$$Y^c = r_{21}X^w + r_{22}Y^w + r_{23}Z^w + T_y$$
$$Z^c = r_{31}X^w + r_{32}Y^w + r_{33}Z^w + T_z. \tag{6.2}$$

☞   Note the slight but important change of notation with respect to Chapter 2. In that chapter, the transformation between the world and camera reference frames was defined by translation *followed* by rotation. Here, the order is reversed and rotation *precedes* translation. While the rotation matrix is the same in both cases, the translation vectors differ (see Question 6.2).

Assuming that radial distortions (section 2.4.3) can be neglected,[2] we can write the image of $[X^c, Y^c, Z^c]^\top$ in the image reference frame as (see (2.14) and (2.20))

$$x_{im} = -\frac{f}{s_x}\frac{X^c}{Z^c} + o_x \tag{6.3}$$

$$y_{im} = -\frac{f}{s_y}\frac{Y^c}{Z^c} + o_y \tag{6.4}$$

For simplicity, and since there is no risk of confusion, we drop the subscript $_{im}$ indicating image (pixel) coordinates, and write (x, y) for (x_{im}, y_{im}). As we know from Chapter 2, (6.3) and (6.4) depend on the five intrinsic parameters f (focal length), s_x and s_y (horizontal and vertical effective pixel size), and o_x and o_y (coordinates of the image center), and, owing to the particular form of (6.3) and (6.4), the five parameters are not independent. However, if we let $f_x = f/s_x$ and $\alpha = s_y/s_x$, we may consider a new set of four intrinsic parameters, o_x, o_y, f_x, and α, all *independent* of one another. The parameter f_x is simply the focal length expressed in the effective horizontal pixel size (the *focal length in horizontal pixels*), while α, usually called *aspect ratio*, specifies the pixel deformation induced by the acquisition process defined in Chapter 2. Let us now summarize all the parameters to be calibrated in a box (see also the discussion in sections 2.4.2 and 2.4.3).

Extrinsic Parameters

- R, the 3×3 rotation matrix
- $\mathbf{T}$, the 3-D translation vector

---

[2] We shall reconsider this assumption in Exercise 6.1, which suggests how to calibrate the parameters of the radial distortion model of Chapter 2.

Intrinsic Parameters

- $f_x = f/s_x$, length in effective horizontal pixel size units
- $\alpha = s_y/s_x$, aspect ratio
- (o_x, o_y), image center coordinates
- k_1, radial distortion coefficient

Plugging (6.2) into (6.3) and (6.4) gives

$$x - o_x = -f_x \frac{r_{11}X^w + r_{12}Y^w + r_{13}Z^w + T_x}{r_{31}X^w + r_{32}Y^w + r_{33}Z^w + T_z} \tag{6.5}$$

$$y - o_y = -f_y \frac{r_{21}X^w + r_{22}Y^w + r_{23}Z^w + T_y}{r_{31}X^w + r_{32}Y^w + r_{33}Z^w + T_z} \tag{6.6}$$

Notice that (6.5) and (6.6) bypass the inaccessible camera reference frame and link *directly* the world coordinates $[X^w, Y^w, Z^w]^\top$ with the coordinates (x, y) of the corresponding image point. If we use a known calibration pattern, both vectors are measurable. This suggests that, given a sufficient number of points on the calibration pattern, we can try to solve (6.5) and (6.6) for the unknown parameters. This is the idea behind the first calibration method, which is articulated in two parts:

1. assuming the coordinates of the image center are known, estimate all the remaining parameters
2. find the coordinates of the image center

6.2.2 Focal Length, Aspect Ratio, and Extrinsic Parameters

We assume that the coordinates of the image center are known. Thus, with no loss of generality, we can consider the translated coordinates $(x, y) = (x - o_x, y - o_y)$. In other words, we assume that the image center is the origin of the image reference frame. As we said in the introduction, the key idea is to exploit the known coordinates of a sufficient number of corresponding image and world points.

Assumptions and Problem Statement

Assuming that the location of the image center (o_x, o_y) is known, and that radial distortion can be neglected, estimate f_x, α, R, and $\mathbf{T}$ from image points (x_i, y_i), $i = 1 \ldots N$, projections of N known world points $[X_i^w, Y_i^w, Z_i^w]^\top$ in the world reference frame.

The key observation is that (6.5) and (6.6) have the same denominator; therefore, from each corresponding pair of points $((X_i^w, Y_i^w, Z_i^w),(x_i, y_i))$ we can write an equation of the form

$$x_i f_y(r_{21}X_i^w + r_{22}Y_i^w + r_{23}Z_i^w + T_y) = y_i f_x(r_{11}X_i^w + r_{12}Y_i^w + r_{13}Z_i^w + T_x). \tag{6.7}$$

Since $\alpha = f_x/f_y$, (6.7) can be thought of as a linear equation for the 8 unknowns $\mathbf{v} = (v_1, v_2, \ldots, v_8)$:

$$x_i X_i^w v_1 + x_i Y_i^w v_2 + x_i Z_i^w v_3 + x_i v_4 - y_i X_i^w v_5 - y_i Y_i^w v_6 - y_i Z_i^w v_7 - y_i v_8 = 0$$

where

$$
\begin{aligned}
v_1 &= r_{21} & v_5 &= \alpha r_{11} \\
v_2 &= r_{22} & v_6 &= \alpha r_{12} \\
v_3 &= r_{23} & v_7 &= \alpha r_{13} \\
v_4 &= T_y & v_8 &= \alpha T_x.
\end{aligned}
$$

Writing the last equation for the N corresponding pairs leads to the homogeneous system of N linear equations

$$A\mathbf{v} = 0 \tag{6.8}$$

where the $N \times 8$ matrix A is given by

$$
A = \begin{bmatrix}
x_1 X_1^w & x_1 Y_1^w & x_1 Z_1^w & x_1 & -y_1 X_1^w & -y_1 Y_1^w & -y_1 Z_1^w & -y_1 \\
x_2 X_2^w & x_2 Y_2^w & x_2 Z_2^w & x_2 & -y_2 X_2^w & -y_2 Y_2^w & -y_2 Z_2^w & -y_2 \\
\cdot & \cdot & \cdot & \cdot & \cdot & \cdot & \cdot & \cdot \\
\cdot & \cdot & \cdot & \cdot & \cdot & \cdot & \cdot & \cdot \\
\cdot & \cdot & \cdot & \cdot & \cdot & \cdot & \cdot & \cdot \\
x_N X_N^w & x_N Y_N^w & x_N Z_N^w & x_N & -y_N X_N^w & -y_N Y_N^w & -y_N Z_N^w & -y_N
\end{bmatrix}.
$$

If $N \geq 7$ and the N points are not coplanar, A has rank 7, and system (6.8) has a nontrivial solution (unique up to an unknown scale factor), which can be determined from the SVD of A, $A = UDV^\top$, as the column of V corresponding to the only null singular value along the diagonal of D (Appendix, section A.6).

☞   The effects of the noise and the inaccurate localization of image and world points make the rank of A likely to be maximum (eight). In this case, the solution is the eigenvector corresponding to the *smallest* eigenvalue.

A rigorous proof of the fact that, in the ideal case (noise-free, perfectly known coordinates) the rank of A is 7 seems too involved to be presented here, and we refer you to the Further Readings section for details. Here, we just observe that, if the effective rank is larger than 7, system (6.8) would only have the trivial solution.

Our next task is to determine the unknown scale factor (and hence the various camera parameters) from the solution vector $\mathbf{v} = \bar{\mathbf{v}}$. If we call γ the scale factor, we have

$$\bar{\mathbf{v}} = \gamma(r_{21}, r_{22}, r_{23}, T_y, \alpha r_{11}, \alpha r_{12}, \alpha r_{13}, \alpha T_x). \tag{6.9}$$

Since $r_{21}^2 + r_{22}^2 + r_{23}^2 = 1$, from the first three components of $\bar{\mathbf{v}}$ we obtain

$$\sqrt{\bar{v}_1^2 + \bar{v}_2^2 + \bar{v}_3^2} = \sqrt{\gamma^2(r_{21}^2 + r_{22}^2 + r_{23}^2)} = |\gamma|. \tag{6.10}$$

Similarly, since $r_{11}^2 + r_{12}^2 + r_{13}^2 = 1$ and $\alpha > 0$, from the fifth, sixth, and seventh component of $\bar{\mathbf{v}}$ we have

$$\sqrt{\bar{v}_5^2 + \bar{v}_6^2 + \bar{v}_7^2} = \sqrt{\gamma^2 \alpha^2 (r_{11}^2 + r_{12}^2 + r_{13}^2)} = \alpha |\gamma|. \tag{6.11}$$

We can solve (6.10) and (6.11) for $|\gamma|$ as well as the aspect ratio α. We observe that the *first two rows* of the rotation matrix, R, and the first two components of the translation vector, $\mathbf{T}$, can now be determined, *up to an unknown common sign*, from (6.9). Furthermore, the *third row* of the matrix R can be obtained as the vector product of the first two estimated rows thought of as 3-D vectors. Interestingly, this implies that the *sign* of the third row is already fixed, as the entries of the third row remain unchanged if the signs of all the entries of the first two rows are reversed.

☞   Since the computation of the estimated rotation matrix, $\hat{R}$, does not take into account explicitly the orthogonality constraints, $\hat{R}$ cannot be expected to be orthogonal ($\hat{R}\hat{R}^\top = I$). In order to enforce orthogonality on $\hat{R}$, one can resort to the ubiquitous SVD decomposition. Assume the SVD of $\hat{R}$ is $\hat{R} = UDV^\top$. Since the three singular values of a 3×3 orthogonal matrix are all 1, we can simply replace D with the 3×3 identity matrix, I, so that the resulting matrix, $UIV^\top$, is exactly orthogonal (see Appendix, section A.6 for details).

Finally, we determine the unknown sign of the scale factor γ, and finalize the estimates of the parameters. To this purpose, we go back to (6.5), for example, with x instead of $x - o_x$, and recall that for every point $Z^c > 0$ and, therefore, x and $r_{11}X^w + r_{12}Y^w + r_{13}Z^w + T_x$ must have opposite sign. Consequently, it is sufficient to check the sign of $x(r_{11}X^w + r_{12}Y^w + r_{13}Z^w + T_x)$ for one of the points. If

$$x(r_{11}X^w + r_{12}Y^w + r_{13}Z^w + T_x) > 0, \tag{6.12}$$

the signs of the first two rows of $\hat{R}$ and of the first two components of the estimated translation vector must be reversed. Otherwise, no further action is required. A similar argument can be applied to y and $r_{21}X^w + r_{22}Y^w + r_{23}Z^w + T_y$ in (6.6).

At this point, we have determined the rotation matrix, R, the first two components of the translation vector, $\mathbf{T}$, and the aspect ratio, α. We are left with two, still undetermined parameters: T_z, the third component of the translation vector, and f_x, the focal length in horizontal pixel units. Both T_z and f_x can be obtained by least squares from a system of equations like (6.5) or (6.6), written for N points. To do this, for each point (x_i, y_i) we can write

$$x_i(r_{31}X_i^w + r_{32}Y_i^w + r_{33}Z_i^w + T_z) = -f_x(r_{11}X_i^w + r_{12}Y_i^w + r_{13}Z_i^w + T_x), \tag{6.13}$$

then solve the overconstrained system of N linear equations

$$A\begin{pmatrix} T_z \\ f_x \end{pmatrix} = \mathbf{b} \tag{6.14}$$

in the two unknowns T_z and f_x, where

$$A = \begin{bmatrix} x_1 & (r_{11}X_1^w + r_{12}Y_1^w + r_{13}Z_1^w + T_x) \\ x_2 & (r_{11}X_1^w + r_{12}Y_1^w + r_{13}Z_1^w + T_x) \\ \cdot & \cdot \\ \cdot & \cdot \\ \cdot & \cdot \\ x_N & (r_{11}X_N^w + r_{12}Y_N^w + r_{13}Z_N^w + T_x) \end{bmatrix}$$

and

$$b = \begin{pmatrix} -x_1(r_{31}X_1^w + r_{32}Y_1^w + r_{33}Z_1^w) \\ \cdot \\ \cdot \\ \cdot \\ -x_N(r_{31}X_N^w + r_{32}Y_N^w + r_{33}Z_N^w) \end{pmatrix}$$

The least squares solution $(\hat{T}_z, \hat{f}_x)$ of system (6.14) is

$$\begin{pmatrix} \hat{T}_z \\ \hat{f}_x \end{pmatrix} = (A^\top A)^{-1} A^\top \mathbf{b}. \tag{6.15}$$

It remains to be discussed how can we actually acquire an image of N points of known *world* coordinates, and locate the N corresponding image points accurately. One possible solution of this problem can be obtained with the pattern shown in Figure 6.1. The pattern consists of two orthogonal grids of equally spaced black squares drawn on white, perpendicular planes. We let the world reference frame be the 3-D reference frame centered in the lower left corner of the left grid and with the axes parallel to the three directions identified by the calibration pattern. If the horizontal and vertical size of the surfaces and the angle between the surfaces are known with high accuracy (from construction), then the 3-D coordinates of the vertices of each of the square in the *world* reference frame can be easily and accurately determined through simple trigonometry. Finally, the location of the vertices on the image plane can be found by intersecting the edge lines of the corresponding square sides. We now summarize the method:

Algorithm EXPL_PARS_CAL

The input is an image of the calibration pattern described in the text (Figure 6.1) and the location of the image center.

1. Measure the 3-D coordinates of each vertex of the n squares on the calibration pattern in the world reference frame. Let $N = 4n$.

2. In order to find the coordinates in the *image* reference frame of each of the N vertices:
   - locate the image lines defined by the sides of the squares (e.g., using procedures EDGE_COMP and HOUGH_LINES of Chapters 3 and 4).
   - estimate the image coordinates of all the vertices of the imaged squares by intersecting the lines found.

3. Having established the N correspondences between image and world points, compute the SVD of A in (6.8). The solution is the column of V corresponding to the smallest singular value of A.

4. Determine $|\gamma|$ and α from (6.10) and (6.11).

5. Recover the first two rows of R and the first two components of $\mathbf{T}$ from (6.9).

6. Compute the third row of R as the vector product of the first two rows estimated in the previous step, and enforce the orthogonality constraint on the estimate of R through SVD decomposition.

7. Pick a point for which $(x - o_x)$ is noticeably different from 0. If inequality (6.12) is satisfied, reverse the sign of the first two rows of R and of the first two components of $\mathbf{T}$.

8. Set up A and $\mathbf{b}$ of system (6.14), and use (6.15) to estimate T_z and f_x.

The output is formed by α, f_x, and the extrinsic parameters of the viewing camera.

☞   When using a calibration pattern like the one in Figure 6.1, the 3-D squares lie on two different planes. The intersections of lines defined by squares from different world planes do not correspond to any image vertices. You must therefore ensure that your implementation considers only the intersections of pairs of lines associated to the *same* plane of the calibration pattern.

☞   The line equations in image coordinates are computed by least squares, using as many collinear edge points as possible. This process improves the accuracy of the estimates of line parameters and vertex location on the image plane.

6.2.3 Estimating the Image Center

In what follows we describe a simple procedure for the computation of the image center. As a preliminary step, we recall the definition of *vanishing points* from projective geometry, and state a simple theorem suggesting how to determine the image center through the orthocenter[3] of a triangle in the image.

Definition: Vanishing Points

Let $L_i, i = 1, \ldots, N$ be parallel lines in 3-D space, and l_i the corresponding image lines. Due to the perspective projection, the lines L_i appear to meet in a point p, called *vanishing point*, defined as the common intersection of all the image lines l_i.

Orthocenter Theorem: Image Center from Vanishing Points

Let T be the triangle on the image plane defined by the three vanishing points of three mutually orthogonal sets of parallel lines in space. The image center is the orthocenter of T.

The proof of the theorem is left as an exercise (Exercise 6.2). The important fact is, the theorem reduces the problem of locating the image center to one of intersecting

---

[3] The orthocenter of a triangle is the common intersection of the three altitudes.

image lines, and can be created easily on a suitable calibration pattern. In fact, we can use the same calibration pattern (actually, the same image!) of Figure 6.1, already used for EXPL_PARS_CAL, so that EXPL_PARS_CAL and the new algorithm, IMAGE_CENTER_CAL, fit nicely together.

Algorithm IMAGE_CENTER_CAL

The input is an image of the calibration pattern in Figure 6.1, and the output of the first two steps of algorithm EXPL_PARS_CAL.

1. Compute the three vanishing points p_1, p_2, and p_3, determined by the three bundles of lines obtained in step 2 of EXPL_PARS_CAL.
2. Compute the orthocenter, O, of the triangle $p_1 p_2 p_3$.

The output are the image coordinates of the image center, O.

☞   It is essential that the calibration pattern is imaged from a viewpoint guaranteeing that no vanishing point lies much farther than the others from the image center; otherwise, the image lines become nearly parallel, and small inaccuracies in the location of the lines result in large errors in the coordinates of the vanishing point. This can happen if one of the three mutually orthogonal directions is nearly parallel to the image plane, a situation to be definitely avoided. Even with a good viewpoint, it is best to determine the vanishing points using several lines and least squares.

☞   To improve the accuracy of the image center estimate, you should run IMAGE_CENTER_CAL with several views of the calibration patterns, and average the results.

Experience shows that an accurate location of the image center is not crucial for obtaining precise estimates of the other camera parameters (see Further Readings). Be careful, however, as accurate knowledge of the image center *is* required to determine the ray in space identified by an image point (as we shall see, for example, in Chapter 8).

6.3   Camera Parameters from the Projection Matrix

We now move on to the description of a second method for camera calibration. The new method consists in two sequential stages:

1. estimate the projection matrix linking world and image coordinates;
2. compute the camera parameters as closed-form functions of the entries of the projection matrix.

6.3.1   Estimation of the Projection Matrix

As we have seen in Chapter 2, the relation between the 3-D coordinates (X_i^w, Y_i^w, Z_i^w) of a point in space and the 2-D coordinates (x, y) of its projection on the image plane can be written by means of a 3×4 projection matrix, M, according to the equation

$$\begin{pmatrix} u_i \\ v_i \\ w_i \end{pmatrix} = M \begin{pmatrix} X_i^w \\ Y_i^w \\ Z_i^w \\ 1 \end{pmatrix},$$

with

$$x = \frac{u_i}{w_i} = \frac{m_{11} X_i^w + m_{12} Y_i^w + m_{13} Z_i^w + m_{14}}{m_{31} X_i^w + m_{32} Y_i^w + m_{33} Z_i^w + m_{34}}$$

$$y = \frac{v_i}{w_i} = \frac{m_{21} X_i^w + m_{22} Y_i^w + m_{23} Z_i^w + m_{24}}{m_{31} X_i^w + m_{32} Y_i^w + m_{33} Z_i^w + m_{34}}$$

(6.16)

The matrix M is defined up to an arbitrary scale factor and has therefore only 11 independent entries, which can be determined through a homogeneous linear system formed by writing (6.16) for at least 6 world-image point matches. However, through the use of calibration patterns like the one in Figure 6.1, many more correspondences and equations can be obtained and M can be estimated through least squares techniques. If we assume we are given N matches for the homogeneous linear system we have

$$A\mathbf{m} = 0,$$

(6.17)

with

$$A = \begin{bmatrix} X_1 & Y_1 & Z_1 & 1 & 0 & 0 & 0 & 0 & -x_1 X_1 & -x_1 Y_1 & -x_1 Z_1 & -x_1 \\ 0 & 0 & 0 & 0 & X_1 & Y_1 & Z_1 & 1 & -y_1 X_1 & -y_1 Y_1 & -y_1 Z_1 & -y_1 \\ X_2 & Y_2 & Z_2 & 1 & 0 & 0 & 0 & 0 & -x_2 X_2 & -x_2 Y_2 & -x_2 Z_2 & -x_2 \\ 0 & 0 & 0 & 0 & X_2 & Y_2 & Z_2 & 1 & -y_2 X_2 & -y_2 Y_2 & -y_2 Z_2 & -y_2 \\ \cdot & \cdot & \cdot & \cdot & \cdot & \cdot & \cdot & \cdot & \cdot & \cdot & \cdot \\ \cdot & \cdot & \cdot & \cdot & \cdot & \cdot & \cdot & \cdot & \cdot & \cdot & \cdot \\ \cdot & \cdot & \cdot & \cdot & \cdot & \cdot & \cdot & \cdot & \cdot & \cdot & \cdot \\ X_N & Y_N & Z_N & 1 & 0 & 0 & 0 & 0 & -x_N X_N & -x_N Y_N & -x_N Z_N & -x_N \\ 0 & 0 & 0 & 0 & X_N & Y_N & Z_N & 1 & -y_N X_N & -y_N Y_N & -y_N Z_N & -y_N \end{bmatrix}$$

and

$$\mathbf{m} = [m_{11}, m_{12}, \ldots, m_{33}, m_{34}]^\top.$$

Since A has rank 11, the vector $\mathbf{m}$ can be recovered from SVD related techniques as the column of V corresponding to the zero (in practice the smallest) singular value of A, with $A = UDV^\top$ (see Appendix, section A.6). In agreement with the above definition of M, this means that the entries of M are obtained up to an unknown scale factor. The following is the detailed implementation of a method for estimating the matrix M:

Algorithm PROJ_MAT_CALIB

The input is an image of the calibration pattern described in the text (see Figure 6.1, for example).

1. Run the first two steps of EXPL_PARS_CAL.

2. Given N world-image matches, compute the SVD of A, system matrix of (6.17), $A = U D V^\top$. The solution **m** is the column of V corresponding to the smallest singular value of A.

   The output is formed by the entries of the projection matrix, determined up to an unknown scale factor.

6.3.2 Computing Camera Parameters

We now want to express the intrinsic and extrinsic camera parameters as functions of the estimated projection matrix. To avoid confusion, we call $\hat{M}$ the projection matrix estimated through PROJ_MAT_CALIB (and hence $\hat{m}_{ij}$ the generic element of $\hat{M}$).

We first rewrite the full expression for the entries of M, or[4]

$$M = \begin{bmatrix} -f_x r_{11} + o_x r_{31} & -f_x r_{12} + o_x r_{32} & -f_x r_{13} + o_x r_{33} & -f_x T_x + o_x T_z \\ -f_y r_{21} + o_y r_{31} & -f_y r_{22} + o_y r_{32} & -f_y r_{23} + o_y r_{33} & -f_y T_y + o_y T_z \\ r_{31} & r_{32} & r_{33} & T_z \end{bmatrix}. \quad (6.18)$$

☞ Notice that we are now using o_x, o_y, f_x, f_y as the four independent parameters, as opposed to $o_x, o_y, f_x, \alpha = s_y/s_x$ used in method 1.

☞ Notice also that not all 3×4 matrices can be written as functions of the extrinsic and intrinsic parameters as per 6.18. For details, see the Further Readings.

In what follows we also need the 3-D vectors

$$\mathbf{q}_1 = [\hat{m}_{11}, \hat{m}_{12}, \hat{m}_{13}]^\top,$$

$$\mathbf{q}_2 = [\hat{m}_{21}, \hat{m}_{22}, \hat{m}_{23}]^\top,$$

$$\mathbf{q}_3 = [\hat{m}_{31}, \hat{m}_{32}, \hat{m}_{33}]^\top,$$

$$\mathbf{q}_4 = [\hat{m}_{14}, \hat{m}_{24}, m_{34}]^\top.$$

Since M is defined up to a scale factor we can write

$$\hat{M} = \gamma M.$$

The absolute value of the scale factor, $|\gamma|$, can be obtained by observing that $\mathbf{q}_3$ is the last row of the rotation matrix, R. Hence,

$$\sqrt{\hat{m}_{31}^2 + \hat{m}_{32}^2 + \hat{m}_{33}^2} = |\gamma| \sqrt{r_{31}^2 + r_{32}^2 + r_{33}^2} = |\gamma|.$$

---

[4] These expressions were already written in Chapter 2 in order to link the perspective and weak-perspective camera models, but scaled by 1)f, and under the simplifying assumptions $s_x = s_y = 1, o_x = o_y = 0$.

We now divide each entry of $\hat{M}$ by $|\gamma|$, and observe that the resulting, normalized projection matrix differs from M by, at most, a sign change. From now on, therefore, we indicate with $\hat{M}$ the normalized matrix. From the last row of (6.18) we have

$$T_z = \sigma \hat{m}_{34}$$

and

$$r_{3i} = \sigma \hat{m}_{3i}, \quad i = 1, 2, 3,$$

with $\sigma = \pm 1$. Taking the dot products of $\mathbf{q}_3$ with $\mathbf{q}_1$ and $\mathbf{q}_2$ we find

$$o_x = \mathbf{q}_1^\top \mathbf{q}_3$$

and

$$o_y = \mathbf{q}_2^\top \mathbf{q}_3.$$

Then we can recover f_x and f_y:

$$f_x = \sqrt{\mathbf{q}_1^\top \mathbf{q}_1 - o_x^2}$$

$$f_y = \sqrt{\mathbf{q}_2^\top \mathbf{q}_2 - o_y^2},$$

(6.19)

We can now compute the remaining extrinsic parameters as

$$
\begin{aligned}
r_{1i} &= \sigma (o_x \hat{m}_{3i} - \hat{m}_{1i})/f_x, & i = 1, 2, 3 \\
r_{2i} &= \sigma (o_y \hat{m}_{3i} - \hat{m}_{2i})/f_y, & i = 1, 2, 3 \\
T_x &= \sigma (o_x T_z - \hat{m}_{14})/f_x \\
T_y &= \sigma (o_y T_z - \hat{m}_{24})/f_y.
\end{aligned}
$$

(6.20)

As usual, the estimated rotation matrix, $\hat{R}$, is not really orthogonal, and we can find the closest orthogonal matrix as done in section 6.2.2.

☞   You may have noticed that M has 11 independent parameters, but there are only 6 extrinsic and 4 intrinsic parameters. The missing parameter is the angle, θ, formed by the axes of the image reference frame. Here we have exploited the fact that this angle is 90° within great accuracy in all commercial cameras, and is therefore not considered explicitly in the intrinsic parameters set. For details of both linear and nonlinear calibration methods estimating θ as well, see the Further Readings.

We are left to discuss how to determine the sign σ. The sign σ can be obtained from $T_z = \sigma \hat{m}_{34}$, because we know whether the origin of the world reference frame is in front of ($T_z > 0$) or behind ($T_z < 0$) the camera. We do not give the usual algorithm, which would merely restate the equations derived in this section.

6.4 Concluding Remarks

Is there any difference between the two calibration methods presented? Obviously, you should expect the same (in practice, very similar) parameter values from both! Method 2 is probably simpler, but the algorithm on which we based EXPL_PARS_CAL (see Further Readings) is a well-known technique in the computer vision community, and has been implemented and used by many groups. PROJ_MAT_CALIB is useful whenever the projection matrix is sufficient to solve a vision problem, and there is no need to make the individual parameters explicit; an example is referenced in the Further Readings.

We said that the precision of calibration depends on how accurately the world and image reference points are located. But *which* accuracy should one pursue? As the errors on the parameter estimates propagate to the results of the application, the answer depends ultimately on the accuracy requirements of the target application. For instance, inspection systems in manufacturing often require submillimeter accuracies, whereas errors of centimeters can be acceptable in vision systems for outdoors navigation. Locating image reference points or lines at pixel precision is generally unsatisfactory in the former case, but acceptable in the latter. A practical guideline is: *the effort going into improving calibration accuracy should be commensurate to the requirements of the application.*

6.5 Summary

After working through this chapter you should be able to:

- ❑ explain what calibration is and why it is necessary
- ❑ calibrate the intrinsic and extrinsic parameters of an intensity camera
- ❑ estimate the entries of the projection matrix
- ❑ design a calibration pattern, motivating your design choices

6.6 Further Readings

EXPL_PARS_CAL has been adapted from a well-known (but rather involved) paper by Tsai [10], which contains a proof that the rank of the matrix A of EXPL_PARS_CAL is 7 in the ideal case. The orthocenter property and the calibration of the image center through vanishing points is due to Caprile and Torre [2], who also suggest a neat method for calibrating the rotation matrix. Direct calibration of the projection matrix is described, for example, by Faugeras and Toscani [5]. The explicit computation of the calibration parameters from the projection matrix is taken from Faugeras's book [4], which also discusses the conditions under which this is possible (Chapter 3, Section 4). Ayache and Lustman [1] describe a stereo system which requires calibration of the projection matrix, but not explicit knowledge of the camera parameters. The influence on vision algorithms of erroneous estimates of the image center and other camera parameters is discussed by [3, 7, 6]. Thacker and Mayhew [9] describe a method for calibrating a stereo system from arbitrary stereo images.

Recent developments on calibration are presented, for instance, by Faugeras and Maybank [8] who introduce an elegant method based solely on point matches and able to obtain camera calibration without the use of calibration patterns.

6.7 Review

Questions

☐ *6.1* Describe the purpose of calibration and name applications in which calibration is necessary.

☐ *6.2* What is the relation between the translation vectors between two reference frames if the transformation between the frames is (a) first rotation and then translation and (b) *vice versa*? Verify your answer in the trivial case in which rotation is described by the identity matrix.

☐ *6.3* Why does algorithm EXPL_PARS_CAL determine only f_x and not f_y?

☐ *6.4* Discuss the construction requirements of a calibration pattern.

☐ *6.5* How would you estimate the accuracy of a calibration program?

Exercises

○ *6.1* In the case of large fields of view, or if pixels very far from the center of the image are considered, the radial distortion cannot be neglected. A possible procedure for calibrating the distortion parameter k_1 is to rewrite (6.13) with $x_i(1 + k_1 r_i^2)$ in place of x_i with $r_i^2 = x_i^2 + \alpha^2 y_i^2$, or

$$x_i(1 + k_1 r_i^2)(r_{31}X_i^w + r_{32}Y_i^w + r_{33}Z_i^w + T_z) = -f_x(r_{11}X_i^w + r_{12}Y_i^w + r_{13}Z_i^w + T_x).$$

The corresponding nonlinear system of equations for the unknowns f_x, T_z, and k_1 can be solved through gradient descent techniques using the output of EXPL_PARS_CAL as initial guess for f_x and T_z and 0 as initial guess for k_1. Write out the complete equations suggested, and devise a calibration algorithm including k_1 in the intrinsic parameters.

○ *6.2* Prove the orthocenter theorem by geometrical arguments. (*Hint*[5]: Let h be the altitude from the vertex (and vanishing point) v to the side s, and O the projection center. Since both the segments h and vO are orthogonal to s, the plane through h and vO is orthogonal to s and hence to the image plane ...)

○ *6.3* Prove that the vanishing points associated to three coplanar bundles of parallel lines are collinear.

○ *6.4* Estimate the theoretical error of the coordinates of the principal point as a function of the error (uncertainty) of the coordinates of the vanishing points. Can you guess the viewpoint that minimizes error propagation?

---

[5] Hint by Francesco Robbiano.

○ *6.5* Explain why errors in the location of the image center do not affect greatly the estimates of the remaining parameters. (*Hint:* Check out (6.18), (6.19) and (6.20).)

○ *6.6* The direct calibration procedure given in Chapter 2 for a range sensor establishes a direct mapping between pixel and world coordinates. Sketch a calibration/acquisition procedure for the same sensor, but based on the estimation/knowledge of the camera parameters.

Projects

● *6.1* Implement the first (EXPL_PARS_CAL and IMAGE_CENTER_CAL) and second (based on PROJ_MAT_CALIB) calibration algorithm in this chapter. Compare the sensitivity of the estimates of the *intrinsic* parameters obtained by the two methods, by repeating the calibration 20 times for different views of the calibration pattern and without moving the camera.

● *6.2* Compare the sensitivity of the estimates of the *extrinsic* parameters obtained by the two methods, by repeating the procedure above without moving the setup.

References

[1]   N. Ayache and F. Lustman, Trinocular Stereo Vision for Robotics, *IEEE Transactions on Pattern Analysis and Machine Intelligence* Vol 13, pp. 73–85 (1991).

[2]   B. Caprile and V. Torre, Using Vanishing Points for Camera Calibration, *International Journal of Computer Vision* Vol 3, pp. 127–140 (1990).

[3]   K. Danillidis and H.-H. Nagel, Analytical Results on Error Sensitivity of Motion Estimation from Two Views, *Image and Vision Computing* Vol 8, pp. 297–303 (1990).

[4]   O.D. Faugeras, *Three-Dimensional Computer Vision: A Geometric Viewpoint*, MIT Press, Cambridge (MA) (1993).

[5]   O.D. Faugeras and G. Toscani, The Calibration Problem for Stereo, in *Proc. IEEE Conference on Computer Vision and Pattern Recognition '86*, Miami Beach (FL), pp. 15-20 (1986).

[6]   B. Kamgar-Parsi and B. Kamgar-Parsi, Evaluation of Quantization Errors in Computer Vision, *IEEE Transactions on Pattern Analysis and Machine Intelligence*, Vol PAMI-11, No 9, pp. 929–939 (1989).

[7]   R.K. Lenz and R.Y. Tsai, Techniques for Calibration of the Scale Factor and Image Center for High-Accuracy 3-D Machine Vision Metrology, *IEEE Transactions on Pattern Analysis and Machine Intelligence*, Vol. PAMI-10, No 10, pp. 713–720 (1988).

[8]   S. Maybank and O.D. Faugeras, A Theory of Self-Calibration of a Moving Camera, *International Journal of Computer Vision* Vol 8, pp. 123–151 (1992).

[9]   N.A. Thacker and J.E.W. Mayhew, Optimal Combination of Stereo Camera Calibration from Arbitrary Stereo Images, *Image and Vision Computing* Vol 9, pp. 27–32 (1991).

[10] R.Y. Tsai, A Versatile Camera Calibration Technique for High Accuracy 3D Machine Vision Metrology Using Off-the-Shelf TV Cameras and Lenses, *IEEE Journal of Robotics and Automation* Vol RA-3, No 4, pp. 323–344 (1987).

7

Stereopsis

> Two are better than one;
> because they have a good reward for their labor.
>
> *Ecclesiastes* 4:9

This chapter is an introduction to *stereopsis*, or simply *stereo*, in computer vision.

Chapter Overview

Section 7.1 is an informal introduction; it subdivides stereo in two subproblems, *correspondence* and *reconstruction*, and analyzes a simple stereo system.

Section 7.2 deals with the problem of establishing correspondences between image elements of a stereo pair.

Section 7.3 is dedicated to the geometry of stereo, the *epipolar geometry*, and how it can be recovered and used.

Section 7.4 discusses three methods for the reconstruction of the 3-D structure of a scene, each assuming a different amount of knowledge on the intrinsic and extrinsic parameters of the cameras.

What You Need to Know to Understand this Chapter

- Working knowledge of Chapters 2, 4, 5, 6.
- Singular value decomposition and constrained optimization (Appendix, section A.6).
- Basic notions of projective geometry (Appendix, section A.4).

7.1   Introduction

Stereo vision refers to the ability to *infer information on the 3-D structure and distance of a scene from two or more images taken from different viewpoints.* You can learn a great deal of the basic principles (and problems) of a stereo system through a simple experiment. Hold one thumb at arm's length and close the right and left eye alternatively. What do you expect to see? With presumably little surprise you find that the relative position of thumb and background appears to change, depending on which eye is open (or closed).[1] It is precisely this difference in retinal location that is used by the brain to reconstruct a 3-D representation of what we see.

7.1.1   The Two Problems of Stereo

From a computational standpoint, a stereo system must solve two problems. The first, known as *correspondence*, consists in determining *which item in the left eye corresponds to which item in the right eye* (Figure 7.1). A rather subtle difficulty here is that some parts of the scene are visible by one eye only. In the thumb experiment, for example, which part of the background is occluded by your thumb depends on which eye is open. Therefore, a stereo system must also be able to determine the image parts that should *not* be matched.

The second problem that a stereo system must solve is *reconstruction*. Our vivid 3-D perception of the world is due to the interpretation that the brain gives of the computed difference in retinal position, named *disparity*, between corresponding items.[2] The disparities of all the image points form the so-called *disparity map*, which can be displayed as an image. If the geometry of the stereo system is known, the disparity map can be converted to a 3-D map of the viewed scene (the reconstruction). Figure 7.2 shows an example of stereo reconstruction with a human face. Figure 7.3 illustrates an application of computational stereopsis in space research: reconstructing the relief of the surface of Venus from two satellite (SAR) images. The images were recorded by the Magellan spacecraft, and cover an area of approximately 120×40 km; each pixel corresponds to 75 m.

Definitions: Stereo Correspondence and Reconstruction

The correspondence problem: Which parts of the left and right images are projections of the same scene element?

The reconstruction problem: Given a number of corresponding parts of the left and right image, and possibly information on the geometry of the stereo system, what can we say about the 3-D location and structure of the observed objects?

---

[1] If you think this is obvious, try to explain why, with both eyes open, you almost invariably see only *one* thumb, well separated in depth from the background.

[2] This is best demonstrated by the now popular *autostereograms*, in which the perception of depth is induced by no cue other than disparity. See the Further Readings for more on autostereograms.

Figure 7.1 An illustration of the correspondence problem. A matching between corresponding points of an image pair is established (only some correspondences are shown).

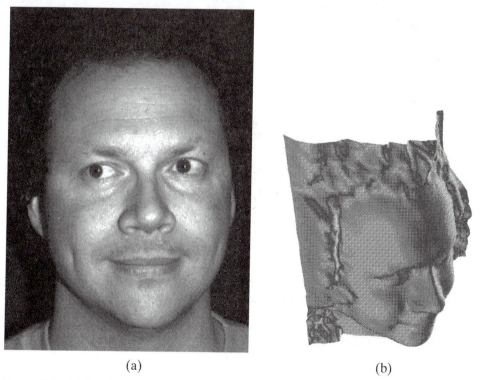

(a)   (b)

Figure 7.2   (a) One image from a stereo pair of Emanuele Trucco's face. (b) 3-D rendering of stereo reconstruction. Courtesy of the Turing Institute, Glasgow (UK).

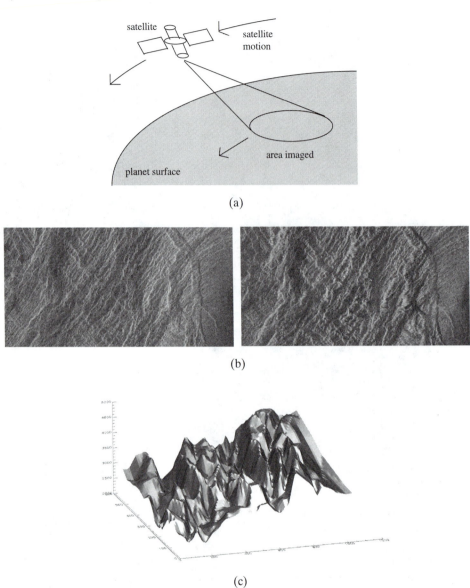

(a)

(b)

(c)

Figure 7.3 Stereo reconstruction of the surface of Venus from a pair of SAR images. (a) Illustration of the application, showing the satellite orbiting around the planet. (b) The stereo pair of the surface of Venus acquired by the Magellan satellite. (c) 3-D rendering of the reconstructed surface. Courtesy of Alois Goller, Institute for Computer Graphics and Vision, Technical University of Graz.

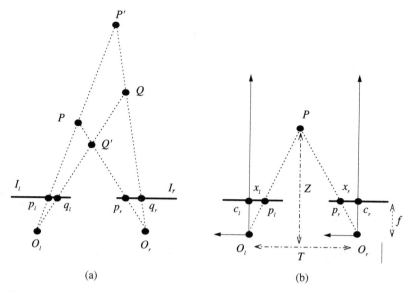

Figure 7.4 A simple stereo system. 3-D reconstruction depends on the solution of the correspondence problem (a); depth is estimated from the disparity of corresponding points (b).

7.1.2 A Simple Stereo System

Before starting our investigation of the correspondence and reconstruction problems with the necessary mathematical machinery, it is useful to learn as much as we can from the very simple model illustrated in Figure 7.4(a). The diagram shows the top view of a stereo system composed of two pinhole cameras. The left and right image planes are coplanar, and represented by the segments I_l and I_r respectively. O_l and O_r are the centers of projection. The optical axes are parallel; for this reason, the *fixation point*, defined as the point of intersection of the optical axes, lies infinitely far from the cameras.

The way in which stereo determines the position in space of P and Q (Figure 7.4(a)) is *triangulation*, that is, by intersecting the rays defined by the centers of projection and the images of P and Q, p_l, p_r, q_l, q_r. Triangulation depends crucially on the solution of the correspondence problem: If (p_l, p_r) and (q_l, q_r) are chosen as pairs of corresponding image points, intersecting the rays $O_l p_l - O_r p_r$ and $O_l q_l - O_r q_r$ leads to interpreting the image points as projections of P and Q; but, if (p_l, q_r) and (q_l, p_r) are the selected pairs of corresponding points, triangulation returns P' and Q'. Note that both interpretations, although dramatically different, stand on an equal footing once we accept the respective correspondences. We will have more to say about the correspondence problem and its solutions in Section 7.2.

Let us now assume that the correspondence problem has been solved, and turn to reconstruction. It is instructive to write the equation underlying the triangulation of Figure 7.4. We concentrate on the recovery of the position of a single point, P, from its projections, p_l and p_r (Figure 7.4(b)). The distance, T, between the centers of projection O_l and O_r, is called the *baseline* of the stereo system. Let x_l and x_r be the coordinates of p_l and p_r with respect to the principal points c_l and c_r, f the common focal length, and Z the distance between P and the baseline. From the similar triangles (p_l, P, p_r) and (O_l, P, O_r) we have

$$\frac{T + x_l - x_r}{Z - f} = \frac{T}{Z}. \tag{7.1}$$

Solving (7.1) for Z we obtain

$$Z = f\frac{T}{d}, \tag{7.2}$$

The larger the disparity, the closer the object.

where $d = x_r - x_l$, the *disparity*, measures the difference in retinal position between the corresponding points in the two images. From (7.2) we see that *depth is inversely proportional to disparity*. You can verify this by looking at moving objects outside: distant objects seem to move more slowly than close ones.

7.1.3 The Parameters of a Stereo System

As shown by (7.2), in our simple example depth depends on the focal length, f, and the stereo baseline, T; the coordinates x_l and x_r are referred to the principal points, c_l and c_r. The quantities f, T, c_l, c_r are the *parameters of the stereo system*, and finding their values is the *stereo calibration* problem. There are two kinds of parameters to be calibrated in a general stereo system.

The Parameters of a Stereo System

The *intrinsic parameters* characterize the transformation mapping an image point from camera to pixel coordinates, in each camera.

The *extrinsic parameters* describe the relative position and orientation of the two cameras.

The intrinsic parameters are the ones introduced in Chapter 2; a minimal set for each camera includes the coordinates of the principal point and the focal lengths in pixel. The extrinsic parameters, instead, are slightly different: they describe the rigid transformation (rotation and translation) that brings the reference frames of the two cameras onto each other.[3]

Since in many cases the intrinsic parameters, or the extrinsic parameters, or both, are unknown, *reconstruction is often a calibration problem*. Rather surprisingly, you will learn in Section 7.4 that a stereo system can compute a great deal of 3-D information without *any* prior knowledge of the stereo parameters (*uncalibrated stereo*). In order

---

[3] In Chapter 2, the same extrinsic parameters were used to define the 3-D rigid motion that brings the camera and the world reference frame onto each other; here, the reference frame of one camera is taken as the world reference frame.

to deal properly with reconstruction, we need to spend some time on the geometry of stereo, the so-called *epipolar geometry* (Section 7.3). As a byproduct, the epipolar geometry will prove useful to get a better understanding of the computational problems of stereo, and to devise more efficient (and effective) correspondence algorithms.

☞   Before starting our investigation of stereo, a word of warning about the validity of the conclusions that can be drawn from the simple stereo model (7.2) referring to Figure 7.4. They illustrate well the main issues of stereo, but are too simple to tell the entire story. In particular, (7.2) may lead you to conclude that the disparity can only decrease as the distance of the object from the cameras increases; instead, in a typical stereo system with converging cameras,[4] disparity actually *increases* with the distance of the objects *from the fixation point*. Clearly, the reason why you cannot infer this property from our example is that the fixation point is at infinity.

7.2   The Correspondence Problem

Let us first discuss the correspondence problem ignoring quantitative knowledge of the cameras' parameters.

7.2.1   Basics

We will start off with the common assumptions underlying most methods for finding correspondences in image pairs.

Assumptions

1. Most scene points are visible from both viewpoints. ✓
2. Corresponding image regions are similar. ✓

These assumptions hold for stereo systems in which the distance of the fixation point from the cameras is much larger than the baseline. In general, however, both assumptions may be false and the correspondence problem becomes considerably more difficult. For the time being, we take the validity of these assumptions for granted and view the correspondence problem as a *search* problem: *given an element in the left image, we search for the corresponding element in the right image.* This involves two decisions:

- which image element to match, and
- which similarity measure to adopt.

☞   We postpone the discussion of the problem arising from the fact that not all the elements of one image have necessarily a corresponding element in the other image.

---

[4] That is, the optical axes intersect in a fixation point at a finite distance from the cameras.

Figure 7.5 An illustration of correlation-based correspondence. We look for the right-image point corresponding to the central pixel of the left-image window. This window is correlated to several windows of the same size in the right image (only a few are shown here). The center of the right-image window producing the highest correlation is the corresponding point sought.

For the sake of convenience, we classify correspondence algorithms in two classes, *correlation-based* and *feature-based* methods, and discuss them separately. Although almost indistinguishable from a conceptual point of view, the two classes lead to quite different implementations: for instance, correlation-based methods apply to the totality of image points; feature-based methods attempt to establish a correspondence between sparse sets of image features.

7.2.2 Correlation-Based Methods

In correlation-based methods, the elements to match are *image windows* of fixed size, and the similarity criterion is a measure of the correlation between windows in the two images. The corresponding element is given by the window that maximizes the similarity criterion within a search region (Figure 7.5). As usual, we give a summary of the algorithm.

Algorithm CORR_MATCHING

The input is a stereo pair of images, I_l (left) and I_r (right).

Let $\mathbf{p}_l$ and $\mathbf{p}_r$ be pixels in the left and right image, $2W + 1$ the width (in pixels) of the correlation window, $R(\mathbf{p}_l)$ the search region in the right image associated with $\mathbf{p}_l$, and $\psi(u, v)$ a function of two pixel values, u, v.

For each pixel $\mathbf{p}_l = [i, j]^\top$ of the left image:

1. for each displacement $\mathbf{d} = [d_1, d_2]^\top \in R(\mathbf{p}_l)$ compute

$$c(\mathbf{d}) = \sum_{k=-W}^{W} \sum_{l=-W}^{W} \psi(I_l(i+k, j+l), I_r(i+k-d_1, j+l-d_2)); \tag{7.3}$$

2. the disparity of $\mathbf{p}_l$ is the vector $\bar{\mathbf{d}} = \left[\bar{d}_1, \bar{d}_2\right]^\top$ that maximizes $c(\mathbf{d})$ over $R(\mathbf{p}_l)$:

$$\bar{\mathbf{d}} = \arg \max_{d \in R} \{c(\mathbf{d})\} .$$

The output is an array of disparities (the *disparity map*), one per each pixel of I_l.

Two widely adopted choices for the function $\psi = \psi(u, v)$ in (7.3) are

$$\psi(u, v) = uv, \tag{7.4}$$

which yields the *cross-correlation* between the window in the left image and the search region in the right image, and

$$\psi(u, v) = -(u - v)^2, \tag{7.5}$$

which performs the so called *SSD (sum of squared differences)* or *block matching*.

☞   The relation between cross-correlation and block matching becomes apparent expanding (7.5) (Exercise 7.3).

Notice that the two choices of ψ in (7.4) and (7.5) lead to the same set of correspondences *if* the *energy* of the right image inside each window, defined as the sum of the squares of the intensity values in the window, is constant across the search region. In many practical situations, this is unlikely to happen, and (7.5) *is usually preferable to (7.4)*. The reason is that SSD, unlike cross-correlation, is not biased by the presence of regions with very small or very large intensity values (see Exercise 7.3 for a quantitative estimation of this effect and the definition of *normalized cross-correlation*).

☞   In an implementation of CORR_MATCHING it is certainly worthwhile to precompute and store the values of the function ψ in a lookup table. For most choices of ψ, this is likely to speed up the algorithm substantially.

We must still discuss how to choose W and R in the implementation of the algorithm. The window width, $2W + 1$, is actually a free parameter, and its choice is left to your ability to grasp the most important spatial scale of the problem from the images you are dealing with.[5]

Fortunately, something more can be said on the initial location and size of the search region, $R(\mathbf{p}_l)$. If the cameras are fixating a common point at a distance much larger than the baseline, the initial location can be chosen to be $\mathbf{p}_r = [i, j]^\top$; that is, choosing the pixel in the right image at exactly the same location of the pixel $\mathbf{p}_l = [i, j]^\top$ in the left image. The size of $R(\mathbf{p}_l)$ can be estimated from the maximum range of

---

[5] See the Further Readings for stereo systems determining the size of the window automatically.

distances that you expect to find in the scene (remember that disparity increases with the inverse of the distance from the fixation point). If nothing can be assumed on the geometry of the two cameras, the initialization of the search region in the right image is more difficult. At first sight, it might even appear that for each point in the left image you have to search over the entire right image. Fortunately, as shown in Section 7.3, *the search region can always be reduced to a 1-D segment*, independent of the relative position of the two cameras.

7.2.3 Feature-based Methods

Feature-based methods restrict the search for correspondences to a sparse set of features. Instead of image windows, they use numerical and symbolic properties of features, available from feature descriptors (Chapters 4 and 5); instead of correlationlike measures, they use a measure of the distance between feature descriptors. Corresponding elements are given by the most similar feature pair, the one associated to the minimum distance.

Most methods narrow down the number of possible matches for each feature by enforcing suitable *constraints* on feasible matches. These constraints can be

- *geometric*, like the *epipolar constraint* discussed in the next section, or
- *analytical*, for instance the *uniqueness constraint* (each feature can at most have one match) or the *continuity constraint* (disparity varies continuously almost everywhere across the image).

Here, we restrict our attention to *unconstrained methods*; pointers to feature-based methods relying on geometric and analytical constraints are given in the Further Readings, and Section 7.3 gives you the basic elements necessary to add the epipolar constraint to our correspondence algorithms.

Typical examples of image features used in stereo are edge points, lines, and corners (either points where edge lines meet, or corners formed by intensity patterns as described in Chapter 4). For example, a feature descriptor for a line could contain values for

- the length, l
- the orientation, o
- the coordinates of the midpoint, $[x, y]^\top$
- the average contrast along the edge line, c

For virtually every possible feature, there is a long list of correspondence algorithms. As you might expect, no single feature type will work at best in all situations. The choice of the feature type (and correspondence method) depends on numerous factors, like the kind of objects you are looking at, the overall conditions of illumination, and the average image contrast. As an example, we suggest a simplified scheme for matching edge lines, assuming the line descriptor above.[6]

An example of similarity criterion between feature descriptors is the inverse of the weighted average, S, of the distances between each of the properties in the descriptors:

$$S = \frac{1}{w_0(l_l - l_r)^2 + w_1(\theta_l - \theta_r)^2 + w_2(m_l - m_r)^2 + w_3(c_l - c_r)^2}, \qquad (7.6)$$

where $w_0, \ldots, w_3$ are weights, and the subscripts l and r refer to the left and right image, respectively. This leaves us with the nontrivial task of determining the weights that yield the best matches. A possible strategy is to determine a working point for the various weights from a subset of easy matches that you believe to be correct. In general, there is not much to say about this problem, except perhaps that if you build a complex and heterogeneous feature descriptor, determining the weights becomes a difficult problem of parameter estimation.

A very simple, feature-based correspondence algorithm is sketched below.

Algorithm FEATURE_MATCHING

The input is a stereo pair of images, I_l and I_r, and two corresponding sets of feature descriptors.

Let $R(f_l)$ be the search region in the right image associated with a feature descriptor f_l, and $\mathbf{d}(f_l, f_r)$ the disparity between two corresponding features, f_l and f_r.

For each f_l in the left image set:

1. Compute the similarity measure between f_l and each image feature in $R(f_l)$.

2. Select the right-image feature, f_r that maximizes the similarity measure.

3. Save the correspondence and the disparity of f_l (the displacement between the points defining the feature's position).

The output is formed by a list of feature correspondences and a disparity map.

☞   The initial location and size of the search region, R, can be determined as in the case of correlation-based methods.

7.2.4 Concluding Remarks

Unfortunately, there is no cut-and-dried correspondence method giving optimal results under all possible circumstances. You have to live with the fact that choosing a method depends on factors like the application, the available hardware, or the software requirements. Having said that, it is useful to keep in mind a few general considerations.

Correlation-based methods are certainly easier to implement (and debug) and provide dense disparity maps. As you may guess, the latter property can be very helpful

---

[6] In reality, feature-based methods can be a great deal more complicated than our solution, especially when enforcing constraints on the search.

for the purpose of reconstructing surfaces. They need textured images to work well. However, due to foreshortening effects and change in illumination direction, they are inadequate for matching image pairs taken from very different viewpoints. Also, the interpolation necessary to refine correspondences from pixel to subpixel precision can make correlation-based matching quite expensive.[7]

Feature-based methods are suitable when *a priori* information is available about the scene, so that optimal features can be used. A typical example is the case of indoor scenes, which usually contain many straight lines but rather untextured surfaces. Feature-based algorithms can also prove faster than correlation-based ones, but any comparison of specific algorithms must take into account the cost of producing the feature descriptors. The sparse disparity maps generated by these methods may look inferior to the dense maps of correlation-based matching, but in some applications (e.g., visual navigation) they may well be all you need in order to perform the required tasks successfully. Another advantage of feature-based techniques is that they are relatively insensitive to illumination changes and highlights.

The performance of any correspondence methods is jeopardised by *occlusions* (points with no counterpart in the other image) and *spurious matches* (false corresponding pairs created by noise). Appropriate constraints reduce the effects of both phenomena: two important ones are the *left-right consistency constraint* (only corresponding pairs found matching left-to-right *and* right-to-left are accepted), and the *epipolar constraint*, explained in the next section.

7.3   Epipolar Geometry

We now move on to study the geometry of stereo in its full generality. This will enable us to clarify what information is needed in order to perform the search for corresponding elements only along image lines. First of all, we need to establish some basic notations.

7.3.1   Notation

The geometry of stereo, known as *epipolar geometry*, is shown in Figure 7.6. The figure shows two pinhole cameras, their projection centers, O_l and O_r, and image planes, π_l and π_r. The focal lengths are denoted by f_l and f_r. As usual, each camera identifies a 3-D reference frame, the origin of which coincides with the projection center, and the Z-axis with the optical axis. The vectors $\mathbf{P}_l = [X_l, Y_l, Z_l]^\top$ and $\mathbf{P}_r = [X_r, Y_r, Z_r]^\top$ refer to the same 3-D point, P, thought of as a vector in the left and right camera reference frames respectively (Figure 7.6). The vectors $\mathbf{p}_l = [x_l, y_l, z_l]^\top$ and $\mathbf{p}_r = [x_r, y_r, z_r]^\top$ refer to the projections of P onto the left and right image plane respectively, and are expressed in the corresponding reference frame (Figure 7.6). Clearly, for all the image points we have $z_l = f_l$ or $z_r = f_r$, according to the image. Since each image plane can be thought of as a subset of the projective space P^2, image points can be equivalently thought of as points of the projective space P^2 (see Appendix, section A.4).

---

[7] One of the projects suggested at the end of this chapter deals with a parsimonious implementation of correlation-based matching.

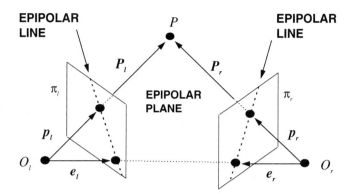

Figure 7.6   The epipolar geometry.

☞   Note that point vectors denoted by the *same* bold capital letter but by *different* subscripts, like $\mathbf{P}_l$ and $\mathbf{P}_r$ identify the *same point* in space. The subscript l or r tells you *the reference frame in which the vectors are expressed* (left or right). Instead, point vectors denoted by the *same* bold small letter but by a *different* subscript, like $\mathbf{p}_l$ and $\mathbf{p}_r$, identify *different points* in space (i.e., belonging to different image planes). In this case, the subscript tells you *also* the image plane on which the vectors lie. This is a slightly unfair but very effective abuse of notation.

7.3.2   Basics

The reference frames of the left and right cameras are related via the extrinsic parameters. These define a rigid transformation in 3-D space, defined by a translation vector, $\mathbf{T} = (O_r - O_l)$, and a rotation matrix, R. Given a point P in space, the relation between $\mathbf{P}_l$ and $\mathbf{P}_r$ is therefore

$$\mathbf{P}_r = R(\mathbf{P}_l - \mathbf{T}). \tag{7.7}$$

The name *epipolar geometry* is used because the points at which the line through the centers of projection intersects the image planes (Figure 7.6) are called *epipoles*. We denote the left and right epipole by $\mathbf{e}_l$ and $\mathbf{e}_r$ respectively. By construction, *the left epipole is the image of the projection center of the right camera and vice versa.*

☞   Notice that, if the line through the centers of projection is parallel to one of the image planes, the corresponding epipole is the point at infinity of that line.

The relation between a point in 3-D space and its projections is described by the usual equations of perspective projection, in vector form:

$$\mathbf{p}_l = \frac{f_l}{Z_l}\mathbf{P}_l \tag{7.8}$$

and

$$\mathbf{p}_r = \frac{f_r}{Z_r}\mathbf{P}_r. \tag{7.9}$$

The practical importance of epipolar geometry stems from the fact that the plane identified by P, O_l, and O_r, called *epipolar plane*, intersects each image in a line, called *epipolar line* (see Figure 7.6). Consider the triplet P, $\mathbf{p}_l$, and $\mathbf{p}_r$. Given $\mathbf{p}_l$, P can lie anywhere on the ray from O_l through $\mathbf{p}_l$. But, since the image of this ray in the right image is the epipolar line through the corresponding point, $\mathbf{p}_r$, *the correct match must lie on the epipolar line*. This important fact is known as the *epipolar constraint*. It establishes a mapping between points in the left image and lines in the right image and *vice versa*.

☞   Incidentally, since all rays include the projection center by construction, this also proves that all the epipolar lines go through the epipole.

So, if we determine the mapping between points on, say, the left image and corresponding epipolar lines on the right image, we can restrict the search for the match of $\mathbf{p}_l$ along the corresponding epipolar line. *The search for correspondences is thus reduced to a 1-D problem.* Alternatively, the same knowledge can be used to verify whether or not a candidate match lies on the corresponding epipolar line. This is usually a most effective procedure to *reject false matches* due to occlusions. Let us now summarize the main ideas encountered in this section:

Definition: Epipolar Geometry

Given a stereo pair of cameras, any point in 3-D space, P, defines a plane, π_P, going through P and the centers of projection of the two cameras. The plane π_P is called *epipolar plane*, and the lines where π_P intersects the image planes *conjugated epipolar lines*. The image in one camera of the projection center of the other is called *epipole*.

Properties of the Epipoles

With the exception of the epipole, only one epipolar line goes through any image point.
    All the epipolar lines of one camera go through the camera's epipole.

Definition: Epipolar Constraint

Corresponding points must lie on conjugated epipolar lines.

The obvious question at this point is, can we estimate the epipolar geometry? Or equivalently, how do we determine the mapping between points in one image and epipolar lines in the other? This is the next problem we consider. Its solution also makes clear the relevance of epipolar geometry for reconstruction.

7.3.3 The Essential Matrix, E

The equation of the epipolar plane through P can be written as the coplanarity condition of the vectors $\mathbf{P}_l$, $\mathbf{T}$, and $\mathbf{P}_l - \mathbf{T}$ (Figure 7.6), or

$$(\mathbf{P}_l - \mathbf{T})^\top \mathbf{T} \times \mathbf{P}_l = 0.$$

Using (7.7), we obtain

$$(R^\top \mathbf{P}_r)^\top \mathbf{T} \times \mathbf{P}_l = 0. \tag{7.10}$$

Recalling that a vector product can be written as a multiplication by a rank-deficient matrix, we can write

$$\mathbf{T} \times \mathbf{P}_l = S\mathbf{P}_l$$

where

$$S = \begin{bmatrix} 0 & -T_z & T_y \\ T_z & 0 & -T_x \\ -T_y & T_x & 0 \end{bmatrix}. \tag{7.11}$$

Using this fact, (7.10) becomes

$$\mathbf{P}_r^\top E\mathbf{P}_l = 0, \tag{7.12}$$

with

$$E = RS. \tag{7.13}$$

Note that, by construction, S has always rank 2. The matrix E is called the *essential matrix* and *establishes a natural link between the epipolar constraint and the extrinsic parameters of the stereo system*. You will learn how to recover the extrinsic parameters from the essential matrix in the next section. In the meantime, observe that, using (7.8) and (7.9), and dividing by $Z_r Z_l$, (7.12) can be rewritten as

$$\mathbf{p}_r^\top E\mathbf{p}_l = 0. \tag{7.14}$$

As already mentioned, the image points $\mathbf{p}_l$ and $\mathbf{p}_r$, which lie on the left and right image planes respectively, can be regarded as points in the projective plane P^2 defined by the left and right image planes respectively (Appendix, section A.4). Consequently, you are entitled to think of $E\mathbf{p}_l$ in (7.14) as the projective line in the right plane, $\mathbf{u}_r$, that goes through $\mathbf{p}_r$ and the epipole $\mathbf{e}_r$:

$$\mathbf{u}_r = E\mathbf{p}_l. \tag{7.15}$$

As shown by (7.14) and (7.15), *the essential matrix is the mapping between points and epipolar lines we were looking for*.

☞   Notice that the whole discussion used coordinates in the *camera* reference frame, but what we actually measure from images are pixel coordinates. Therefore, in order to be able to make profitable use of the essential matrix, we need to know the transformation from *camera coordinates* to *pixel coordinates*, that is, the intrinsic parameters. This limitation is removed in the next section, but at a price.

7.3.4 The Fundamental Matrix, F

We now show that the mapping between points and epipolar lines can be obtained from corresponding points only, *with no prior information on the stereo system.*

Let M_l and M_r be the matrices of the intrinsic parameters (Chapter 2) of the left and right camera respectively. If $\bar{\mathbf{p}}_l$ and $\bar{\mathbf{p}}_r$ are the points in *pixel* coordinates corresponding to $\mathbf{p}_l$ and $\mathbf{p}_r$ in camera coordinates, we have

$$\mathbf{p}_l = M_l^{-1}\bar{\mathbf{p}}_l \tag{7.16}$$

and

$$\mathbf{p}_r = M_r^{-1}\bar{\mathbf{p}}_r. \tag{7.17}$$

By substituting (7.16) and (7.17) into (7.14), we have

$$\bar{\mathbf{p}}_r^\top F \bar{\mathbf{p}}_l = 0, \tag{7.18}$$

where

$$F = M_r^{-\top} E M_l^{-1}. \tag{7.19}$$

F is named *fundamental matrix*. The essential and fundamental matrix, as well as (7.14) and (7.18), are formally very similar. As with $E\mathbf{p}_l$ in (7.14), $F\bar{\mathbf{p}}_l$ in (7.18) can be thought of as the equation of the projective epipolar line, $\bar{\mathbf{u}}_r$, that correspond to the point $\bar{\mathbf{p}}_l$, or

$$\bar{\mathbf{u}}_r = F\bar{\mathbf{p}}_l. \tag{7.20}$$

The most important difference between (7.15) and (7.20), and between the essential and fundamental matrices, is that *the fundamental matrix is defined in terms of pixel coordinates, the essential matrix in terms of camera coordinates.* Consequently, if you can estimate the fundamental matrix from a number of point matches in pixel coordinates, *you can reconstruct the epipolar geometry with no information at all on the intrinsic or extrinsic parameters.*

☞   This indicates that the epipolar constraint, as the mapping between points and corresponding epipolar lines, can be established with *no* prior knowledge of the stereo parameters.

The definitions and basic mathematical properties of these two important matrices are worth a summary.

Definition: Essential and Fundamental Matrices

For each pair of corresponding points $\mathbf{p}_l$ and $\mathbf{p}_r$ in *camera* coordinates, the *essential matrix* satisfies the equation

$$\mathbf{p}_r^\top E \mathbf{p}_l = 0.$$

For each pair of corresponding points $\bar{\mathbf{p}}_l$ and $\bar{\mathbf{p}}_r$ in *pixel* coordinates, the *fundamental matrix* satisfies the equation

$$\bar{\mathbf{p}}_r^\top F \bar{\mathbf{p}}_l = 0.$$

Properties

Both matrices enable full reconstruction of the epipolar geometry.

If M_l and M_r are the matrices of the intrinsic parameters, the relation between the essential and fundamental matrices is given by

$$F = M_r^{-T} E M_l^{-1}.$$

The essential matrix:

1. encodes information on the extrinsic parameters only (see (7.13))
2. has rank 2, since S in (7.11) has rank 2 and R full rank
3. its two nonzero singular values are equal

The fundamental matrix:

1. encodes information on both the intrinsic and extrinsic parameters
2. has rank 2, since T_l and T_r have full rank and E has rank 2

7.3.5   Computing E and F: The Eight-point Algorithm

How do we compute the essential and fundamental matrices? Of the various methods possible, the eight-point algorithm is by far the simplest and definitely the one you cannot ignore (if you are curious about other techniques, look into the Further Readings). We consider here the fundamental matrix only, and leave it to you to work out the straightforward modification needed to recover the essential matrix.

The idea behind the eight-point algorithm is very simple. Assume that you have been able to establish n point correspondences between the images. Each correspondence gives you a homogeneous linear equation like (7.18) for the nine entries of F; these equations form a homogeneous linear system. If you have at least eight correspondences (i.e., $n \geq 8$) and the n points do not form degenerate configurations,[8] the nine entries of F can be determined as the nontrivial solution of the system. Since the system is homogeneous, the solution is unique up to a signed scaling factor. If one uses more than eight points, so that the system is overdetermined, the solution can once again be obtained by means of SVD related techniques. If A is the system's matrix and $A = UDV^\top$, the solution is the column of V corresponding to the only null singular value of A (see Appendix, section A.6),

☞   Because of noise, numerical errors and inaccurate correspondences, A is more likely to be full rank, and the solution is the column of V associated with the *least* singular value of A.

☞   The estimated fundamental matrix is almost certainly nonsingular. We can enforce the singularity constraint by adjusting the entries of the estimated matrix F as done in Chapter 6

---

[8] For a thorough discussion of the degenerate configurations of eight or more points, as well as of the instabilities in the estimation of the essential and fundamental matrices, see the Further Readings.

for rotation matrices: we compute the singular value decomposition of the estimated matrix, $\hat{F} = UDV^\top$, and set the smallest singular value on the diagonal of the matrix D equal to 0. If D' is the corrected D matrix, the corrected estimate, F' is given by $F' = UD'V^\top$ (see Appendix, section A.6).

The following is the basic structure of the eight-point algorithm:

Algorithm EIGHT_POINT

The input is formed by n point correspondences, with $n \geq 8$.

1. Construct system (7.18) from n correspondences. Let A be the $n \times 9$ matrix of the coefficients of the system and $A = UDV^\top$ the SVD of A.

2. The entries of F (up to an unknown, signed scale factor) are the components of the column of V corresponding to the least singular value of A.

3. To enforce the singularity constraint, compute the singular value decomposition of F:

$$F = UDV^\top.$$

4. Set the smallest singular value in the diagonal of D equal to 0; let D' be the corrected matrix.

5. The corrected estimate of F, F', is finally given by

$$F' = UD'V^\top.$$

The output is the estimate of the fundamental matrix, F'.

☞ In order to avoid numerical instabilities, the eight-point algorithm should be implemented with care. The most important action to take is *to normalize the coordinates of the corresponding points so that the entries of A are of comparable size*. Typically, the first two coordinates (in pixels) of an image point are referred to the top left corner of the image, and can vary between a few pixels to a few hundreds; the differences can make A seriously ill-conditioned (Appendix, section A.6). To make things worse, the third (homogeneous) coordinate of image points is usually set to one. A simple procedure to avoid numerical instability is to translate the first two coordinates of each point to the centroid of each data set, and scale the norm of each point so that the average norm over the data set is 1. This can be accomplished by multiplying each left (right) point by two suitable 3×3 matrices, H_l and H_r (see Exercise 7.6 for details on how to compute both H_l and H_r). The algorithm EIGHT_POINT is then used to estimate the matrix $\bar{F} = H_r F H_l$, and F obtained as $H_r^{-1} \bar{F} H_l^{-1}$.

7.3.6 Locating the Epipoles from E and F

We can now establish the relation between the epipoles and the two matrices E *and* F. Consider for example the fundamental matrix, F. Since $\bar{e}_l$ lies on all the epipolar lines of the left image, we can rewrite (7.18) as

$$\bar{p}_r^\top F \bar{e}_l = 0$$

for every $\bar{\mathbf{p}}_r$. But since F is not identically zero, this is possible if and only if

$$F\bar{\mathbf{e}}_l = 0. \tag{7.21}$$

From (7.21) and the fact that F has rank 2, it follows that *the epipole, $\bar{\mathbf{e}}_l$, is the null space of F*; similarly, $\bar{\mathbf{e}}_r$ is the null space of $F^\top$.

We are now in a position to present an algorithm for finding the epipoles. Accurate epipole localization is helpful for refining the location of corresponding epipolar lines, checking the geometric consistency of the entire construction, simplifying the stereo geometry, and recovering 3-D structure in the case of uncalibrated stereo.

Again we present the algorithm in the case of the fundamental matrix. The adaptation to the case of the essential matrix is even simpler than before. The algorithm follows easily from (7.21): To determine the location of the epipoles, it is sufficient to find the null spaces of F and $F^\top$.

☞   These can be determined, for instance, from the singular value decomposition $F = UDV^\top$ and $F^\top = VDU^\top$ as column of V and U respectively corresponding to the null singular value in the diagonal matrix D.

Algorithm EPIPOLES_LOCATION

The input is the fundamental matrix F.

1. Find the SVD of F, that is, $F = UDV^\top$.
2. The epipole $\mathbf{e}_l$ is the column of V corresponding to the null singular value.
3. The epipole $\mathbf{e}_r$ is the column of U corresponding to the null singular value.

The output are the epipoles, $\mathbf{e}_l$ and $\mathbf{e}_r$.

☞   Notice that we can safely assume that there is exactly one singular value equal to 0 because algorithm EIGHT_POINT enforces the singularity constraint explicitly.

It has to be noticed that there are alternative methods to locate the epipoles, not based on the fundamental matrix and requiring as few as 6 point correspondences. More about them in the Further Readings.

7.3.7 Rectification

Before moving on to the problem of 3-D reconstruction, we want to address the issue of *rectification*. Given a pair of stereo images, rectification determines a transformation (or *warping*) of each image such that *pairs of conjugate epipolar lines become collinear and parallel to one of the image axes*, usually the horizontal one. Figure 7.7 shows an example. The importance of rectification is that the correspondence problem, which involves 2-D search in general, *is reduced to a 1-D search on a scanline identified trivially*. In other words, to find the point corresponding to (i_l, j_l) of the left image, we just look along the scanline $j = j_l$ in the right image.

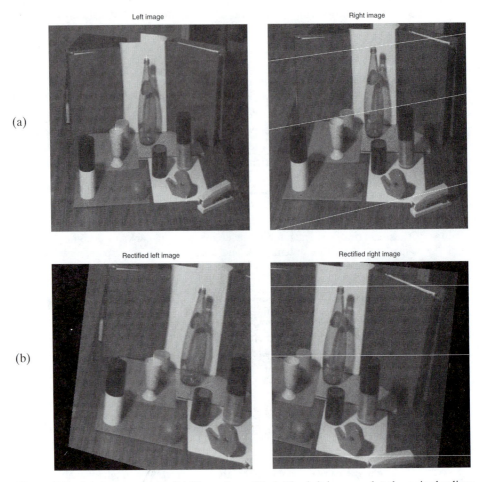

Figure 7.7 (a) A stereo pair. (b) The pair rectified. The left images plot the epipolar lines corresponding to the points marked in the right pictures. Stereo pair courtesy of INRIA (France).

Let us begin by stating the problem and our assumptions.

Assumptions and Problem Statement

Given a stereo pair of images, the intrinsic parameters of each camera, and the extrinsic parameters of the system, R and $\mathbf{T}$, compute the image transformation that makes conjugated epipolar lines collinear and parallel to the horizontal image axis.

The assumption of knowing the intrinsic and extrinsic parameters is not strictly necessary (see Further Readings) but leads to a very simple technique. How do we

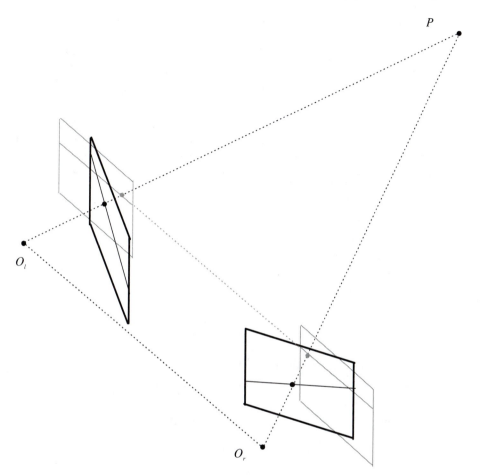

Figure 7.8 Rectification of a stereo pair. The epipolar lines associated to a 3-D point *P* in the original cameras (black lines) become collinear in the rectified cameras (light grey). Notice that the original cameras can be in any position, and the optical axes may not intersect.

go about computing the rectifying image transformation? The rectified images can be thought of as acquired by a new stereo rig, obtained by rotating the original cameras around their optical centers. This is illustrated in Figure 7.8, which shows also how the points of the rectified images are determined from the points of the original images and their corresponding projection rays.

We proceed to describe a rectification algorithm assuming, without losing generality, that in both cameras

1. the origin of the image reference frame is the principal point;
2. the focal length is equal to f.

The algorithm consists of four steps:

- Rotate the left camera so that the epipole goes to infinity along the horizontal axis.
- Apply the same rotation to the right camera to recover the original geometry.
- Rotate the right camera by R.
- Adjust the scale in both camera reference frames.

To carry out this method, we construct a triple of mutually orthogonal unit vectors $\mathbf{e}_1$, $\mathbf{e}_2$, and $\mathbf{e}_3$. Since the problem is underconstrained, we are going to make an arbitrary choice. The first vector, $\mathbf{e}_1$, is given by the epipole; since the image center is in the origin, $\mathbf{e}_1$ coincides with the direction of translation, or

$$\mathbf{e}_1 = \frac{\mathbf{T}}{\|\mathbf{T}\|}.$$

The only constraint we have on the second vector, $\mathbf{e}_2$, is that it must be orthogonal to $\mathbf{e}_1$. To this purpose, we compute and normalize the cross product of $\mathbf{e}_1$ with the direction vector of the optical axis, to obtain

$$\mathbf{e}_2 = \frac{1}{\sqrt{T_x^2 + T_y^2}} \left[-T_y, T_x, 0 \right]^\top.$$

The third unit vector is unambiguously determined as

$$\mathbf{e}_3 = \mathbf{e}_1 \times \mathbf{e}_2.$$

It is easy to check that the orthogonal matrix defined as

$$R_{rect} = \begin{pmatrix} \mathbf{e}_1^\top \\ \mathbf{e}_2^\top \\ \mathbf{e}_3^\top \end{pmatrix} \tag{7.22}$$

rotates the left camera about the projection center in such a way that the epipolar lines become parallel to the horizontal axis. This implements the first step of the algorithm. Since the remaining steps are straightforward, we proceed to give the customary algorithm:

Algorithm RECTIFICATION

The input is formed by the intrinsic and extrinsic parameters of a stereo system and a set of points in each camera to be rectified (which could be the whole images). In addition, Assumptions 1 and 2 above hold.

1. Build the matrix R_{rect} as in (7.22);
2. Set $R_l = R_{rect}$ and $R_r = R R_{rect}$;

3. For each left-camera point, $\mathbf{p}_l = [x, y, f]^\top$ compute

$$R_l\mathbf{p}_l = [x', y', z']$$

and the coordinates of the corresponding rectified point, $\mathbf{p}'_l$, as

$$\mathbf{p}'_l = \frac{f}{z'}[x', y', z'].$$

4. Repeat the previous step for the right camera using R_r and $\mathbf{p}_r$.

The output is the pair of transformations to be applied to the two cameras in order to rectify the two input point sets, as well as the rectified sets of points.

Notice that the rectified coordinates are in general not integer. Therefore, if you want to obtain integer coordinates (for instance if you are rectifying the whole images), you should implement RECTIFICATION backwards, that is, starting from the *new* image plane and applying the *inverse* transformations, so that the pixel values in the *new* image plane can be computed as a bilinear interpolation of the pixel values in the *old* image plane.

☞   A rectified image is not in general contained in the same region of the image plane as the original image. You may have to alter the focal lengths of the rectified cameras to keep all the points within images of the same size as the original.

We are now fully equipped to deal with the reconstruction problem of stereo.

7.4 3-D Reconstruction

We have learned methods for solving the correspondence problem and determining the epipolar geometry from at least eight point correspondences. At this point, the 3-D reconstruction that can be obtained depends on the amount of *a priori* knowledge available on the parameters of the stereo system; we can identify three cases.[9] First, if both intrinsic and extrinsic parameters are known, you can solve the reconstruction problem unambiguously by triangulation, as detailed in section 7.1. Second, if only the intrinsic parameters are known, you can still solve the problem and, at the same time, estimate the extrinsic parameters of the system, but only *up to an unknown scaling factor*. Third, if the pixel correspondences are the only information available, and neither the intrinsic nor the extrinsic parameters are known, you can still obtain a reconstruction of the environment, but only *up to an unknown, global projective transformation*. Here is a visual summary.

---

[9] In reality there are several intermediate cases, but we concentrate on these three for simplicity.

A Priori Knowledge	3-D Reconstruction from Two Views
Intrinsic and extrinsic parameters	Unambiguous (absolute coordinates)
Intrinsic parameters only	Up to an unknown scaling factor
No information on parameters	Up to an unknown projective transformation of the environment

We now consider these three cases in turn.

7.4.1 Reconstruction by Triangulation

This is the simplest case. If you know both the intrinsic and the extrinsic parameters of your stereo system, reconstruction is straightforward.

Assumptions and Problem Statement

Under the assumption that the intrinsic and extrinsic parameters are known, compute the 3-D location of the points from their projections, $\mathbf{p}_l$ and $\mathbf{p}_r$.

As shown in Figure 7.6, the point P, projected into the pair of corresponding points $\mathbf{p}_l$ and $\mathbf{p}_r$, lies at the intersection of the two rays from O_l through $\mathbf{p}_l$ and from O_r through $\mathbf{p}_r$ respectively. In our assumptions, the rays are known and the intersection can be computed. The problem is, since parameters and image locations are known only approximately, *the two rays will not actually intersect in space*; their intersection can only be estimated as the point of minimum distance from both rays. This is what we set off to do.

Let $a\mathbf{p}_l$ ($a \in \mathbb{R}$) be the ray, l, through O_l and $\mathbf{p}_l$. Let $\mathbf{T} + bR^\top\mathbf{p}_r$ ($b \in \mathbb{R}$) be the ray, r, through O_r and $\mathbf{p}_r$ expressed in the left reference frame. Let $\mathbf{w}$ be a vector orthogonal to both l and r. Our problem reduces to determining the midpoint, P', of the segment parallel to $\mathbf{w}$ that joins l and r (Figure 7.9).

This is very simple because the endpoints of the segment, say $a_0\mathbf{p}_l$ and $\mathbf{T} + b_0R^\top\mathbf{p}_r$, can be computed solving the linear system of equations

$$a\mathbf{p}_l - bR^\top\mathbf{p}_r + c(\mathbf{p}_l \times R^\top\mathbf{p}_r) = \mathbf{T} \tag{7.23}$$

for a_0, b_0, and c_0. We summarize this simple method below:

Algorithm TRIANG

All vectors and coordinates are referred to the left camera reference frame. The input is formed by a set of corresponding points; let $\mathbf{p}_l$ and $\mathbf{p}_r$ be a generic pair.

Let $a\mathbf{p}_l$, $a \in \mathbb{R}$, be the ray l through O_l ($a = 0$) and $\mathbf{p}_l$ ($a = 1$). Let $\mathbf{T} + bR^\top\mathbf{p}_r$, $b \in \mathbb{R}$, the ray r through O_r ($b = 0$) and $\mathbf{p}_r$ ($b = 1$). Let $\mathbf{w} = \mathbf{p}_l \times R^\top\mathbf{p}_r$ the vector orthogonal to both l and r, and $a\mathbf{p}_l + c\mathbf{w}$, $c \in \mathbb{R}$, the line w through $a\mathbf{p}_l$ (for some fixed a) and parallel to $\mathbf{w}$.

1. Determine the endpoints of the segment, s, belonging to the line parallel to $\mathbf{w}$ that joins l and r, $a_0\mathbf{p}_l$ and $\mathbf{T} + b_0R^\top\mathbf{p}_r$, by solving (7.23).
2. The triangulated point, P', is the midpoint of the segment s.

The output is the set of reconstructed 3-D points.

☞   The determinant of the coefficients of system (7.23) is the triple product of $\mathbf{p}_l$, $R^\top\mathbf{p}_r$, and $\mathbf{p}_l \times R^\top\mathbf{p}_r$. Therefore, as expected from geometric considerations, the system has a unique solution if and only if the two rays l and r are not parallel.

☞   Reconstruction can be performed from rectified images directly; that is, without going back to the coordinate frames of the original pair (Exercise 7.7).

How often can we assume to know the intrinsic and extrinsic parameters of a stereo system? If the geometry of the system does not change with time, the intrinsic and extrinsic parameters of each camera can be estimated through the procedures of Chapter 6. If $\mathbf{T}_l$, R_l, and $\mathbf{T}_r$, R_r are the extrinsic parameters of the two cameras in the world reference frame, it is not difficult to show that the extrinsic parameters of the stereo system, $\mathbf{T}$ and R, are

$$R = R_r R_l^\top$$
$$\mathbf{T} = \mathbf{T}_l - R^\top\mathbf{T}_r. \tag{7.24}$$

Try to derive (7.24) yourself. If you need help, see Exercise 7.10.

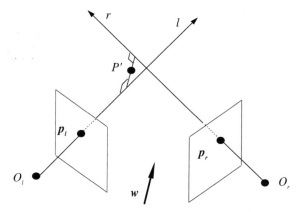

Figure 7.9 Triangulation with nonintersecting rays.

7.4.2 Reconstruction up to a Scale Factor

We now consider the case in which *only the intrinsic parameters of both the cameras are known* and derive a method to estimate the extrinsic parameters of the stereo system as well as the 3-D structure of the scene. Since the method makes use of the essential matrix, we must assume that at least eight point correspondences have been established.

Assumptions and Problem Statement

Assuming that only the intrinsic parameters and n point correspondences are given, with $n \geq 8$, compute the location of the 3-D points from their projections, $\mathbf{p}_l$ and $\mathbf{p}_r$.

Unlike triangulation, in which the geometry of the stereo system was fully known, the solution cannot rely on sufficient information to locate the 3-D points unambiguously. Intuitively, *since we do not know the baseline of the system, we cannot recover the true scale of the viewed scene.* Consequently, the reconstruction is unique only up to an unknown scaling factor. This factor can be determined if we know the distance between two points in the observed scene.

The origin of this ambiguity is quite clear in the method that we now present. The first step requires estimation of the essential matrix, E, which can only be known up to an arbitrary scale factor; therefore, we look for a convenient normalization of E. From the definition of the essential matrix, (7.13), we have

$$E^\top E = S^\top R^\top R S = S^\top S,$$

or

$$E^\top E = \begin{bmatrix} T_y^2 + T_z^2 & -T_x T_y & -T_x T_z \\ -T_y T_x & T_z^2 + T_x^2 & -T_y T_z \\ -T_z T_x & -T_z T_y & T_x^2 + T_y^2 \end{bmatrix}. \tag{7.25}$$

From (7.25) we have that the trace of $EE^\top$ is

$$Tr(E^\top E) = 2\|\mathbf{T}\|^2,$$

so that dividing the entries of the essential matrix by

$$N = \sqrt{Tr(E^\top E)/2}$$

is equivalent to normalizing the length of the translation vector to unit.

☞   Notice that, by effect of this normalization, the difference between the true essential matrix and the one estimated through the eight-point algorithm is, at most, a global sign change.

Using this normalization, (7.25) can be rewritten as

$$\hat{E}^{\top}\hat{E} = \begin{bmatrix} 1 - \hat{T}_x^2 & -\hat{T}_x\hat{T}_y & -\hat{T}_x\hat{T}_z \\ -\hat{T}_y\hat{T}_x & 1 - \hat{T}_y^2 & -\hat{T}_y\hat{T}_z \\ -\hat{T}_z\hat{T}_x & -\hat{T}_z\hat{T}_y & 1 - \hat{T}_z^2 \end{bmatrix}, \tag{7.26}$$

where $\hat{E}$ is the normalized essential matrix and $\hat{\mathbf{T}} = \mathbf{T}/\|\mathbf{T}\|$ the normalized translation vector. Recovering the components of $\hat{\mathbf{T}}$ from any row or column of the matrix $\hat{E}^{\top}\hat{E}$ is now a simple matter. However, since each entry of the matrix $\hat{E}^{\top}\hat{E}$ in (7.26) is quadratic in the components of $\hat{\mathbf{T}}$, the estimated components might differ from the true components by a global sign change. Let us assume, for the time being, that $\hat{\mathbf{T}}$ has been recovered with the proper global sign; then the rotation matrix can be obtained by simple algebraic computations. We define

$$\mathbf{w}_i = \hat{\mathbf{E}}_i \times \hat{\mathbf{T}}, \tag{7.27}$$

with $i = 1, 2, 3$ and $\hat{\mathbf{E}}_i$ the three rows of the normalized essential matrix $\hat{E}$, thought of as 3-D vectors. If $\mathbf{R}_i$ are the rows of the rotation matrix R, again thought of as 3-D vectors, easy but rather lengthy algebraic calculations yield

$$\mathbf{R}_i = \mathbf{w}_i + \mathbf{w}_j \times \mathbf{w}_k \tag{7.28}$$

with the triplet (i, j, k) spanning all cyclic permutations of $(1, 2, 3)$.

  In summary, given an estimated, normalized essential matrix, we end up with four different estimates for the pair $(\hat{\mathbf{T}}, R)$. These four estimates are generated by the twofold ambiguity in the sign of $\hat{E}$ and $\hat{\mathbf{T}}$. The 3-D reconstruction of the viewed points resolves the ambiguity and finds the only correct estimate. For each of the four pairs $(\hat{\mathbf{T}}, R)$, we compute the third component of each point in the left camera reference frame. From (7.7) and (7.9), and since $Z_r = \mathbf{R}_3^{\top}(\mathbf{P}_l - \hat{\mathbf{T}})$, we obtain

$$\mathbf{p}_r = \frac{f_r R(\mathbf{P}_l - \hat{\mathbf{T}})}{\mathbf{R}_3^{\top}(\mathbf{P}_l - \hat{\mathbf{T}})}.$$

Thus, for the first component of $\mathbf{p}_r$ we have

$$x_r = \frac{f_r \mathbf{R}_1^{\top}(\mathbf{P}_l - \hat{\mathbf{T}})}{\mathbf{R}_3^{\top}(\mathbf{P}_l - \hat{\mathbf{T}})}. \tag{7.29}$$

Finally, plugging (7.8) into (7.29) with $\mathbf{T} = \hat{\mathbf{T}}$, and solving for Z_l,

$$Z_l = f_l \frac{(f_r \mathbf{R}_1 - x_r \mathbf{R}_3)^{\top} \hat{\mathbf{T}}}{(f_r \mathbf{R}_1 - x_r \mathbf{R}_3)^{\top} \mathbf{p}_l}. \tag{7.30}$$

We can recover the other coordinates of $\mathbf{P}_l$ from (7.8), and the coordinates of $\mathbf{P}_r$ from the relation

$$\mathbf{P}_r = R(\mathbf{P}_l - \hat{\mathbf{T}}). \tag{7.31}$$

It turns out that only one of the four estimates of $(\hat{\mathbf{T}}, R)$ yields geometrically consistent (i.e., positive) Z_l and Z_r coordinates for *all* the points. The actions to take in order to determine the correct solution are detailed in the box below, which summarizes the entire algorithm.

Algorithm EUCLID_REC

The input is formed by a set of corresponding image points in camera coordinates, with $\mathbf{p}_l$ and $\mathbf{p}_r$ a generic pair, and an estimate of the normalized essential matrix, $\hat{E}$.

1. Recover $\hat{\mathbf{T}}$ from (7.26).

2. Construct the vectors $\mathbf{w}$ from (7.27), and compute the rows of the matrix R through (7.28).

3. Reconstruct the Z_l and Z_r coordinates of each point using (7.30), (7.8) and (7.31).

4. If the signs of Z_l and Z_r of the reconstructed points are

   (a) both negative for some point, change the sign of $\hat{\mathbf{T}}$ and go to step 3;
   (b) one negative, one positive for some point, change the sign of each entry of $\hat{E}$ and go to step 2;
   (c) both positive for all points, exit.

The output is the set of reconstructed 3-D points (up to a scale factor).

☞   When implementing EUCLID_REC, make sure that the algorithm does not go through more than 4 iterations of steps 2-4 (since there are only 4 possible combinations for the unknown signs of $\hat{\mathbf{T}}$ and $\hat{E}$). Keep in mind that, in the case of very small displacements, the errors in the disparity estimates may be sufficient to make the 3-D reconstruction inconsistent; when this happens, the algorithm keeps going through steps 2-4.

7.4.3 Reconstruction up to a Projective Transformation

The aim of this section is to show that you can compute a 3-D reconstruction even in the absence of *any* information on the intrinsic and extrinsic parameters. The price to pay is that *the reconstruction is unique only up to an unknown projective transformation of the world*. The Further Readings point you to methods for determining this transformation.

Assumptions and Problem Statement

Assuming that only n point correspondences are given, with $n \geq 8$ (and therefore the location of the epipoles, $\mathbf{e}$ and $\mathbf{e}'$), compute the location of the 3-D points from their projections, $\mathbf{p}_l$ and $\mathbf{p}_r$.

☞   It is worth noticing that, if no estimates of the intrinsic and extrinsic parameters are available and nonlinear deformations can be neglected, the accuracy of the reconstruction is only affected by that of the algorithms computing the disparities, not by calibration.

The plan for this section is as follows. We show that, mapping five arbitrary scene points into the standard projective basis of P^3, and using the epipoles, the projection

matrix of each camera can be explicitly recovered up to an unknown projective transformation (the one associating the standard basis to the five points selected, which is
unknown as we do not know the location of the five 3-D points in camera coordinates).[10]
Once the projection matrices are determined, the 3-D location of an arbitrary point in
space is obtained by triangulation in projective space. You can find the essential notions
of projective geometry needed to cope with all this in the Appendix, section A.4.

Determining the Projection Matrices.   In order to carry out our plan, we introduce a slight change of notation. In what follows, we drop the l and r subscripts and
adopt the unprimed and primed letters to indicate points in the left and right images
respectively. In addition, capital letters now denote points in the projective space P^3
(four coordinates), while small letters points in P^2 (three coordinates). The 3-D space
is regarded as a subset of P^3, and each image plane as a subset of P^2. This means
that we regard the 3-D point $[X, Y, Z]^\top$ of $\mathbb{R}^3$ as the point $[X, Y, Z, 1]^\top$ of P^3, and a
point $[x, y]^\top$ of $\mathbb{R}^2$ as the point $[x, y, 1]^\top$ of P^2. Let $\mathbf{O}$ and $\mathbf{O}'$ denote the projection
centers.

We let $\mathbf{P}_1, \ldots, \mathbf{P}_n$ be the points in P^3 to be recovered from their left and right
images, $\mathbf{p}_1, \ldots, \mathbf{p}_n$ and $\mathbf{p}'_i, \ldots, \mathbf{p}'_n$, and assume that, of the first five $\mathbf{P}_i$ ($\mathbf{P}_1, \mathbf{P}_2, \ldots \mathbf{P}_5$),
no three are collinear and no four are coplanar.

We first show that, if we choose $\mathbf{P}_1, \mathbf{P}_2, \ldots, \mathbf{P}_5$ as the standard projective basis of P^3
(see Appendix, section A.4), *each projection matrix can be determined up to a projective
factor that depends on the location of the epipoles.* Since a *spatial projective transformation* is fixed if the destiny of five points is known, we can, without losing generality,
set up a projective transformation that sends $\mathbf{P}_1, \mathbf{P}_2, \ldots \mathbf{P}_5$ into the standard projective
basis of P^3, $\mathbf{P}_1 = [1, 0, 0, 0]^\top$, $\mathbf{P}_2 = [0, 1, 0, 0]^\top$, $\mathbf{P}_3 = [0, 0, 1, 0]^\top$, $\mathbf{P}_4 = [0, 0, 0, 1]^\top$, and
$\mathbf{P}_5 = [1, 1, 1, 1]^\top$.

For the corresponding image points $\mathbf{p}_i$ in the left camera, we can write

$$M\mathbf{P}_i = \rho_i \mathbf{p}_i, \tag{7.32}$$

where M is the projection matrix and $\rho_i \neq 0$. Similarly, since a *planar projective transformation* is fixed if the destiny of four points is known, we can also set up a projective
transformation that sends the first four $\mathbf{p}_i$ into the standard projective basis of P^2, that
is, $\mathbf{p}_1 = (1, 0, 0)^\top$, $\mathbf{p}_2 = (0, 1, 0)^\top$, $\mathbf{p}_3 = (0, 0, 1)^\top$, and $\mathbf{p}_4 = (1, 1, 1)^\top$. In what follows, it
is assumed that the coordinates of the fifth point, $\mathbf{p}_5$, of the epipole $\mathbf{e}$, and of any other
image point, $\mathbf{p}_i$, are obtained applying this transformation to their *old* coordinates.

The purpose of all this is to simplify the expression of the projection matrix:
substituting $\mathbf{P}_1, \ldots, \mathbf{P}_4$ and $\mathbf{p}_1, \ldots, \mathbf{p}_4$ into (7.32), we see that the matrix M can be
rewritten as

$$M = \begin{bmatrix} \rho_1 & 0 & 0 & \rho_4 \\ 0 & \rho_2 & 0 & \rho_4 \\ 0 & 0 & \rho_3 & \rho_4 \end{bmatrix}. \tag{7.33}$$

---

[10] You should convince yourself that knowing the locations of the five points *in the camera reference frame*
amounts to camera calibration, which rather defeats the point of uncalibrated stereo.

Let $[\alpha, \beta, \gamma]^{\top}$ be the coordinates of $\mathbf{p}_5$ in the standard basis; (7.32) with $i = 5$ makes it possible to eliminate ρ_1, ρ_2 and ρ_3 from (7.33), obtaining

$$M = \begin{bmatrix} \alpha\rho_5 - \rho_4 & 0 & 0 & \rho_4 \\ 0 & \beta\rho_5 - \rho_4 & 0 & \rho_4 \\ 0 & 0 & \gamma\rho_5 - \rho_4 & \rho_4 \end{bmatrix}. \tag{7.34}$$

Finally, since a projection matrix is defined only up to a scale factor, we can divide each entry of matrix (7.34) by ρ_4, obtaining

$$M = \begin{bmatrix} \alpha x - 1 & 0 & 0 & 1 \\ 0 & \beta x - 1 & 0 & 1 \\ 0 & 0 & \gamma x - 1 & 1 \end{bmatrix}. \tag{7.35}$$

where $x = \rho_5/\rho_4$. *The projection matrix of the left camera has been determined up to the unknown projective parameter x.*

In order to determine x, it is useful to relate the entries of M to the coordinates of the projection center, $\mathbf{O}$. This can be done by observing that M models a perspective projection with $\mathbf{O}$ as projection center. Therefore, M *projects* every point of P^3, with the exception of $\mathbf{O}$, into a point of P^2. Since M has rank 3, the null space of M is nontrivial and consists necessarily of $\mathbf{O}$:

$$M\mathbf{O} = 0. \tag{7.36}$$

Equation (7.36) can be solved for O_x, O_y and O_z:

$$\mathbf{O} = \left[\frac{1}{1 - \alpha x}, \frac{1}{1 - \beta x}, \frac{1}{1 - \gamma x}, 1 \right]^{\top}. \tag{7.37}$$

Corresponding relations and results can be obtained for the right camera (in the primed reference frame). In particular, we can write

$$M' = \begin{bmatrix} \alpha' x' - 1 & 0 & 0 & 1 \\ 0 & \beta' x' - 1 & 0 & 1 \\ 0 & 0 & \gamma' x' - 1 & 1 \end{bmatrix}.$$

and

$$\mathbf{O}' = \left[\frac{1}{1 - \alpha' x'}, \frac{1}{1 - \beta' x'}, \frac{1}{1 - \gamma' x'}, 1 \right]^{\top}. \tag{7.38}$$

Since the location of the epipoles is known, x and x' (and hence the full projection matrices and the centers of projection) can be determined from

$$M\mathbf{O}' = \sigma\mathbf{e} \tag{7.39}$$

and

$$M'\mathbf{O} = \sigma'\mathbf{e}' \tag{7.40}$$

with $\sigma \neq 0$ and $\sigma' \neq 0$.[11]

Let's see first what we can recover from (7.39). Substituting (7.35) and (7.38) into (7.39), we obtain the following system of equations

$$\begin{bmatrix} \alpha & -\alpha' & \alpha'e_x \\ \beta & -\beta' & \beta'e_y \\ \gamma & -\gamma' & \gamma'e_z \end{bmatrix} \begin{pmatrix} x \\ x' \\ \sigma x' \end{pmatrix} = \begin{pmatrix} \sigma e_x \\ \sigma e_y \\ \sigma e_z \end{pmatrix} \tag{7.41}$$

Since σ is unknown, system (7.41) is homogeneous and nonlinear in the three unknown x, x', and σ. However, we can regard it as a linear system in the unknown x, x' and $\sigma x'$, so that solving for $\sigma x'$ we have

$$\sigma x' = \sigma \frac{\mathbf{e}^\top(\mathbf{p}_5 \times \mathbf{p}'_5)}{\mathbf{v}^\top(\mathbf{p}_5 \times \mathbf{p}'_5)} \tag{7.42}$$

with $\mathbf{v} = (\alpha'e_x, \beta'e_y, \gamma'e_z)$. Since $\mathbf{e}$, $\mathbf{p}_5$, $\mathbf{p}'_5$, and $\mathbf{v}$ are known and the unknown factor σ cancels out, (7.42) actually determines x'.

A similar derivation applied to (7.40) yields

$$x = \frac{\mathbf{e}'^\top(\mathbf{p}_5 \times \mathbf{p}'_5)}{\mathbf{v}'^\top(\mathbf{p}_5 \times \mathbf{p}'_5)} \tag{7.43}$$

with $\mathbf{v}' = (\alpha e'_x, \beta e'_y, \gamma e'_z)$. Having determined both x and x' we can regard both the projection matrices and the centers of projections as completely determined.

Computing the Projective Reconstruction. We are now in a position to reconstruct *any* point in P^3 given its corresponding image points, $\mathbf{p} = \begin{bmatrix} p_x, p_y, p_z \end{bmatrix}^\top$ and $\mathbf{p}' = \begin{bmatrix} p'_x, p'_y, p'_z \end{bmatrix}^\top$. The reconstruction is unique up to the unknown projective transformation fixed by the choice of $\mathbf{P}_1, \ldots, \mathbf{P}_5$ as the standard basis for P^3. Observe that the projective line l defined by

$$\lambda\mathbf{O} + \mu \begin{bmatrix} O_x p_x, O_y p_y, O_z p_z, 0 \end{bmatrix}^\top, \tag{7.44}$$

with $\lambda, \mu \in \mathbb{R}$ and not both 0, goes through $\mathbf{O}$ (for $\lambda = 1$ and $\mu = 0$) and also through $\mathbf{p}$, since

$$M \begin{pmatrix} O_x p_x \\ O_y p_y \\ O_z p_z \\ 0 \end{pmatrix} = \mathbf{p}.$$

---

[11] Since the epipoles and the centers of projection lie on a straight line, (7.39) and (7.40) are not independent. For the purpose of this brief introduction, however, this can be safely ignored.

Similarly, the projective line l'

$$\lambda' \mathbf{O}' + \mu' \left[O'_x p'_x, O'_y p'_y, O'_z p'_z, 0 \right]^\top ,$$

with $\lambda', \mu' \in \mathbb{R}$ and not both 0, goes through $\mathbf{O}'$ and $\mathbf{p}'$. *The projective point $\mathbf{P}$ can thus be obtained by intersecting the two projective lines l and l'.* This amounts to looking for the non-trivial solution of the homogeneous system of linear equations

$$\begin{bmatrix} O_x & O_x p_x & -O'_x & -O'_x p'_x \\ O_y & O_y p_y & -O'_y & -O'_y p'_y \\ O_z & O_z p_z & -O'_z & -O'_z p'_z \\ 1 & 0 & -1 & 0 \end{bmatrix} \begin{pmatrix} \lambda \\ \mu \\ \lambda' \\ \mu' \end{pmatrix} = 0. \qquad (7.45)$$

☞   Once again, singular value decomposition $UDV^\top$ of the system matrix of (7.45) provides a numerically stable procedure for solving this linear system: The solution is given by the column of V associated with the smallest singular value along the diagonal of D.

Algorithm UNCAL_STEREO

The input is formed by n pairs of corresponding points, $\mathbf{p}_i$ and $\mathbf{p}'_i$, with $i =, \ldots, n$ and $n \geq 5$, images of n points, $\mathbf{P}_1, \ldots, \mathbf{P}_n$. We assume that, of the first five $\mathbf{P}_i$ ($\mathbf{P}_1, \mathbf{P}_2, \ldots \mathbf{P}_5$), no three are collinear and no four are coplanar.

We assume to have estimated the location of the epipoles, $\mathbf{e}$ and $\mathbf{e}'$, using EPIPOLES_ LOCATION. Let $\mathbf{P}_1, \ldots, \mathbf{P}_5$ be the standard projective basis of P^3. We assume the same notation used throughout the section.

1. Determine the planar projective transformations T and T' that map the $\mathbf{p}_i$ and $\mathbf{p}'_i$ ($i = 1, \ldots, 4$) into the standard projective basis of P^2 on each image plane. Apply T to the $\mathbf{p}_i$ and the epipole $\mathbf{e}$, and T' to the $\mathbf{p}'_i$ and the epipole $\mathbf{e}'$. Let (α, β, γ) and $(\alpha', \beta', \gamma')$ be the new coordinates of $\mathbf{p}_5$ and $\mathbf{p}'_5$.

2. Determine x and x' from (7.42) and (7.43).

3. Determine $\mathbf{O}$ and $\mathbf{O}'$ from (7.37) and (7.38).

4. Given a pair of corresponding points $\mathbf{p}$ and $\mathbf{p}'$, reconstruct the location of the point $\mathbf{P}$ in the standard projective basis of P^3 using (7.44) with λ and μ nontrivial solution of (7.45).

The output is formed by the coordinates of $\mathbf{P}_1, \ldots, \mathbf{P}_n$ in the standard projective basis.

Having found a projective reconstruction of our points, *how do we go back to Euclidean coordinates?* If we know the location of $\mathbf{P}_1, \ldots, \mathbf{P}_5$ in the world frame, we can determine the projective transformation introduced at the beginning of this section that mapped these five points, thought of as points of P^3, into the standard projective basis (see the Appendix, section A.4 for details on how to do it). The Further Readings point to (nontrivial) algorithms for Euclidean reconstruction which relax this assumption, but need more than two images.

7.5   Summary

After working through this chapter you should be able to:

- ❏ explain the fundamental concepts and problems of stereo
- ❏ solve the correspondence problem by means of correlation-based and feature-based techniques
- ❏ estimate the fundamental (and, if possible, the essential) matrix from point correspondences
- ❏ determine the epipolar geometry and rectifying a stereo pair
- ❏ recover 3-D structure from image correspondences when *(a)* both intrinsic and extrinsic parameters are known, *(b)* only the intrinsic parameters are known, and *(c)* both intrinsic and extrinsic parameters are unknown

7.6   Further Readings

The literature on stereo is immense. You may wish to start with the two classic correspondence algorithms by Marr and Poggio [14, 15]. Among the multitude of correspondence algorithms proposed in the last two decades, we suggest the methods proposed in [2, 9, 12, 17]. A way to adapt the shape and size of SSD windows to different image parts is described in Kanade and Okutomi [10].

Rectification is discussed by Ayache [1] and Faugeras [5]. A MATLAB implementation of a rectification algorithm based on projection matrices can be downloaded from `ftp://taras.dimi.uniud.it/pub/sources/rectif_m.tar.gz`. The algorithm and the implementation are due to Andrea Fusiello. For uncalibrated rectification, see Robert *et al.* [20].

The eight-point algorithm is due to Longuet-Higgins [11]; the normalization procedure to avoid numerical instabilities (discussed at length in Exercise 7.6) is due to Hartley [7]. Linear and nonlinear methods for determining the fundamental matrix, as well as stability issues and critical configurations, are studied by Luong and Faugeras [13]. Shashua [21] proposed a method for locating the epipoles and achieving projective reconstruction that require only six point matches. The recovery of the extrinsic parameters from the essential matrix described in this chapter is again due to Longuet-Higgins [11]. An alternative method for the calibration of the extrinsic parameters can be found in [8]. We have largely based the introduction to uncalibrated stereo on the seminal paper by Faugeras [4]. Similar results are discussed by Sparr [22] and Mohr and Arbogast [18]. More recently a number of methods for the recovery of Euclidean structure from the projective reconstruction have been proposed (see [3, 6, 19] for example).

If you are curious about understanding and creating autostereograms, check out the Web site `http://www.ccc.nottingham.ac.uk/~etzpc/sirds.html`. The face reconstruction in Figure 7.2 was computed by a stereo system commercialized by the Turing Institute (`http://www.turing.gla.ac.uk`), which maintains an interesting Web site on stereopsis. The INRIA-Syntim Web site contains useful test data, including calibrated stereo pairs (please notice the copyright attached!): `http://www-syntim.inria.fr/syntim/analyze/paires-eng.html`.

To conclude, we observe that stereo-like visual systems can also be built taking two pictures of the same scene from the same viewpoint, but under different illuminations. This is the so-called *photometric stereo*, first proposed by Woodham [23].

7.7 Review

Questions

- **7.1** What are the intrinsic and extrinsic parameters of a stereo system?
- **7.2** What are the main properties of correlation-based and feature-based methods for finding correspondences?
- **7.3** What is the epipolar constraint and how could you use it to speed up the search for corresponding points?
- **7.4** What are the main properties of the essential and fundamental matrices?
- **7.5** What is the purpose of rectification?
- **7.6** What happens in rectification if the focal lengths of the two original cameras are not equal?
- **7.7** What sort of 3-D reconstruction can be obtained if all the parameters, only the intrinsic parameters, or no parameters can be assumed to be known?
- **7.8** What is the purpose of step 4 in algorithm EUCLID_REC?
- **7.9** Is the reconstruction of the fundamental matrix necessary for uncalibrated stereo?
- **7.10** How can (7.23) reconstruct depths in millimeters if the focal length in millimeters is not known?

Exercises

- **7.1** Estimate the accuracy of the simple stereo system of Figure 7.4 assuming that the only source of noise is the localization of corresponding points in the two images. (*Hint:* Take the partial derivatives of Z with respect to x, T, f.) Discuss the dependence of the error in depth estimation as a function of the baseline width and the focal length.
- **7.2** Using your solution to Exercise 7.1, estimate the accuracy with which features should be localized in the two images in order to reconstruct depth with a relative error smaller than 1%.
- **7.3** Check what happens if you compute SSD and cross-correlation between an arbitrary pattern and a perfectly black pattern over a window W. Discuss the effect of replacing the definition of cross-correlation with the *normalized cross-correlation*

$$\psi(x, y) = \frac{(x - \bar{x})(y - \bar{y})}{N_x N_y} \tag{7.46}$$

where $\bar{x} = \sum_W x$, $\bar{y} = \sum_W y$, $N_x = \sqrt{\sum_W x^2}$, and $N_y = \sqrt{\sum_W y^2}$. Can you precompute and store the possible values of ψ if you are using (7.46)?

○ *7.4* Discuss strategies for estimating correspondences at subpixel precision using correlation-based methods. (*Hint:* Keep track of the values of ψ in the neighborhood of the maximum, and take a weighted average of every integer location in that neighborhood.) Make sure that the weights are positive and sum to 1.

○ *7.5* Design a correlation-based method that can be used to match edge points.

○ *7.6* Determine the matrices H_l and H_r needed to normalize the entries of the fundamental matrix before applying algorithm EIGHT_POINT. (*Hint:* Given a set of points $\mathbf{p}_i = [x_i, y_i, 1]^\top$ with $i = 1, \ldots, n$, define $\bar{x} = \sum_i x_i/n$, $\bar{y} = \sum_i y_i/n$, and

$$\bar{d} = \frac{\sum_i \sqrt{(x_i - \bar{x})^2 + (y_i - \bar{y})^2}}{n\sqrt{2}}.$$

Then, find the 3×3 matrix H such that

$$H\mathbf{p}_i = \hat{\mathbf{p}}_i$$

with $\mathbf{p}_i = [(x_i - \bar{x})/d, (y_i - \bar{y})/d, 1]^\top$.) Verify that the average length of each component of $\hat{\mathbf{p}}_i$ equals 1.

○ *7.7* Write an algorithm reconstructing a scene from a rectified stereo pair using rectified images coordinates. (*Hint:* Use the simultaneous projection equations associated to a 3-D point in the two cameras.)

○ *7.8* Verify that (7.2) can be derived from (7.23) in the special case of the stereo system of Figure 7.4.

○ *7.9* In analogy with the case of point matching, compute the solution to the triangulation problem in the case of line matching. If l_l and l_r are the matched lines, this amounts to find the 3-D line intersection of the planes through O_l and l_l, and O_r and l_r respectively. Why is the triangulation based on lines computationally easier than the triangulation based on points?

○ *7.10* Assume $\mathbf{T}_l$ and R_l, and $\mathbf{T}_r$ and R_r are the extrinsic parameters of two cameras with respect to the same world reference frame. Show that the translation vector, $\mathbf{T}$, and rotation matrix, R, which define the extrinsic parameters of the stereo system composed of the two cameras are given by (7.24). (*Hint:* For a point $\mathbf{P}$ in the world reference frame we have $\mathbf{P}_r = R_r\mathbf{P} + \mathbf{T}_r$ and $\mathbf{P}_l = R_l\mathbf{P} + \mathbf{T}_l$. But the relation between $\mathbf{P}_l$ and $\mathbf{P}_r$ is given by $\mathbf{P}_r = R(\mathbf{P}_l - \mathbf{T})$.)

○ *7.11* In the same notation of Section 7.4, let $\mathbf{w}_i = \hat{\mathbf{E}}_i \times \hat{\mathbf{T}}_i$ with $\hat{\mathbf{E}}_i = \hat{\mathbf{T}}_i \times \mathbf{R}_i$. Prove that

$$(\hat{\mathbf{T}}^\top \mathbf{R}_i)\hat{\mathbf{T}} = \mathbf{w}_j \times \mathbf{w}_k$$

for every triplet (i, j, k) which is a cyclic permutation of $(1, 2, 3)$ and use the result to derive (7.28). (*Hint:* Make use of the vector identity $\mathbf{A} \times (\mathbf{B} \times \mathbf{C}) = (\mathbf{A}^\top \mathbf{C})\mathbf{B} - (\mathbf{A}^\top \mathbf{B})\mathbf{C}$.)

Projects

● *7.1* In a typical implementation of CORR-MATCHING, one computes $c(d) = \sum_W \psi$ at each pixel for each possible shift (where W is the window over which $c(d)$ is evaluated and ψ is the pixelwise cross-correlation measure), and stores the shift for which $c(d)$ is maximum. If the size of W is $n \times n$, this implementation requires $O(n^2)$ additions. However, if n is larger than a few units, the overlap between the correlation windows centered at neighbor pixels can be exploited to obtain a more efficient implementation that requires $O(2n)$ additions. The key idea is to compute $c(d)$ for each possible shift at all pixels first. This makes it possible to use the result of the computation of $c(d)$ for some d at one pixel to evaluate $c(d)$ at the neighboring pixel. Here is a simple way to do it. For each possible shift, evaluate ψ over the entire image. Once you have obtained $c(d)$ at some pixel p over the window W, you can compute $c(d)$ for the pixel immediately to the right of p, for example, by simply subtracting the contribution to $c(d)$ from the leftmost column of W and adding the contribution from the column immediately to the right of W. The memory requirement is not much different, as for each shift you do not need to save the value of ψ over the entire image, but only the intermediate maximum of $c(d)$ (and corresponding shift) for all pixels. Implement this version of CORR-MATCHING and compare it with the standard implementation.

● *7.2* Design and implement a program that, given a stereo pair, determines at least eight point matches, then recovers the fundamental matrix and the location of the epipoles. Check the accuracy of the result by measuring the distance between the estimated epipolar lines and image points not used by the matrix estimation.

References

[1] N. Ayache, *Artificial Vision for Mobile Robots: Stereo Vision and Multisensory Perception*, MIT Press, Cambridge (MA) (1991).

[2] N. Ayache and B. Faverjon, Efficient Registration of Stereo Images by Matching Graph Descriptions of Edge Segments, *International Journal of Computer Vision*, Vol. 1, no. 2 (1987).

[3] F. Devernay and O.D. Faugeras, From Projective to Euclidean Reconstruction, *Technical Report 2725*, INRIA (1995) (available from http://www.inria.fr).

[4] O.D. Faugeras, What Can Be Seen in Three Dimensions with an Uncalibrated Stereo Rig?, *Proc. 2nd European Conference on Computer Vision*, Santa Margherita (Italy), pp. 563-578 (1992).

[5] O.D. Faugeras, *Three-Dimensional Computer Vision: A Geometric Viewpoint*, MIT Press, Cambridge (MA) (1993).

[6] R.I. Hartley, Estimation of Relative Camera Positions for Uncalibrated Cameras, *Proc. 2nd European Conference on Computer Vision*, Santa Margherita (Italy), pp. 579-587 (1992).

[7] R.I. Hartley, In Defence of the 8-Point Algorithm, *Proc. 5th International Conference on Computer Vision*, Cambridge (MA), pp. 1064-1070 (1995).

[8] B.K.P. Horn, Relative Orientation, *International Journal of Computer Vision*, Vol. 4, pp. 59-78 (1990).

[9] D.G. Jones and J. Malik, Determining 3-D Shape from Orientation and Spatial Frequency Disparities, *Proc. 2nd European Conference on Computer Vision*, Santa Margherita (Italy), pp. 661-669 (1992).

[10] T. Kanade and M. Okutomi, A Stereo Matching Algorithm with an Adaptive Window: Theory and Experiments, *IEEE Transactions on Pattern Analysis and Machine Intelligence*, Vol. PAMI-16, pp. 920-932 (1994).

[11] H.C. Longuet-Higgins, A Computer Algorithm for Reconstructing a Scene from Two Projections, *Nature*, Vol. 293, no. 10, pp. 133-135 (1981).

[12] B.D. Lucas and T. Kanade, An Iterative Image Registration Technique with an Application to Stereo Vision, *Proc. International Joint Conference on Artificial Intelligence*, pp. 674-679 (1981).

[13] Q.-T. Luong and O.D. Faugeras, The Fundamental Matrix: Theory, Algorithms, and Stability Analysis, *International Journal of Computer Vision*, Vol. 17, pp. 43-75 (1996).

[14] D. Marr and T. Poggio, Cooperative Computation of Stereo Disparity, *Science* Vol. 194, pp. 283-287 (1976).

[15] D. Marr and T. Poggio, A Computational Theory of Human Stereo Vision, *Proc. R. Soc. Lond. B* Vol. 204, pp. 301-328 (1979).

[16] L. Matthies, T. Kanade and R. Szeliski, Kalman Filter-Based Algorithms for Estimating Depth from Image Sequences, *International Journal of Computer Vision*, Vol. 3, pp. 209-236 (1989).

[17] J.E.W. Mayhew and J.P. Frisby, Psychophysical and Computational Studies Towards a Theory of Human Stereopsis, *Artificial Intelligence*, Vol. 17, pp. 349-385 (1981).

[18] R. Mohr and E. Arbogast, It Can Be Done without Camera Calibration, *Pattern Recognition Letters*, Vol. 12, pp. 39-43 (1990).

[19] M. Pollefeys, L. Van Gool and M. Proesmans, Euclidean 3-D Reconstruction from Image Sequences with Variable Focal Lenghts, *Proc. European Conference on Computer Vision*, Cambridge (UK), pp. 31-42 (1996).

[20] L. Robert, C. Zeller, O.D. Faugeras and M. Hebert, Applications of Non-Metric Vision to Some Visually Guided Robotic Tasks, *Technical Report 2584*, INRIA (1995) (available from `http://www.inria.fr`).

[21] A. Shashua, Projective Depth: a Geometric Invariant for 3-D Reconstruction from Two Perspective/Orthographic Views and for Visual Recognition, *Proc. IEEE Int. Conf. on Computer Vision*, Berlin (Germany), pp. 583-590 (1993).

[22] G. Sparr, An Algebraic-Analytic Method for Reconstruction from Image Correspondences, in *Proc. 7th Scandinavian Conference on Image Analysis*, pp. 274-281 (1991).

[23] R. J. Woodham, Photometric Stereo: a Reflectance Map Technique for Determining Surface Orientation from a Single View, *Proc. SPIE Technical Symposium on Image Understanding Systems and Industrial Applications*, Vol. 155, pp. 136-143 (1978).

8

Motion

Eppur si muove.[1]

Galileo

This chapter concerns the analysis of the *visual motion* observed in time-varying image sequences.

Chapter Overview

Section 8.1 presents the basic concepts, importance and problems of visual motion.

Section 8.2 introduces the notions of *motion field* and *motion parallax*, and their fundamental equations.

Section 8.3 discusses the *image brightness constancy equation* and the *optical flow*, the approximation of the motion field which can be computed from the changing image brightness pattern.

Section 8.4 presents methods for estimating the motion field, divided in *differential* and *feature-matching/tracking* methods.

Section 8.5 deals with the *reconstruction of 3-D motion and structure*.

Section 8.6 discusses *motion-based segmentation* based on change detection.

What You Need to Know to Understand this Chapter

- Working knowledge of Chapters 2 and 7.
- Eigenvalues and eigenvectors of a matrix.
- Least squares and SVD (Appendix, section A.6).
- The basics of Kalman filtering (Appendix, section A.8).

---

[1] And yet it is moving.

8.1 Introduction

Until now, we have studied visual computations on single images, or two images acquired simultaneously. In this chapter, we broaden our perspective and focus on the
processing of images *over time*. More precisely, we are interested in the visual information that can be extracted from the spatial and temporal changes occurring in an *image
sequence*.

Definition: Image Sequence

An image sequence is a series of N images, or *frames*, acquired at discrete time instants $t_k = t_0 + k\Delta t$, where Δt is a fixed time interval, and $k = 0, 1, \ldots, N - 1$.

☞   In order to acquire an image sequence, you need a frame grabber capable of storing frames
at a fast rate. Typical rates are the so called *frame rate* and *field rate*, corresponding to a time
interval Δt of 1/24sec and 1/30sec respectively. If you are allowed to choose a different time
interval, or simply want to subsample an image sequence, make sure that Δt is small enough
to guarantee that the discrete sequence is a representative sampling of the continuous image
evolving over time; as a rule of thumb, this means that the apparent displacements over the
image plane between frames should be at most a few pixels.

Assuming the illumination conditions do not vary, image changes are caused by a
relative motion between camera and scene: the viewing camera could move in front of
a static scene, or parts of the scene could move in front of a stationary camera, or, in
general, both camera and objects could be moving with different motions.

8.1.1 The Importance of Visual Motion

The temporal dimension in visual processing is important primarily for two reasons.
First, the apparent motion of objects onto the image plane is a strong visual cue for
understanding structure and 3-D motion. Second, biological visual systems use visual
motion to infer properties of the 3-D world with little *a priori* knowledge of it. Two
simple examples may be useful to illustrate these points.

Example 1: Random Dot Sequences.   Consider an image of *random dots*, generated by assigning to each pixel a random grey level. Consider a second image obtained
by shifting a squared, central region of the first image by a few pixels, say, to the right,
and filling the gap thus created with more random dots. Two such images are shown in
Figure 8.1. If you display the two images in sequence on a computer screen, in the same
window and one after the other at a sufficiently fast rate, you will unmistakably *see* a
square moving sideways back and forth against a steady background. Notice that the

Figure 8.1   A sequence of two random dot images: a square has been displaced between the two frames.

visual system bases its judgement on the only information available in the sequence; that is, the displacement of the square in the two images.[2]

Example 2: Computing Time-to-Impact.   Visual motion allows us to compute useful properties of the observed 3-D world with very little knowledge about it. Consider a planar version of the usual pinhole camera model, and a vertical bar perpendicular to the optical axis, travelling towards the camera with constant velocity as shown in Figure 8.2. We want to prove a simple but very important fact: *It is possible to compute the time, τ, taken by the bar to reach the camera only from image information;* that is, without knowing either the real size of the bar or its velocity in 3-D space.[3]

As shown in Figure 8.2, we denote with L the real size of the bar, with V its constant velocity, and with f the focal length of the camera. The origin of the reference frame is the projection center. If the position of the bar on the optical axis is $D(0) = D_0$ at time $t = 0$, its position at a later time t will be $D = D_0 - Vt$. Note that L, V, f, D_0, and the choice of the time origin are all unknown, but that τ can be written as

$$\tau = \frac{V}{D}. \tag{8.1}$$

From Figure 8.2, we see that $l(t)$, the *apparent* size of the bar at time t on the image plane, is given by

$$l(t) = f\frac{L}{D}.$$

---

[2] Incidentally, you can look at the two images of Figure 8.1 as a *random-dot stereogram* to perceive a square floating in the background. Stand a diskette (or a sheet of paper of the same size) between the two images and touch your nose against the diskette, so that each eye can see only one image. Focus your eyes *behind* the page. After a while, the two images should fuse and produce the impression of a square floating against the background.

[3] In the biologically-oriented community of computer vision, τ is called, rather pessimistically, *time-to-collision* or even *time-to-crash!*

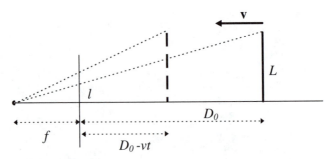

Figure 8.2   How long before the bar reaches the camera?

If we now compute the time derivative of $l(t)$,

$$l'(t) = \frac{dl(t)}{dt} = -f\frac{L}{D^2}\frac{dD}{dt} = f\frac{LV}{D^2},$$

take the ratio between $l(t)$ and $l'(t)$, and use (8.1), we obtain

$$\frac{l(t)}{l'(t)} = \tau. \tag{8.2}$$

This is the equation we were after: since both the apparent size of the bar, $l(t)$, and its time derivative, $l'(t)$, are measured from the images, (8.2) allows us to compute τ in the absence of *any* 3-D information, like the size of the bar and its velocity.

8.1.2   The Problems of Motion Analysis

It is now time to state the main problems of motion analysis. The analogies with stereo suggest to begin by dividing the motion problem into two subproblems.

Two Subproblems of Motion

Correspondence: Which elements of a frame correspond to which elements of the next frame of the sequence?
Reconstruction: Given a number of corresponding elements, and possibly knowledge of the camera's intrinsic parameters, what can we say about the 3-D motion and structure of the observed world?

Main Differences between Motion and Stereo

Correspondence: As image sequences are sampled temporally at usually high rates, the spatial differences (disparities) between consecutive frames are, on average, much smaller than those of typical stereo pairs.
Reconstruction: Unlike stereo, in motion the relative 3-D displacement between the viewing camera and the scene is not necessarily caused by a single 3-D rigid transformation.

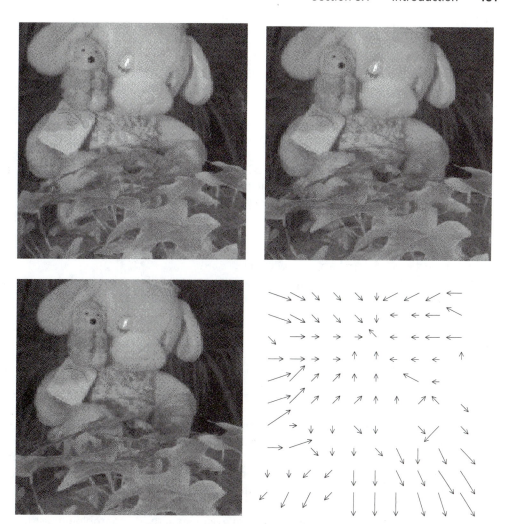

Figure 8.3   Three frames from a long image sequence (left to right and top to bottom) and the optical flow computed from the sequence, showing that the plant in the foreground is moving towards the camera, and the soft toys away from it.

Regarding correspondence, the fact that motion sequences make many, closely sampled frames available for analysis is an advantage over the stereo case for at least two reasons. First, feature-based approaches can be made more effective by *tracking* techniques, which exploit the past history of the features' motion to predict disparities in the next frame. Second, due to the generally small spatial and temporal differences between consecutive frames, the correspondence problem can also be cast as the problem of *estimating the apparent motion of the image brightness pattern*, usually called *optical flow* (see Figure 8.3).

We shall use two strategies for solving the correspondence problem.

Differential methods (section 8.4.1) lead to *dense* measures; that is, computed at each image pixel. They use estimates of time derivatives, and require therefore image sequences sampled closely.

Matching methods (section 8.4.2) lead to *sparse* measures; that is, computed only at a subset of image points. We shall place emphasis on *Kalman filtering* as a technique for matching and tracking efficiently sparse image features over time.

Unlike correspondence, and perhaps not surprisingly, reconstruction is more difficult in motion than in stereo. Even in the presence of only one 3-D motion between the viewing camera and the scene, frame-by-frame recovery of motion and structure turns out to be more sensitive to noise. The reason is that the baseline between consecutive frames, regarded as a stereo pair, is very small (see Chapter 7). 3-D motion and structure estimation from both sparse and dense estimates of the image motion is discussed in sections 8.5.1 and 8.5.2, respectively.

This chapter discusses and motivates methods for solving correspondence and reconstruction under the following simplifying assumption.

Assumption

There is only one, rigid, relative motion between the camera and the observed scene, and the illumination conditions do not change.

This assumption of single, rigid motion implies that *the 3-D objects observed cannot move of different motions.* This assumption is violated, for example, by sequences of football matches, motorway traffic or busy streets, but satisfied by, say, the sequence of a building viewed by a moving observer. The assumption also rules out flexible (nonrigid) objects: deformable objects like clothes or moving human bodies are excluded.

If the camera is looking at more than one moving object, or you simply cannot assume a moving camera in a static environment, a third subproblem must be added.

The Third Subproblem of Motion

The segmentation problem: What are the regions of the image plane which correspond to different moving objects?

The main difficulty here is a chicken and egg problem: should we first solve the matching problem and then determine the regions corresponding to the different moving objects, or find the regions first, and then look for correspondences? This question is addressed in section 8.6 in the hypothesis that the viewing camera is not moving. Pointers to solutions to this difficult problem in more general cases are given in the Further Readings.

We now begin by establishing some basic facts.

8.2 The Motion Field of Rigid Objects

Definition: Motion Field

The motion field is the 2-D vector field of velocities of the image points, induced by the relative motion between the viewing camera and the observed scene.

The motion field can be thought of as *the projection of the 3-D velocity field on the image plane* (to visualize this vector field, imagine to project the 3-D velocity vectors on the image). The purpose of this section is to get acquainted with the theory and geometrical properties of the motion field. *We shall work in the camera reference frame*, ignoring the image reference frame and the pixelization.[4] The issue of camera calibration will be raised in due time.

This section presents some essential facts of motion fields, compares disparity representations in motion and stereo, analyzes two special cases of rigid motion leading to generally useful facts, and introduces the concept of motion parallax.

8.2.1 Basics

Notation. We let $\mathbf{P} = [X, Y, Z]^\top$ be a 3-D point in the usual camera reference frame: The projection center is in the origin, the optical axis is the Z axis, and f denotes the focal length. The image of a scene point, $\mathbf{P}$, is the point $\mathbf{p}$ given by

$$\mathbf{p} = f\frac{\mathbf{P}}{Z}. \tag{8.3}$$

As usual (see Chapter 2), since the third coordinate of $\mathbf{p}$ is always equal to f, we write $\mathbf{p} = [x, y]^\top$ instead of $\mathbf{p} = [x, y, f]^\top$. The relative motion between $\mathbf{P}$ and the camera can be described as

$$\mathbf{V} = -\mathbf{T} - \boldsymbol{\omega} \times \mathbf{P}, \tag{8.4}$$

where $\mathbf{T}$ is the translational component of the motion,[5] and $\boldsymbol{\omega}$ the angular velocity. As the motion is rigid, $\mathbf{T}$ and $\boldsymbol{\omega}$ are the same for any $\mathbf{P}$. In components, (8.4) reads

$$V_x = -T_x - \omega_y Z + \omega_z Y$$
$$V_y = -T_y - \omega_z X + \omega_x Z \tag{8.5}$$
$$V_z = -T_z - \omega_x Y + \omega_y X.$$

---

[4] Remember, this means that we consider the intrinsic parameters known.

[5] Note that $\mathbf{T}$ denotes a velocity vector only in this chapter, not a displacement vector as in the rest of the book.

The Basic Equations of the Motion Field. To obtain the relation between the velocity of **P** in space and the corresponding velocity of **p** on the image plane, we take the time derivative of both sides of (8.3), which gives an important set of equations.

The Basic Equations of the Motion Field

The motion field, **v**, is given by

$$\mathbf{v} = f \frac{Z\mathbf{V} - V_z\mathbf{P}}{Z^2}.$$

(8.6)

In components, and using (8.5), (8.6) read

$$v_x = \frac{T_z x - T_x f}{Z} - \omega_y f + \omega_z y + \frac{\omega_x xy}{f} - \frac{\omega_y x^2}{f}$$

$$v_y = \frac{T_z y - T_y f}{Z} + \omega_x f - \omega_z x - \frac{\omega_y xy}{f} + \frac{\omega_x y^2}{f}.$$

(8.7)

Notice that *the motion field is the sum of two components, one of which depends on translation only, the other on rotation only*. In particular, the translational components of the motion field are

$$v_x^T = \frac{T_z x - T_x f}{Z}$$

$$v_y^T = \frac{T_z y - T_y f}{Z},$$

and the rotational components are

$$v_x^\omega = -\omega_y f + \omega_z y + \frac{\omega_x xy}{f} - \frac{\omega_y x^2}{f}$$

$$v_y^\omega = \omega_x f - \omega_z x - \frac{\omega_y xy}{f} + \frac{\omega_x y^2}{f}.$$

Since the component of the motion field along the optical axis is always equal to 0, we shall write $\mathbf{v} = [v_x, v_y]^\top$ instead of $\mathbf{v} = [v_x, v_y, 0]^\top$. Notice that, in the last two pairs of equations, the terms depending on the angular velocity, ω, and depth, Z, are decoupled. This discloses an important property of the motion field: *the part of the motion field that depends on angular velocity does not carry information on depth.*

Comparing Disparity Representations in Stereo and Motion. As we said before, stereo and motion pose similar computation problems, and one of these is correspondence. Point displacements are represented by disparity maps in stereo, and by motion fields in motion. An obvious question is, how similar are disparity maps and motion fields? The key difference is that *the motion field is a differential concept, stereo disparity is not*. The motion field is based on velocity, and therefore on time derivatives:

Consecutive frames must be as close as possible to guarantee good discrete approximations of the continuous time derivatives. In stereo, there is no such constraint on the two images, and the disparities can take, in principle, any value.

Stereo Disparity Map and Motion Field

The spatial displacements of corresponding points between the images of a stereo pair (forming the stereo disparity map) are *finite*, and, in principle, unconstrained.

The spatial displacements of corresponding points between consecutive frames of a motion sequence (forming the motion field) are discrete approximations of time-varying derivatives, and must therefore be suitably small.

The motion field coincides with the stereo disparity map *only if* spatial and temporal differences between frames are sufficiently small.

8.2.2 Special Case 1: Pure Translation

We now analyze the case in which the relative motion between the viewing camera and the scene has no rotational component. The resulting motion field has a peculiar spatial structure, and its analysis leads to concepts very useful in general.

Since $\omega = 0$, (8.7) read

$$v_x = \frac{T_z x - T_x f}{Z}$$
$$v_y = \frac{T_z y - T_y f}{Z} \tag{8.8}$$

We first consider the general case in which $T_z \neq 0$. Introducing a point $\mathbf{p}_0 = [x_0, y_0]^\top$ such that

$$x_0 = f T_x / T_z$$
$$y_0 = f T_y / T_z, \tag{8.9}$$

(8.8) become

$$v_x = (x - x_0) \frac{T_z}{Z}$$
$$v_y = (y - y_0) \frac{T_z}{Z}. \tag{8.10}$$

Equation (8.10) say that *the motion field of a pure translation is radial*: It consists of vectors radiating from a common origin, the point $\mathbf{p}_0$, which is therefore the *vanishing point* of the translation direction. In particular, if $T_z < 0$, the vectors point away from $\mathbf{p}_0$, and $\mathbf{p}_0$ is called the *focus of expansion* (Figure 8.4 (a)); if $T_z > 0$, the motion field vectors point towards $\mathbf{p}_0$, and $\mathbf{p}_0$ is called the *focus of contraction* (Figure 8.4 (b)). In addition, the length of $\mathbf{v} = \mathbf{v}(\mathbf{p})$ is proportional to the distance between $\mathbf{p}$ and $\mathbf{p}_0$, and inversely proportional to the depth of the 3-D point $\mathbf{P}$.

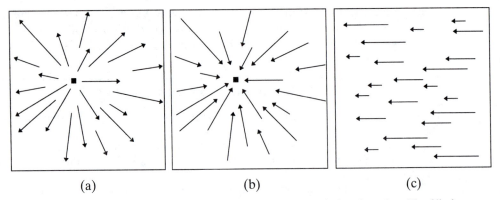

Figure 8.4   The three types of motion field generated by translational motion. The filled square marks the instantaneous epipole.

☞   Notice that the point p_0 retains its significance and many of its properties even in the presence of a rotational component of 3-D motion (section 8.5.2).

If T_z vanishes (a rather special case), (8.8) become

$$v_x = -f \frac{T_x}{Z}$$

$$v_y = -f \frac{T_y}{Z}.$$

Therefore, *if $T_z = 0$, all the motion field vectors are parallel* (see Figure 8.4 (c)) and their lengths are inversely proportional to the depth of the corresponding 3-D points.

☞   In homogeneous coordinates, there would be no need to distinguish between the two cases $T_z \neq 0$ and $T_z = 0$: For *all* possible values of T_z, including $T_z = 0$, p_0 is the vanishing point of the direction in 3-D space of the translation vector **T**, and the 3-D line through the center of projection and p_0 is parallel to **T**.

Following is a summary of the main properties of the motion field of a purely translational motion.

Pure Translation: Properties of the Motion Field

1. If $T_z \neq 0$, the motion field is *radial* (see (8.10)), and all vectors point towards (or away from) a single point, p_0, given by (8.8). If $T_z = 0$, the motion field is *parallel*.

2. The length of motion field vectors is inversely proportional to the depth Z; if $T_z \neq 0$, it is also inversely proportional to the distance from **p** to p_0.

3. p_0 is the vanishing point of the direction of translation (see (8.10)).

4. p_0 is the intersection of the ray parallel to the translation vector with the image plane.

8.2.3 Special Case 2: Moving Plane

Planes are common surfaces in man-made objects and environments, so it is useful to investigate the properties of the motion field of a moving plane. Assume that the camera is observing a planar surface, π, of equation

$$\mathbf{n}^\top \mathbf{P} = d \tag{8.11}$$

where $\mathbf{n} = \begin{bmatrix} n_x, n_y, n_z \end{bmatrix}^\top$ is the unit vector normal to π, and d the distance between π and the origin (the center of projection). Let π be moving in space with translational velocity $\mathbf{T}$ and angular velocity ω, so that both $\mathbf{n}$ and d in (8.11) are functions of time. By means of (8.3), (8.11) can be rewritten as

$$\frac{n_x x + n_y y + n_z f}{f} Z = d. \tag{8.12}$$

Solving for Z in (8.12), and plugging the resulting expression into (8.7), we have

$$v_x = \frac{1}{fd}(a_1 x^2 + a_2 xy + a_3 fx + a_4 fy + a_5 f^2)$$

$$v_y = \frac{1}{fd}(a_1 xy + a_2 y^2 + a_6 fy + a_7 fx + a_8 f^2) \tag{8.13}$$

where

$$a_1 = -d\omega_y + T_z n_x, \qquad a_2 = d\omega_x + T_z n_y,$$
$$a_3 = T_z n_z - T_x n_x, \qquad a_4 = d\omega_z - T_x n_y,$$
$$a_5 = -d\omega_y - T_x n_z, \qquad a_6 = T_z n_z - T_y n_y,$$
$$a_7 = -d\omega_z - T_y n_x, \qquad a_8 = d\omega_x - T_y n_z.$$

The (8.13) states, interestingly, that *the motion field of a moving planar surface, at any instant t, is a quadratic polynomial in the coordinates (x, y, f) of the image points.*

The remarkable symmetry of the time-dependent coefficients $a_1 \ldots a_8$ is not coincidental. You can easily verify that the a_i remain unchanged if d, $\mathbf{n}$, $\mathbf{T}$, and ω are replaced by

$$d' = d$$
$$\mathbf{n}' = \mathbf{T}/\|\mathbf{T}\|$$
$$\mathbf{T}' = \|\mathbf{T}\|\mathbf{n}$$
$$\omega' = \omega + \mathbf{n} \times \mathbf{T}/d.$$

This means that, apart from the special case in which **n** and **T** are parallel, *the same motion field can be produced by two different planes undergoing two different 3-D motions.*[6]

☞   The practical consequence is that it is usually impossible to recover uniquely the 3-D structure parameters, **n** and d, and motion parameters, **T** and ω, of a planar set of points from the motion field *alone*.

You might be tempted to regard this discussion on the motion field of a planar surface as a mere mathematical curiosity. On the contrary, we can draw at least two important and general conclusions from it.

1. Since the motion field of a planar surface is described *exactly and globally* by a polynomial of second degree (see (8.13)), *the motion field of any smooth surface is likely to be approximated well by a low-order polynomial even over relatively large regions of the image plane* (Exercise 8.1). The useful consequence is that very simple parametric models enable a quite accurate estimation of the motion field in rather general circumstances (section 8.4.1).

2. As algorithms recovering 3-D motion and structure cannot be based on motion estimates produced by coplanar points, measurements must be made at many different locations of the image plane in order to minimize the probability of looking at points that lie on planar or nearly planar surfaces.[7] We will return to this point in sections 8.5.1 and 8.5.2.

We conclude this section with a summary of the main properties of the motion field of a planar surface.

Moving Plane: Properties of Motion Field

1. The motion field of a planar surface is, at any time instant, a quadratic polynomial in the image coordinates.

2. Due to the special symmetry of the polynomial coefficients, the same motion field can be produced by two different planar surfaces undergoing different 3-D motions.

8.2.4 Motion Parallax

The decoupling of rotational parameters and depth in the (8.7) is responsible for what is called *motion parallax*. Informally, motion parallax refers to the fact that *the relative motion field of two instantaneously coincident points does not depend on the rotational*

---

[6] This result should not surprise you. Planar surfaces lack generality: The eight-point algorithm (Chapter 7), for example, fails to yield a unique solution if the points are all coplanar in 3-D space.

[7] A "nearly planar" surface is a surface that can be approximated by a plane within a given tolerance, which is typically proportional to the distance of the surface from the image plane.

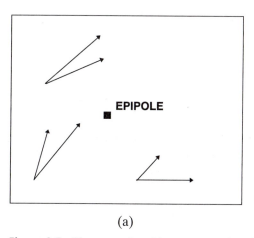
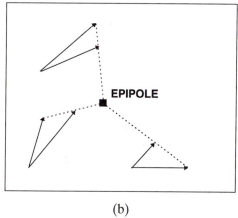

(a)   (b)

Figure 8.5 Three couples of instantaneously coincident image points and their flow vectors (a); the difference vectors point towards the instantaneous epipole (b).

component of motion in 3-D space; this section makes this statement more precise. Motion parallax will be used in section 8.5.2 to compute structure and motion from optical flow.

Let two points $\mathbf{P} = [X, Y, Z]^\top$ and $\bar{\mathbf{P}} = [\bar{X}, \bar{Y}, \bar{Z}]^\top$ be projected into the image points $\mathbf{p}$ and $\bar{\mathbf{p}}$, respectively. We know that the corresponding motion field vectors can be written as

$$v_x = v_x^T + v_x^\omega$$
$$v_y = v_y^T + v_y^\omega$$

and

$$\bar{v}_x = \bar{v}_x^T + \bar{v}_x^\omega$$
$$\bar{v}_y = \bar{v}_y^T + \bar{v}_y^\omega.$$

If, at some instant t, the points $\mathbf{p}$ and $\bar{\mathbf{p}}$ happen to be coincident (Figure 8.5(a)), we have

$$\mathbf{p} = \bar{\mathbf{p}} = [x, y]^\top,$$

and the rotational components of the observed motion, (v_x^ω, v_y^ω) and $(\bar{v}_x^\omega, \bar{v}_y^\omega)$, become

$$v_x^\omega = \bar{v}_x^\omega = -\omega_y f + \omega_z y + \frac{\omega_x xy}{f} - \frac{\omega_y x^2}{f}$$

$$v_y^\omega = \bar{v}_y^\omega = \omega_x f - \omega_z x - \frac{\omega_y xy}{f} + \frac{\omega_x y^2}{f}.$$

$$(8.14)$$

Therefore, by taking the difference between $\mathbf{v}$ and $\bar{\mathbf{v}}$, the rotational components cancel out, and we obtain

$$\Delta v_x = v_x^T - \bar{v}_x^T = (T_z x - T_x f)(\frac{1}{Z} - \frac{1}{\bar{Z}})$$

$$\Delta v_y = v_y^T - \bar{v}_y^T = (T_z y - T_y f)(\frac{1}{Z} - \frac{1}{\bar{Z}}).$$

The vector $(\Delta v_x, \Delta v_y)$ can be thought of as the *relative motion field*. Other factors being equal, Δv_x and Δv_y increase with the separation in depth between $\mathbf{P}$ and $\bar{\mathbf{P}}$.

Notice that the ratio between Δv_y and Δv_x can be written as

$$\frac{\Delta v_y}{\Delta v_x} = \frac{y - y_0}{x - x_0}$$

with $[x_0, y_0]^\top$ image coordinates of $\mathbf{p}_0$, the vanishing point of the translation direction (Figure 8.5(b)).[8] Hence, *for all possible rotational motions, the vector* $(\Delta v_x^T, \Delta v_y^T)$ *points in the direction of* $\mathbf{p}_0$. Consequently, the dot product between the motion field, $\mathbf{v}$, and the vector $[y - y_0, -(x - x_0)]^\top$, which is perpendicular to $\mathbf{p} - \mathbf{p}_0$, depends neither on the 3-D structure of the scene nor on the translational component of motion, and can be written as

$$v_\perp = (y - y_0)v_x^\omega - (x - x_0)v_y^\omega.$$

We will make use of this result in section 8.5.2, where we will learn how to compute motion and structure from dense estimates of the motion field.

☞ Be aware that the vanishing point of translation, $\mathbf{p}_0$, and the point at which $\mathbf{v}$ vanishes, call it $\mathbf{q}$, are in general *different*; they coincide only if the motion is purely translational. Any rotational component about an axis not perpendicular to the image plane shifts the position of $\mathbf{q}$, whereas the position of $\mathbf{p}_0$ remains unchanged, as it is determined by the translational component only. Somewhat deceptively, the flow field in the neighborhood of $\mathbf{q}$ might still look very much like a focus of expansion or contraction (see Figure 8.3).

And here is the customary summary of the main ideas.

Motion Parallax

The relative motion field of two instantaneously coincident points:

1. does not depend on the rotational component of motion
2. points towards (away from) the point $\mathbf{p}_0$, the vanishing point of the translation direction

---

[8] Section 8.2.5 makes it clear that this point can be regarded as an *instantaneous epipole*.

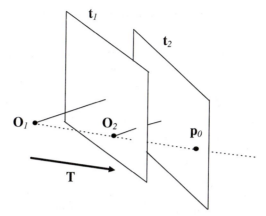

Figure 8.6 The point $\mathbf{p}_0$ as instantaneous epipole.

8.2.5   The Instantaneous Epipole

We close this introductory section with an important remark. The point $\mathbf{p}_0$, being the intersection of the image plane with the direction of translation of the center of projection, can be regarded as the *instantaneous epipole* between pairs of consecutive frames in the sequence (Figure 8.6). The main consequence of this property is that *it is possible to locate* $\mathbf{p}_0$ *without prior knowledge of the camera intrinsic parameters* (section 8.5.2).

☞   Notice that, as in the case of stereo, knowing the epipole's location in image coordinates is *not* equivalent to knowing the direction of translation (the baseline vector for stereo). The relation between epipole location and translation direction is specified by (8.9), which is written in the camera (not image) frame, and contains the focal length f. Therefore, *the epipole's location gives the direction of translation only if the intrinsic parameters of the viewing camera are known.*

8.3   The Notion of Optical Flow

We now move to the problem of *estimating the motion field from image sequences*, that is, from the spatial and temporal variations of the image brightness. To do this, we must model the link between brightness variations and motion field, and arrive at a fundamental equation of motion analysis, the *image brightness constancy equation*. We want also to analyze the power and validity of this equation, that is, understand how much and how well it can help us to estimate the motion field. For simplicity, we will assume that *the image brightness is continuous and differentiable as many times as needed in both the spatial and temporal domain.*

8.3.1 The Image Brightness Constancy Equation

It is common experience that, under most circumstances, the apparent brightness of moving objects remains constant. We have seen in Chapter 2 that the image irradiance is proportional to the scene radiance in the direction of the optical axis of the camera; if we assume that the proportionality factor is the same across the entire image plane, the constancy of the apparent brightness of the observed scene can be written as the stationarity of the image brightness E over time:

$$\frac{dE}{dt} = 0. \tag{8.15}$$

☞   In (8.15), the image brightness, E, should be regarded as a function of both the spatial coordinates of the image plane, x and y, and of time, that is, $E = E(x, y, t)$. Since x and y are in turn functions of t, the *total* derivative in (8.15) should not be confused with the *partial* derivative $\partial E / \partial t$.

Via the chain rule of differentiation, the total temporal derivative reads

$$\frac{dE(x(t), y(t), t)}{dt} = \frac{\partial E}{\partial x}\frac{dx}{dt} + \frac{\partial E}{\partial y}\frac{dy}{dt} + \frac{\partial E}{\partial t} = 0. \tag{8.16}$$

The partial spatial derivatives of the image brightness are simply the components of the spatial image gradient, ∇E, and the temporal derivatives, dx/dt and dy/dt, the components of the motion field, $\mathbf{v}$. Using these facts, we can rewrite (8.16) as the image brightness constancy equation.

The Image Brightness Constancy Equation

Given the image brightness, $E = E(x, y, t)$, and the motion field, $\mathbf{v}$,

$$(\nabla E)^\top \mathbf{v} + E_t = 0. \tag{8.17}$$

The subscript t denotes partial differentiation with respect to time.

We shall now discuss the relevance and applicability of this equation for the estimation of the motion field.

8.3.2 The Aperture Problem

How much of the motion field can be determined through (8.17)? *Only its component in the direction of the spatial image gradient,*[9] v_n. We can see this analytically by isolating the measurable quantities in (8.17):

$$-\frac{E_t}{\|\nabla E\|} = \frac{(\nabla E)^\top \mathbf{v}}{\|\nabla E\|} = v_n \tag{8.18}$$

---

[9] This component is sometimes called the *normal component*, because the spatial image gradient is normal to the spatial direction along which image intensity remains constant.

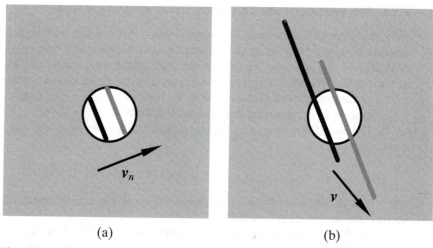

(a)   (b)

Figure 8.7   The aperture problem: the black and grey lines show two positions of the same image line in two consecutive frames. The image velocity perceived in (a) through the small aperture, v_n, is only the component parallel to the image gradient of the true image velocity, v, revealed in (b).

The Aperture Problem

The component of the motion field in the direction *orthogonal* to the spatial image gradient is not constrained by the image brightness constancy equation.

The aperture problem can be visualized as follows. Imagine to observe a thin, black rectangle moving against a white background through a small aperture. "Small" means that the corners of the rectangle are not visible through the aperture (Figure 8.7(a)); the small aperture simulates the narrow support of a differential method. Clearly, there are many, actually infinite, motions of the rectangle compatible with what you see through the aperture (Figure 8.7(b)); the visual information available is only sufficient to determine the velocity in the direction *orthogonal* to the visible side of the rectangle; the velocity in the *parallel* direction cannot be estimated.

☞   Notice that the parallel between (8.17) and Figure 8.7 is not perfect. Equation (8.17) relates the image gradient and the motion field at the *same* image point, thereby establishing a constraint on an *infinitely small* spatial support; instead, Figure 8.7 describes a state of affairs over a *small but finite* spatial region. This immediately suggests that a possible strategy for solving the aperture problem is to look at the spatial and temporal variations of the image brightness over a neighborhood of each point.[10]

---

[10] Incidentally, this strategy appears to be adopted by the visual system of primates.

8.3.3   The Validity of the Constancy Equation: Optical Flow

How well does (8.17) estimate the normal component of the motion field? To answer this question, we can look at the difference, Δv, between the true value and the one estimated by the equation. To do this, we must introduce a model of image formation, accounting for the reflectance of the surfaces and the illumination of the scene.

For the purposes of this discussion, we restrict ourselves to a Lambertian surface, S, illuminated by a pointwise light source infinitely far away from the camera (Chapter 2). Therefore, ignoring photometric distorsion, we can write the image brightness, E, as

$$E = \rho \mathbf{I}^\top \mathbf{n}, \tag{8.19}$$

where ρ is the surface albedo, $\mathbf{I}$ identifies the direction and intensity of illumination, and $\mathbf{n}$ is the unit normal to S at $\mathbf{P}$.

Let us now compute the total temporal derivative of both sides of (8.19). The only quantity that depends on time on the right hand side is the normal to the surface. If the surface is moving relative to the camera with translational velocity $\mathbf{T}$ and angular velocity $\boldsymbol{\omega}$, the orientation of the normal vector $\mathbf{n}$ will change according to

$$\frac{d\mathbf{n}}{dt} = \boldsymbol{\omega} \times \mathbf{n}, \tag{8.20}$$

where $\times$ indicates vector product. Therefore, taking the total temporal derivative of both sides of (8.19), and using (8.17) and (8.20), we have

$$\nabla E^\top \mathbf{v} + E_t = \rho \mathbf{I}^\top (\boldsymbol{\omega} \times \mathbf{n}). \tag{8.21}$$

We can obtain the desired expression for Δv from (8.18) and (8.21):

$$|\Delta v| = \rho \frac{|\mathbf{I}^\top \boldsymbol{\omega} \times \mathbf{n}|}{\|\nabla E\|}.$$

We conclude that, even under the simplifying assumption of Lambertian reflectance, the image brightness constancy equation yields the true normal component of the motion field (that is, $|\Delta v|$ is identically 0 for every possible surface) only for (a) purely translational motion, or (b) for any rigid motion such that the illumination direction is parallel to the angular velocity.

Other factors being equal, the difference Δv decreases as the magnitude of the spatial gradient increases; this suggests that *points with high spatial image gradient are the locations at which the motion field can be best estimated by the image brightness constancy equation.*

In general, $|\Delta v|$ is unlikely to be identically zero, and *the apparent motion of the image brightness is almost always different from the motion field.* For this reason, to avoid confusion, we call the apparent motion an *optical flow*, and refer to techniques estimating the motion field from the image brightness constancy equation as *optical flow techniques.* Here is a summary of similarities and differences between motion field and optical flow.

Definition: Optical Flow

The *optical flow* is a vector field subject to the constraint (8.17), and loosely defined as the *apparent motion* of the image brightness pattern.

Optical Flow and Motion Field

The optical flow is the *approximation of the motion field* which can be computed from time-varying image sequences. Under the simplifying assumptions of

- Lambertian surfaces
- pointwise light source at infinity
- no photometric distortion

the *error* of this approximation is

- *small* at points with high spatial gradient
- *exactly zero* only for translational motion or for any rigid motion such that the illumination direction is parallel to the angular velocity

---

We are now ready to learn algorithms estimating the motion field.

8.4   Estimating the Motion Field

The estimation of the motion field is a useful starting point for the solution of many motion problems. The many techniques devised by the computer vision community can be roughly divided into two major classes: *differential techniques* and *matching techniques*. Differential techniques are based on the spatial and temporal variations of the image brightness at all pixels, and can be regarded as methods for computing optical flow. Matching techniques, instead, estimate the disparity of special image points (features) between frames. We examine differential techniques in section 8.4.1; matching is the theme of section 8.4.2.

8.4.1   Differential Techniques

In recent (and not so recent) years a large number of differential techniques for computing optical flow have been proposed. Some of them require the solution of a system of partial differential equations, others the computation of second and higher-order derivatives of the image brightness, others again least-squares estimates of the parameters characterizing the optical flow. Methods in the latter class have at least two advantages over those in the first two:

- They are not iterative; therefore, they are genuinely local, and less biased than iterative methods by possible discontinuities of the motion field.
- They do not involve derivatives of order higher than the first; therefore, they are less sensitive to noise than methods requiring higher-order derivatives.

We describe a differential technique that gives good results. The basic assumption is that the motion field is well approximated by a *constant* vector field, **v**, within any small region of the image plane.[11]

Assumptions

1. The image brightness constancy equation yields a good approximation of the normal component of the motion field.
2. The motion field is well approximated by a *constant* vector field within any small patch of the image plane.

An Optical Flow Algorithm.   Given Assumption 1, for each point $\mathbf{p}_i$ within a small, $N \times N$ patch, Q, we can write

$$(\nabla E)^{\top} \mathbf{v} + E_t = 0$$

where the spatial and temporal derivatives of the image brightness are computed at $\mathbf{p}_1, \mathbf{p}_2 \cdots \mathbf{p}_{N^2}$.

☞   A typical size of the "small patch" is 5×5.

Therefore, the optical flow can be estimated within Q as the constant vector, $\bar{\mathbf{v}}$, that minimizes the functional

$$\Psi[\mathbf{v}] = \sum_{\mathbf{p}_i \in Q} \left[(\nabla E)^{\top} \mathbf{v} + E_t \right]^2.$$

The solution to this least squares problem can be found by solving the linear system

$$A^{\top} A \mathbf{v} = A^{\top} \mathbf{b}. \tag{8.22}$$

The i-th row of the $N^2 \times 2$ matrix A is the spatial image gradient evaluated at point $\mathbf{p}_i$:

$$A = \begin{bmatrix} \nabla E(\mathbf{p}_1) \\ \nabla E(\mathbf{p}_2) \\ \cdot \\ \cdot \\ \cdot \\ \nabla E(\mathbf{p}_{N \times N}) \end{bmatrix}, \tag{8.23}$$

and **b** is the N^2-dimensional vector of the partial temporal derivatives of the image brightness, evaluated at $\mathbf{p}_1, \ldots \mathbf{p}_{N^2}$, after a sign change:

$$\mathbf{b} = - \left[E_t(\mathbf{p}_1), \ldots, E_t(\mathbf{p}_{N \times N}) \right]^{\top}. \tag{8.24}$$

---

[11] Notice that this is in agreement with the first conclusion of section 8.2.3 (motion field of moving planes) regarding the approximation of smooth motion fields.

The least squares solution of the overconstrained system (8.22) can be obtained as[12]

$$\bar{\mathbf{v}} = (A^{\top}A)^{-1}A^{\top}\mathbf{b}. \tag{8.25}$$

$\bar{\mathbf{v}}$ is the optical flow (the estimate of the motion field) at the center of patch Q; repeating this procedure for all image points, we obtain a dense optical flow. We summarize the algorithm as follows:

Algorithm CONSTANT_FLOW

The input is a time-varying sequence of n images, $E_1, E_2, \ldots E_n$. Let Q be a square region of $N \times N$ pixels (typically, $N = 5$).

1. Filter each image of the sequence with a Gaussian filter of standard deviation equal to σ_s (typically $\sigma_s = 1.5$ pixels) along each spatial dimension.

2. Filter each image of the sequence along the temporal dimension with a Gaussian filter of standard deviation σ_t (typically $\sigma_t = 1.5$ frames). If $2k + 1$ is the size of the temporal filter, leave out the first and last k images.

3. For each pixel of each image of the sequence:

   (a) compute the matrix A and the vector $\mathbf{b}$ using (8.23) and (8.24)
   (b) compute the optical flow using (8.25)

The output is the optical flow computed in the last step.

☞   The purpose of spatial filtering is to attenuate noise in the estimation of the spatial image gradient; temporal filtering prevents aliasing in the time domain. For the implementation of the temporal filtering, imagine to stack the images one on top of the other, and filter sequences of pixels having the same coordinates. Note that the size of the temporal filter is linked to the maximum speed that can be "measured" by the algorithm.

An Improved Optical Flow Algorithm.   We can improve CONSTANT_FLOW by observing that the error made by approximating the motion field at $\mathbf{p}$ with its estimate at the center of a patch increases with the distance of $\mathbf{p}$ from the center itself. This suggests a *weighted* least-square algorithm, in which the points close to the center of the patch are given more weight than those at the periphery. If W is the weight matrix, the solution, $\bar{\mathbf{v}}_w$, is given by

$$\bar{\mathbf{v}}_w = (A^{\top}W^2A)^{-1}A^{\top}W^2\mathbf{b}.$$

Concluding Remarks on Optical Flow Methods.   It is instructive to examine the image locations at which CONSTANT_FLOW fails. As we have seen in Chapter 4, the 2×2 matrix

$$A^{\top}A = \begin{pmatrix} \sum E_x^2 & \sum E_x E_y \\ \sum E_x E_y & \sum E_y^2 \end{pmatrix}, \tag{8.26}$$

---

[12] See Appendix, section A.6 for alternative ways of solving overconstrained linear systems.

computed over an image region Q, is singular if and only if all the spatial gradients in Q are null or parallel. In this case the aperture problem cannot be solved, and the only possibility is to pick the solution of minimum norm, that is, the normal flow. The fact that we have already met the matrix $A^\top A$ in Chapter 4 is not a coincidence; the next section tells you why.

Notice that CONSTANT_FLOW gives good results because the spatial structure of the motion field of a rigid motion is well described by a low-degree polynomial in the image coordinates (as shown in section 8.2.3). For this reason, the assumption of local constancy of the motion field over small image patches is quite effective.

8.4.2 Feature-based Techniques

The second class of methods for estimating the motion field is formed by so-called *matching techniques*, which estimate the motion field at feature points only. The result is a sparse motion field. We start with a two-frame analysis (finding feature disparities between consecutive frames), then illustrate how *tracking* the motion of a feature across a long image sequence can improve the robustness of frame-to-frame matching.

Two-Frame Methods: Feature Matching.   If motion analysis is restricted to two consecutive frames, the same matching methods can be used for stereo and motion.[13] This is true for both correlation-based and feature-based methods (Chapter 7). Here we concentrate on *matching feature points*. You can easily adapt this method for the stereo case too.

The point-matching method we describe is reminiscent of the CONSTANT_FLOW algorithm, and based on the features we met in Chapter 4. There, we looked at the matrix $A^\top A$ of (8.26), computed over small, square image regions: the features were the centers of those regions for which the smallest eigenvalue of $A^\top A$ was larger than a threshold. The idea of our matching method is simple: compute the displacement of such feature points by iterating algorithm CONSTANT_FLOW.

The procedure consists of three steps. First, the uniform displacement of the square region Q is estimated through CONSTANT_FLOW, and added to the current displacement estimate (initially set to 0). Second, the patch Q is *warped* according to the estimated flow. This means that Q is displaced according to the estimated flow, and the resulting patch, Q', is resampled in the pixel grid of frame I_2. If the estimated flow equals (v_x, v_y), the gray value at pixel (i, j) of Q' can be obtained from the gray values of the pixels of Q close to $(i - v_y, j - v_x)$. For our purpose, bilinear interpolation[14] is sufficient. Third, the first and second steps are iterated until a stopping criterion is met. Here is the usual algorithm box, containing an example of stopping criterion.

---

[13] But keep in mind the discussion of section 8.2.1 on the differences between stereo and motion disparities.

[14] Bilinear interpolation means that the interpolation is linear in each of the four pixels closest to $(i - v_y, j - v_x)$.

Algorithm FEATURE_POINT_MATCHING

The input is formed by I_1 and I_2, two frames of an image sequence, and a set of corresponding feature points in the two frames.

Let Q_1, Q_2, and Q' be three $N \times N$ image regions, and τ a fixed, positive real number. Let $\mathbf{d}$ be the unknown displacement between I_1 and I_2 of a feature point $\mathbf{p}$ on which Q_1 is centered.

For all feature points $\mathbf{p}$:

1. Set $\mathbf{d} = 0$ and center Q_1 on $\mathbf{p}$.
2. Estimate the displacement $\mathbf{d}_0$ of $\mathbf{p}$, center of Q_1, through (8.25) and let $\mathbf{d} = \mathbf{d} + \mathbf{d}_0$.
3. Let Q' be the patch obtained by warping Q_1 according to $\mathbf{d}_0$. Compute S, the sum of the squared differences (the SSD of Chapter 7) between the new patch Q' and the corresponding patch Q_2 in the frame I_2.
4. If $S > \tau$, set $Q_1 = Q'$ and go to step 1; otherwise exit.

The output is an estimate of $\mathbf{d}$ for all feature points.

☞   In both the smoothing stage, necessary to compute the derivatives in (8.25), and the warping stage of steps 2 and 3 respectively, you should consider a region actually larger than Q_1 (say by a factor 2). This enables you to iterate the procedure without introducing boundary effects.

☞   An alternative stopping criterion is to control the relative variation of the estimated flow at each iteration and exit the loop if the relative variation falls below a fixed threshold.

Multiple-Frame Methods: Feature Tracking. As we assume to analyze long image sequences, not just pairs of frames, we can improve on two-frame feature matching. We start with an intuitive fact: if the motion of the observed scene is continuous, as it nearly always is, we should be able to make *predictions* on the motion of the image points, at any instant, on the basis of their previous trajectories. In other words, we expect the motion of image points to be continuous, and therefore predictable, in most cases: we should be able to use the disparities computed between frames I_{i-1} and I_{i-2}, I_{i-2} and I_{i-3}, and so on, to make predictions on the disparities between I_{i-1} and I_i, *before* observing frame I_i.

Definition: Feature Tracking

Feature tracking is the problem of matching features from frame to frame in long sequences of images.

We approach tracking in the general framework of *optimal estimation theory*; our solution is the *Kalman filter*. For our purposes, a Kalman filter is a recursive algorithm which estimates the *position* and *uncertainty* of a moving feature point in the next frame, that is, where to look for the feature, and how large a region should be searched in the next frame, around the predicted position, to be sure to find the feature within a certain confidence. An introduction to the basic elements of Kalman filter theory is necessary to understand this section, and can be found in the Appendix, section A.8. Read it now if you are not familiar with the Kalman filter.

Let us formalize the tracking problem. A new frame of the image sequence is acquired and processed at each instant, $t_k = t_0 + k$, where k is a natural number. The sampling interval is assumed 1 for simplicity, and, more importantly, small enough to consider the motion of feature points from frame to frame linear.

We consider *only one* feature point, $\mathbf{p}_k = [x_k, y_k]^\top$, in the frame acquired at instant t_k, moving with velocity $\mathbf{v}_k = [v_{x,k}, v_{y,k}]^\top$. We describe the motion on the image plane with the state vector $\mathbf{x} = [x_k, y_k, v_{x,k}, v_{y,k}]^\top$. Assuming a sufficiently small sampling interval (and therefore constant feature velocity between frames), we write the system model of the linear Kalman filter as

$$\mathbf{p}_k = \mathbf{p}_{k-1} + \mathbf{v}_{k-1} + \boldsymbol{\xi}_{k-1}$$
$$\mathbf{v}_k = \mathbf{v}_{k-1} + \boldsymbol{\eta}_{k-1}, \tag{8.27}$$

where $\boldsymbol{\xi}_{k-1}$ and $\boldsymbol{\eta}_{k-1}$ are zero-mean, white, Gaussian random processes modelling the system noise. In terms of the state vector $\mathbf{x}_k$, (8.27) rewrites

$$\mathbf{x}_k = \Phi_{k-1}\mathbf{x}_{k-1} + \mathbf{w}_{k-1},$$

with

$$\Phi_{k-1} = \begin{bmatrix} 1 & 0 & 1 & 0 \\ 0 & 1 & 0 & 1 \\ 0 & 0 & 1 & 0 \\ 0 & 0 & 0 & 1 \end{bmatrix}$$

and

$$\mathbf{w}_{k-1} = \begin{bmatrix} \boldsymbol{\xi}_{k-1} \\ \boldsymbol{\eta}_{k-1} \end{bmatrix}.$$

As to measurements, we assume that a fast feature extractor estimates $\mathbf{z}_k$, the position of the feature point $\mathbf{p}_k$, at every frame of a sequence. Therefore, the measurement model of the Kalman filter becomes

$$\mathbf{z}_k = \begin{bmatrix} 1 & 0 & 0 & 0 \\ 0 & 1 & 0 & 0 \end{bmatrix} \begin{bmatrix} \mathbf{p}_k \\ \mathbf{v}_k \end{bmatrix} + \boldsymbol{\mu}_k,$$

with $\boldsymbol{\mu}_k$ a zero-mean, white, Gaussian random processes modelling the measurement noise.

Assumptions and Problem Statement

In the assumptions of the linear Kalman filter (Appendix, section A.8), and given the noisy observations $\mathbf{z}_k$, compute the best estimate of the feature's position and velocity at instant t_k and their uncertainties.

Figure 8.8   An example of feature tracking over three frames of a traffic sequence. The feature tracked is the centroid of the car, marked with a cross.

The Kalman filter algorithm is summarized in the following equations, repeated here from the Appendix, section A.8 for completeness.

Algorithm KALMAN_TRACKING

The input is formed, at instant t_k, by the covariance matrices of system and measurement noise at time t_{k-1}, Q_{k-1} and R_{k-1} respectively, the time-invariant state matrix, Φ, the time-invariant measurement matrix, H, and the position measurement at time t_k, $\mathbf{z}_k$. The entries of P_0 are set to high, arbitrary values.

$$P'_k = \Phi_{k-1} P_{k-1} \Phi_{k-1}^\top + Q_{k-1}$$

$$K_k = P'_k H_k^\top (H_k P'_k H_k^\top + R_k)^{-1}$$

$$\hat{\mathbf{x}}_k = \Phi_{k-1}\hat{\mathbf{x}}_{k-1} + K_k(\mathbf{z}_k - H_k \Phi_{k-1}\hat{\mathbf{x}}_{k-1})$$

$$P_k = (I - K_k)P'_k(I - K_k)^\top + K_k R_k K_k^\top.$$

The output is the optimal estimation of the position and velocity at time t_k, $\hat{\mathbf{x}}_k$, and their uncertainties, given by the diagonal elements of P_k.

Two things are worth noticing here. First, we do not just believe the noisy measurements of $\mathbf{p}_k$ of the feature detector; the filter integrates them with model predictions to obtain optimal estimates. Second, *the filter quantifies the uncertainty on the state estimate*, in the form of the diagonal elements of the state covariance matrix. This information allows the feature detector to dimension automatically the image region to be searched to find the feature point in the next frame. The search region is centered on the best position estimate, and is larger the larger the uncertainty. The elements of the state covariance matrix are usually initialized to very large values; in a well-designed filter, they decrease and reach a steady state rapidly, thereby restricting the search region of an image feature within a few frames. An example of tracking with Kalman filtering is shown in Figure 8.8. The centroid of the car in the image (indicated by the white cross) is tracked over time. The size of the cross is proportional to the uncertainty in the system's state.

Two problems arise in the implementation of this algorithm.

Missing Information

Kalman filtering is based on the knowledge of the following:

1. the system model and the corresponding noise covariance matrix, Q_k
2. the measurement model and the corresponding noise covariance matrix, R_k
3. the initial (time t_0) system state, $\hat{x}_0$, and state covariance matrix, P_0

However, several of these quantities are usually unknown.

Data Association

In the presence of several image features and multiple measurements, which observed measurements should be associated with which feature?

Missing Information. Fortunately, this problem is not as bad as you could expect. The system model is usually unknown, but is assumed linear if the time sampling is fast enough. The measurement model is available, as we assume that feature positions are computed at each frame. A really critical parameter, instead, is *the relative weight of model prediction and measurements* expressed by the filter's *gain*, K_k. From the equation of K_k, we see that the gain depends on the covariance matrices of system and measurement noise. In particular, if the entries of R_k are much smaller than those of Q_k (that is, the system model is much noisier, and therefore more uncertain, than the measurements), the Kalman filter ignores the prediction of the system model and relies almost entirely on measurements. Conversely, if the entries of Q_k are much smaller than the entries of R_k (that is, the measurements are much more uncertain than the prediction), the filter ignores the measurements and relies almost entirely on the prediction of the system model. Clearly, one aims at a balanced situation to achieve the greatest benefit from the integration of measurements and prediction. To achieve a balance, one can estimate R_k on the basis of the information available on the measuring process, then scale the entries of Q_k, making them comparable with those of R_k.

Finally, the state and its covariance can be initialized far off their asymptotic values with no risk of compromising the filter convergence.[15]

Data Association. This is a nontrivial problem in general, as there may be many features to be tracked. You should look into the Further Readings for a detailed analysis of techniques dealing with it. Here, we just consider briefly the case of *low clutter* and multiple but *noninterfering targets*. Low clutter means that the likelihood of noisy features at each frame (e.g., false features, features appearing for one frame only) is low. Noninterfering targets means that feature paths do not intersect. In this case, the technique known as *nearest neighbor data association* (NNDA) is the most effective. NNDA just selects the measurement associated with the updated state nearest to the

---

[15] If the filter's assumptions are satisfied, of course.

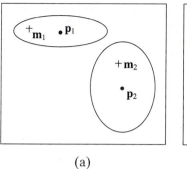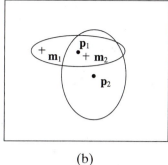

(a)   (b)

Figure 8.9   (a) Disjoint search regions of two features, centered around the best position estimates $\mathbf{p}_1$, $\mathbf{p}_2$; the measurements $\mathbf{m}_1$, $\mathbf{m}_2$ are associated to the closest estimates. (b) If the search regions intersect, the minimum-distance criterion fails.

predicted state (see Figure 8.9(a)). It is good practice to measure the distance between states by means of the inverse of the state covariance matrix (Exercise 8.5). NNDA is clearly suboptimal (Figure 8.9(b)) and more sophisticated methods are required to deal with high clutter and interfering targets.

8.5   Using the Motion Field

Now that we have various ways to estimate the motion field, what do we do with it? We target two tasks of practical importance, the reconstruction of 3-D motion and structure.

Problem Statement

Given the motion field estimated from an image sequence, compute the shape, or *structure*, of the visible objects, and their *motion* with respect to the viewing camera.

Once again we distinguish between methods using dense and sparse estimates of the motion field.

8.5.1   3-D Motion and Structure from a Sparse Motion Field

In this section we estimate 3-D motion and structure from a sparse set of matched image features. If the average disparity between consecutive frames is small, the reconstruction can gain in stability and robustness from the time integration of long sequences of frames. If, on the contrary, the average disparity between frames is large, this problem can be dealt with in a stereo-like fashion, for example by means of the eight-point algorithm of Chapter 7 applied to a pair of frames. Of the many methods proposed in the literature for the former case, we have chosen the *factorization method*, which is simple to implement and gives very good (and numerically stable) results for objects viewed from rather large distances. The necessary assumptions are summarized below.

Assumptions: Factorization Method

1. The camera model is orthographic.

2. The position of n image points, corresponding to the scene points $\mathbf{P}_1, \mathbf{P}_2 \ldots \mathbf{P}_n$, not all coplanar, have been tracked in N frames, with $N \geq 3$.

☞   Note that Assumption 2 is equivalent to acquiring the *entire* sequence before starting any processing. This may or may not be acceptable depending on the application. Notice also that, since the camera model is orthographic, camera calibration can be altogether ignored *if* we accept to reconstruct the 3-D points only up to a scale factor.

The remainder of this section introduces the necessary notations, discusses the *rank theorem*, on which the whole method is based, and states the complete factorization algorithm.

Notation.   We let $\mathbf{p}_{ij} = \left[x_{ij}, y_{ij}\right]^{\top}$ denote the jth image point ($j = 1, \ldots n$) at the i-th frame ($i = 1, \ldots N$), and think of the x_{ij} and y_{ij} as entries of two $N \times n$ matrices, X and Y, respectively. We then form the $2N \times n$ *measurement matrix*

$$W = \begin{bmatrix} X \\ Y \end{bmatrix}.$$

For reasons that will be clear shortly, we subtract the mean of the entries on the same row from each x_{ij} and y_{ij}:

$$\tilde{x}_{ij} = x_{ij} - \bar{x}_i$$
$$\tilde{y}_{ij} = y_{ij} - \bar{y}_i, \tag{8.28}$$

where

$$\bar{x}_i = \frac{1}{n} \sum_{j=1}^{n} x_{ij}$$

$$\bar{y}_i = \frac{1}{n} \sum_{j=1}^{n} y_{ij} \tag{8.29}$$

are the coordinates of $\bar{\mathbf{p}}_i$, the centroid of the image points in the i-th frame. Again, we think of the $\tilde{x}_{ij}$ and $\tilde{y}_{ij}$ as entries of two $N \times n$ matrices, $\tilde{X}$ and $\tilde{Y}$, and form the $2N \times n$ matrix $\tilde{W}$, called the *registered measurement matrix*.

$$\tilde{W} = \begin{bmatrix} \tilde{X} \\ \tilde{Y} \end{bmatrix}. \tag{8.30}$$

The Rank Theorem.   The factorization method is based on the proof of a simple but fundamental result.

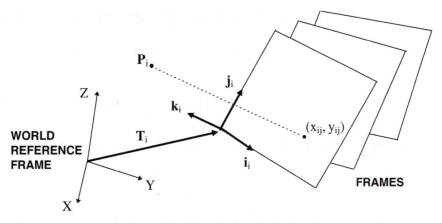

Figure 8.10   The geometry of the factorization method.

Rank Theorem

The registered measurement matrix (without noise) has at most rank 3.

The proof is based on the decomposition (*factorization*) of $\tilde{W}$ of (8.30) into the product of a $2N \times 3$ matrix, R, and a $3 \times n$ matrix, S. R describes the frame-to-frame rotation of the camera with respect to the points $\mathbf{P}_j$. S describes the points' structure (coordinates). The proof is essential for the actual algorithm, so we will go through it in detail.

We consider all quantities expressed in an object-centered reference frame with the origin in the centroid of $\mathbf{P}_1, \ldots \mathbf{P}_n$ (Figure 8.10), and let $\mathbf{i}_i$ and $\mathbf{j}_i$ denote the unit vectors of the image reference frame, expressed in the world reference frame and at time instant i. Thus, the direction of the optical axis is given by the cross product of $\mathbf{i}_i$ and $\mathbf{j}_i$,

$$\mathbf{k}_i = \mathbf{i}_i \times \mathbf{j}_i.$$

It can be seen from Figure 8.10 that

$$x_{ij} = \mathbf{i}_i^\top (\mathbf{P}_j - \mathbf{T}_i) \tag{8.31}$$

$$y_{ij} = \mathbf{j}_i^\top (\mathbf{P}_j - \mathbf{T}_i), \tag{8.32}$$

where $\mathbf{T}_i$ is the vector from the world origin to the origin of the i-th image frame; moreover, as the origin is in the centroid of the points,

$$\frac{1}{n} \sum_{j=1}^{n} \mathbf{P}_j = 0. \tag{8.33}$$

Now, plugging (8.31) and (8.32) into (8.28), and using (8.29), we obtain

$$\tilde{x}_{ij} = \mathbf{i}_i^\top (\mathbf{P}_j - \mathbf{T}_i) - \frac{1}{n} \sum_{m=1}^{n} \mathbf{i}_i^\top (\mathbf{P}_m - \mathbf{T}_i)$$

$$\tilde{y}_{ij} = \mathbf{j}_i^\top (\mathbf{P}_j - \mathbf{T}_i) - \frac{1}{n} \sum_{m=1}^{n} \mathbf{j}_i^\top (\mathbf{P}_m - \mathbf{T}_i). \tag{8.34}$$

But due to (8.33), and to the fact that the index i is not summed, (8.34) become

$$\tilde{x}_{ij} = \mathbf{i}_i^\top \mathbf{P}_j$$

$$\tilde{y}_{ij} = \mathbf{j}_i^\top \mathbf{P}_j$$

Therefore, if we define the $2N \times 3$ *rotation matrix* R as

$$R = \begin{bmatrix} \mathbf{i}_1^\top \\ \mathbf{i}_2^\top \\ \cdot \\ \cdot \\ \cdot \\ \mathbf{i}_N^\top \\ \mathbf{j}_1^\top \\ \mathbf{j}_2^\top \\ \cdot \\ \cdot \\ \cdot \\ \mathbf{j}_N^\top \end{bmatrix}, \tag{8.35}$$

and a $3 \times n$ *shape matrix* S as

$$S = [\, \mathbf{P}_1 \quad \mathbf{P}_2 \quad \cdots \quad \mathbf{P}_n \,], \tag{8.36}$$

we can write

$$\tilde{W} = RS.$$

Since the rank of R is 3 because $N \geq 3$, and the rank of S is also 3 because the N points in 3-D space are not all coplanar, the theorem is proved.

☞ Notice the importance of the assumption of noncoplanar points.

The importance of the rank theorem is twofold. First, it tells you that there is a great deal of redundancy in the image data: no matter how many points and views you are considering, the rank of the registered measurement matrix does not exceed three. Second, and most importantly, the factorization of the registered measurement matrix, $\tilde{W}$, as the product of **R** and **S** suggests a method for reconstructing structure and motion from a sequence of tracked image points.

The Factorization Algorithm. The factorization of $\tilde{W}$ is relatively straightforward. First of all, note that this factorization is not unique: if R and S factorize $\tilde{W}$, and Q is any invertible 3×3 matrix, then RQ and $Q^{-1}S$ also factorize $\tilde{W}$. The proof is simple:

$$(RQ)(Q^{-1}S) = R(QQ^{-1})S = RS = W.$$

Fortunately, we can add two constraints:

1. the rows of R, thought of as 3-D vectors, must have unit norm;
2. the first n rows of R (the $\mathbf{i}_i^\top$) must be orthogonal to the corresponding last n rows (the $\mathbf{j}_i^\top$).

Our last effort before reaching an algorithm box is to show that these constraints allow us to compute a factorization of $\tilde{W}$ which is unique up to an unknown initial orientation of the world reference frame with respect to the camera frame (Figure 8.10). At the same time, we also show how to extend the method to the case in which, due to noise or imperfect matching, the rank of the matrix $\tilde{W}$ is greater than 3. Here is the proof.

First, consider the singular value decomposition (Appendix, section A.6) of $\tilde{W}$,

$$\tilde{W} = UDV^\top. \tag{8.37}$$

The fact that the rank of $\tilde{W}$ is greater than 3 means that more than 3 singular values along the diagonal of D will not be zero. The rank theorem can be enforced simply by setting all but the three largest singular values in D to zero, and recomputing the corrected matrix $\tilde{W}$ from (8.37).

☞   By now, this should not surprise you. We used the same method elsewhere; e.g., to compute the closest rotation matrix to a numerical estimate in Chapter 6. Notice that, if the ratio between the third and fourth singular value is not large, as expected, the SVD warns you about the consistency of the data.

Then let D' be the 3×3 top left submatrix of D corresponding to the three largest singular values, σ_1, σ_2, and σ_3, and U' and V' the $2N \times 3$ and $n \times 3$ submatrices of U and V formed by the columns corresponding to σ_1, σ_2, and σ_3.

$$\hat{R} = U'D'^{1/2}$$
$$\hat{S} = D'^{1/2}V'^\top. \tag{8.38}$$

In general, the rows $\hat{\mathbf{i}}_i^\top$ and $\hat{\mathbf{j}}_i^\top$ of the matrix $\hat{R}$ will not satisfy the constraints mentioned above; however, if we look for a matrix Q such that

$$\hat{\mathbf{i}}_i^\top Q Q^\top \hat{\mathbf{i}}_i = 1$$
$$\hat{\mathbf{j}}_i^\top Q Q^\top \hat{\mathbf{j}}_i = 1 \tag{8.39}$$
$$\hat{\mathbf{i}}_i^\top Q Q^\top \hat{\mathbf{j}}_i = 0,$$

then the new matrices $R = \hat{R}Q$ and $S = Q^{-1}R$ still factorize $\tilde{W}$, and the rows of R satisfy the constraints. The obtained factorization is now clearly unique up to an arbitrary rotation. One possible choice is to assume that at time $t = 0$ the world and camera reference frame coincide.

Here is a concise description of the entire method. A method for determining Q from (8.39) is discussed in Exercise 8.8.

Algorithm MOTSTRUCT_FROM_FEATS

The input is the registered measurement matrix $\tilde{W}$, computed from n features tracked over N consecutive frames.

1. Compute the SVD of $\tilde{W}$,

$$\tilde{W} = UDV^{\top},$$

   where U is a $2N \times 2N$ matrix, V $n \times n$, and D $2N \times n$; $U^{\top}U = I$, $V^{\top}V = I$; and D is the diagonal matrix of the singular values.

2. Set to zero all but the three largest singular values in D.

3. Define $\hat{R}$ and $\hat{S}$ as in (8.5.1).

4. Solve (8.39) for Q, for example by means of Newton's method (Exercise 8.8).

The output are the rotation and shape matrices, given by

$$R = \hat{R}Q \quad \text{and} \quad S = Q^{-1}\hat{S}.$$

The algorithm determines the rotation of a set of 3-D points with respect to the camera, but how about their translation? The component of the translation parallel to the image plane is simply proportional to the frame-by-frame motion of the centroid of the data points on the image plane. However, because of the orthographic assumption, *the component of the translation along the optical axis cannot be determined*.

8.5.2 3-D Motion and Structure from a Dense Motion Field

We now discuss the reconstruction of 3-D motion and structure from optical flow. The two major differences with the previous section are that

- optical flow provides dense but often inaccurate estimates of the motion field;
- the analysis is instantaneous, not integrated over many frames.

Problem Statement

Given an optical flow and the intrinsic parameters of the viewing camera, recover the 3-D motion and structure of the observed scene with respect to the camera reference frame.

We have chosen a method that represents a good compromise between ease of implementation and quality of results. The method consists of two stages:

1. determine the direction of translation through approximate motion parallax;
2. determine a least-squares approximation of the rotational component of the optical flow, and use it in the motion field equations to compute depth.

Stage 1: Translation Direction.   The first stage is rather complex. We start by explaining the solution in the ideal case of *exact* motion parallax, then move to the case of approximate parallax. We learned in section 8.2 that the relative motion field of two *instantaneously coincident* image points, $[\Delta v_x, \Delta v_y]^\top$, is directed towards (or away from) the vanishing point of the translation direction, $\mathbf{p}_0$ (the instantaneous epipole), according to

$$\Delta v_x = (T_z x - T_x f) \left(\frac{1}{Z} - \frac{1}{\bar{Z}} \right)$$

$$\Delta v_y = (T_z y - T_y f) \left(\frac{1}{Z} - \frac{1}{\bar{Z}} \right)$$

(8.40)

where Z and $\bar{Z}$ are the depths of the 3-D points $P = (X, Y, Z)$ and $\bar{P} = (\bar{X}, \bar{Y}, \bar{Z})$, which project onto the same image point, $\mathbf{p} = [x, y]^\top$, in the frame considered. If (8.40) can be written for two different image points, we can locate the epipole, $\mathbf{p}_0$, as the intersection of the estimated, relative motion fields. Once the epipole is known, it is straightforward to get the direction of translation from (8.9).

☞   If (8.40) can be written for more than two points, we can resort to least squares to obtain a better estimate of the epipole's location.

This solution can be extended to the more realistic case of *approximate* motion parallax, in which the estimates of the relative motion field are available only for *almost coincident* image points. The key observation is that *the differences between the optical flow vectors at an image point* $\mathbf{p}$ *and at any point close to* $\mathbf{p}$ *can be regarded as noisy estimates of the motion parallax at* $\mathbf{p}$ (section 8.2.4).

We must now rewrite the (8.40) for the case of approximate parallax. We begin by writing the translational and rotational components of the relative motion field, $\left[\Delta v_x^T, \Delta v_y^T\right]^\top$ and $\left[\Delta v_x^\omega, \Delta v_y^\omega\right]^\top$ respectively, for two almost coincident image points, $\mathbf{p}$ and $\bar{\mathbf{p}}$. These are

$$\Delta v_x^T = \frac{T_z x - T_x f}{Z} - \frac{T_z \bar{x} - T_x f}{\bar{Z}}$$

$$\Delta v_y^T = \frac{T_z y - T_y f}{Z} - \frac{T_z \bar{y} - T_y f}{\bar{Z}}$$

(8.41)

and

$$\Delta v_x^\omega = \omega_z(y - \bar{y}) + \frac{\omega_x}{f}(xy - \bar{x}\bar{y}) - \frac{\omega_y}{f}(x^2 - \bar{x}^2)$$

$$\Delta v_y^\omega = -\omega_z(x - \bar{x}) - \frac{\omega_y}{f}(xy - \bar{x}\bar{y}) + \frac{\omega_x}{f}(y^2 - \bar{y}^2).$$

(8.42)

From the rotation equations we notice that $\Delta v_x^\omega \to 0$ and $\Delta v_y^\omega \to 0$ for $\bar{p} \to p$. As to the translation equations, we can rewrite them as

$$\Delta v_x^T = (T_z x - T_x f)\left(\frac{1}{Z} - \frac{1}{\bar{Z}}\right) + \frac{T_z}{\bar{Z}}(x - \bar{x})$$
$$\Delta v_y^T = (T_z y - T_y f)\left(\frac{1}{Z} - \frac{1}{\bar{Z}}\right) + \frac{T_z}{\bar{Z}}(y - \bar{y}). \tag{8.43}$$

The second terms of the right-hand side of the (8.43) tend to zero for $\bar{p} \to p$, while the first terms tend to the expression obtained for the exact motion parallax, (8.40). We can therefore write the relative motion field of two almost coincident points concisely as

$$\Delta v_x = (T_z x - T_x f)(\frac{1}{Z} - \frac{1}{\bar{Z}}) + e_x(\mathbf{p} - \bar{\mathbf{p}})$$
$$\Delta v_y = (T_z y - T_y f)(\frac{1}{Z} - \frac{1}{\bar{Z}}) + e_y(\mathbf{p} - \bar{\mathbf{p}}), \tag{8.44}$$

with e_x and e_y smooth functions of the difference between $\mathbf{p}$ and $\bar{\mathbf{p}}$, and $e_x(0) = e_y(0) = 0$.

Equations (8.44) show that, if $\mathbf{p}$ and $\bar{\mathbf{p}}$ are close enough, a *large* relative motion field can only be due to a *large* difference in depth between the 3-D points $\mathbf{P}$ and $\bar{\mathbf{P}}$. This observation suggests a relatively simple algorithm for locating the instantaneous epipole (and therefore the direction of translation) from a number of approximate motion parallax estimates. We compute the flow differences $(\Delta v_x, \Delta v_y)$ between a point $\mathbf{p}_i$ and all its neighbors within a small patch Q_i, then determine the eigenvalues and eigenvectors of the matrix

$$A_i = \begin{bmatrix} \sum \Delta^2 v_x & \sum \Delta v_x \Delta v_y \\ \sum \Delta v_x \Delta v_y & \sum \Delta^2 v_y \end{bmatrix}, \tag{8.45}$$

where the sums are taken over the Q_i; the eigenvector corresponding to λ_i, the greater eigenvalue, identifies the direction of the line $\hat{\mathbf{l}}_i$ through $\mathbf{p}_i$ which minimizes the sum of the squared distances to the set of difference vectors (Appendix, section A.6). This direction is taken to be the optimal estimate of motion parallax within the patch Q_i.

Moreover, λ_i itself can be regarded as a measure of the estimate's *reliability*. If λ_i is large,[16] the underlying distribution of the flow differences has a peak in the direction of $\hat{\mathbf{l}}_i$. This is likely to be due to the presence of considerable differences in depth within Q_i. Instead, if λ_i is small, the underlying distribution of the flow differences is flatter, and almost certainly created by the flow field of a surface that does not vary much in depth within Q_i.

---

[16] One might argue that what really counts should be the ratio between the smaller and greater eigenvalues. However, since the range of Δv_x and Δv_y is finite, the greater eigenvalue is *large* in absolute terms.

We can now formulate a weighted least squares scheme to compute the intersection of the several lines $\hat{\mathbf{l}}_i$, that is, the epipole $\mathbf{p}_0$. Since $\hat{\mathbf{l}}_i$, $\mathbf{p}_0$, and $\mathbf{p}_i$ are coplanar, for each patch Q_i we can write

$$(\hat{\mathbf{l}}_i \times \mathbf{p}_i)^\top \mathbf{p}_0 = 0. \tag{8.46}$$

If there are N patches, we can write N simultaneous instances of (8.46), that is, in matrix notation,

$$B\mathbf{p}_0 = 0 \tag{8.47}$$

with

$$B = \begin{bmatrix} \hat{\mathbf{l}}_1 \times \mathbf{p}_1^\top \\ \hat{\mathbf{l}}_2 \times \mathbf{p}_2^\top \\ \cdot \\ \cdot \\ \cdot \\ \hat{\mathbf{l}}_N \times \mathbf{p}_N^\top \end{bmatrix} \tag{8.48}$$

The problem of determining a least-squares estimate of $\mathbf{p}_0$ is thus reduced to the problem of solving the overconstrained homogeneous system (8.47). As customary, the solution can be found from the SVD of B, $B = UDV^\top$ (Appendix, section A.6), as the column of V corresponding to the null (in practice, the smallest) singular value of B.

☞   In order to give appropriate weights to the different estimates, it is better to use a weighted least-square scheme and consider the matrix WB, where the entries of the diagonal matrix W are the larger eigenvalues of A_i.

Stage Two: Rotational Flow and Depth.   The rest of the algorithm is straightforward. We simply form the pointwise dot product, $v_\perp$, between the optical flow at point $\mathbf{p}_i = [x_i, y_i]^\top$ and the vector $[y_i - y_0, -(x_i - x_0)]^\top$. As we know from section 8.2, $v_\perp$ depends only on the rotational component of motion; therefore, at each point $\mathbf{p}_i$ of the image plane we have

$$v_\perp = v_x^\omega(y_i - y_0) - v_y^\omega(x_i - x_0) \tag{8.49}$$

with v_x^ω and v_y^ω as in (8.42). If the intrinsic parameters of the camera are known, we can write a linear system of N simultaneous instances of (8.49) in the image reference frame by using (8.14), and solve for the three components of the angular velocity using least squares. Finally, we recover the translational direction from the epipole coordinates by means of (8.9), and solve (8.7) for the depth Z of each image point.

It is now time to summarize the method.

Algorithm MOTSTRUCT_FROM_FLOW

The input quantities are the intrinsic parameters of the viewing camera, and a dense optical flow field, **v**, produced by a single rigid motion.

1. Write (8.7) in the image reference frame, using the knowledge of the intrinsic parameters.

2. For each image point $\mathbf{p}_i, i = 1, \ldots N$:

   (a) compute the flow differences Δv_x and Δv_y between the optical flow at $\mathbf{p}_i$ and at all the points $\mathbf{p}$ in a neighborhood of $\mathbf{p}_i$, Q_i;

   (b) compute the eigenvalues and eigenvectors of the matrix A_i of (8.45); let λ_i be the greater eigenvalue, and $\hat{\mathbf{l}}_i$ the unit eigenvector corresponding to λ_i.

3. Compute the SVD of WB, $WB = UDV^\top$, with B as in (8.48) and W a diagonal matrix such that $W_{ii} = \lambda_i$. Estimate the epipole $\mathbf{p}_0$ as the column of V corresponding to the smallest singular value.

4. Form the dot product of (8.49) for $i = 1, \ldots, N$ and rewrite the equations obtained in the image reference frame.

5. Determine the angular velocity components as the least-squares solution of a system of N simultaneous instances of (8.49).

6. Determine the translational direction from the epipole coordinates and the knowledge of the intrinsic parameters (see (8.9)).

7. Solve (8.7) for the depth Z of each image point.

The output quantities are the direction of translation, the angular velocity, and the 3-D coordinates of the scene points.

☞   Notice that, as discussed in section 8.2.5, the epipole can be estimated *without prior knowledge of the camera parameters*, that is, with an uncalibrated camera. The direction of translation, instead, can be obtained from the epipole *only if the intrinsic parameters of the camera are known.*

MOTSTRUCT_FROM_FLOW is not as accurate as MOTSTRUCT_FROM_ FEATS. This is not surprising, as MOTSRUCT_FROM_FLOW is an instantaneous method, which relies on local approximations of the observed motion, on the assumption of large variation in depth in the observed scene, and on the accuracy of camera calibration.

8.6 Motion-based Segmentation

In this final section, we relax the assumption that the motion between the camera and the scene is described by a single 3-D motion to deal with the problem of *multiple motions*. For the sake of simplicity, we restrict the analysis to the case in which the camera is fixed. If you are interested in motion segmentation in the presence of camera motion, a problem which is still waiting for a general and satisfactory solution, see the Further Readings.

If the camera is fixed, identifying moving objects can be seen as a problem of *detecting changes against a fixed background.*

Problem Statement

Given a sequence of images taken by a fixed camera, find the regions of the image, if any, corresponding to the different moving objects.

This problem can be thought of as a classification problem. One has to classify the pixels of each frame of a sequence as either *moving* or *fixed* with respect to the camera. On the basis of what we have seen so far, a possible procedure seems to be the computation of optical flow followed by a thresholding procedure. The pixels for which the norm of the optical flow is large enough are labelled as "moving", the others as "fixed". Two criticisms can be raised against this approach. First, the estimation of optical flow inside patches that contain pixels from independently moving objects is usually rather poor. Second, in many applications (surveillance tasks for example) the detection of motion or changes in the scene is all that is needed.

Presumably the simplest strategy for detecting changes in an image sequence is *image differencing*: *(i)* take the pointwise difference between consecutive frames, and *(ii)* label as "moving" the pixels for which this difference exceeds a predetermined threshold, t.

Algorithm CHANGE_DETECTION

The input is an image sequence, $I_1, I_2, \ldots, I_n$, and a positive real number, τ; for each image pair (I_k, I_{k+1})

1. Compute the pointwise image difference $\Delta_k(i, j) = I_{k+1}(i, j) - I_k(i, j)$;
2. if $|\Delta_k(i, j)| > \tau$, label pixel (i, j) of the frame k as moving.

The output is a map of the moving image regions.

The threshold must be chosen so that the probability of mistaking differences created by image noise for real motion is very small. A simple way to do this is to acquire two images of a static scene, with no illumination changes, and look at the histogram of the difference image. In agreement with the assumption that this difference is mainly due to the camera noise (see Chapter 2), the histogram should look like a zero-mean Gaussian function (see Figure 8.11). The threshold can thus be estimated as a multiple of the standard deviation of the computed distribution.

☞   Of course, the more different the images used to estimate the standard deviation, the more accurately the histogram reflects the distribution of the noise.

Better results can be obtained in a number of different ways. One possibility is to resort to statistical tests (see Further Readings). A second possibility is to take *motion measures* more sophisticated than image differencing. For instance, an alternative to image differencing is the weighted average of the normal flow magnitude, $|E_t|/\|\nabla E\|$,

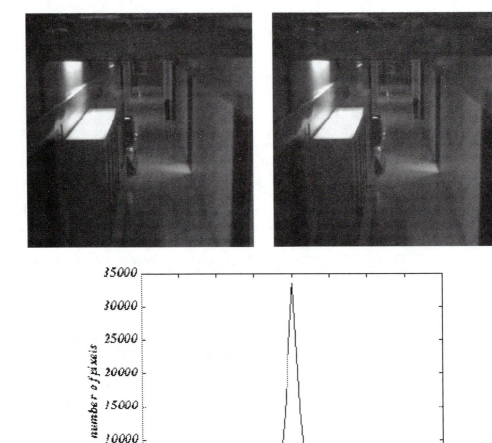

Figure 8.11 Two sample snapshots, taken at close instants, from an overnight, indoors surveillance sequence, and the histogram of the differences of the gray values between the two images.

over a small patch Q_i centered at point $\mathbf{p}_i$. The weights are taken to be the square of the norm of the spatial image gradient, so that this motion measure can be written as

$$M = \frac{|E_t|\|\nabla E\|}{C + \sum \|\nabla E\|^2},$$

where the sum are taken over Q_i, and the temporal and spatial derivatives are meant to be computed at each point of Q_i. The purpose of the constant C is to remove the instability that may be caused by uniform patches. The choice of the weights agrees with the results of section 8.3, which proved that the difference between the true motion field and the apparent motion of the image brightness is smaller at the locations where the norm of the spatial image gradient is larger. The conversion of this idea into an algorithm for change detection is straightforward, and left as an exercise.

8.7 Summary

After working through this chapter you should be able to:

❑ explain the fundamental problems of motion analysis in computer vision

❑ estimate the optical flow from a sequence of images

❑ match and tracking image features over time

❑ estimate 3-D motion and structure from both dense and sparse estimates of the motion field

❑ detect changes in image sequences taken from a fixed camera

8.8 Further Readings

The discussion on the difference between motion field and optical flow is taken from Verri and Poggio [19]. The method described here for estimating optical flow is due to Lucas and Kanade [12]. A variational approach to the computation of optical flow was first proposed by Horn and Schunck [8]. Of the many parametric methods proposed since see, for example, Campani and Verri [3] (in which you can also find a discussion on the spatial properties of the motion field). For a strictly local computation of optical flow you should start with Nagel's work (see [14], for example). A correlation-based method is described by Poggio, Little, and Gamble [15]. Finally, the paper by Barron, Fleet, and Beauchemin [2] is an excellent review of many methods for estimating optical flow, and includes Internet sites with public-domain code.

The literature on feature matching for motion is also vast. Among the many techniques available, you may want to look at the classic books by Ullman [18] and Hildreth [7]. A nice and simple account of Kalman filtering is the paper by Cooper [4]. Alternatively, you can look at the very clear and complete book by Maybeck [13]. Data association is discussed, for example, in the book edited by Bar-Shalom [1].

The factorization method for 3-D motion and structure from sparse features is due to Tomasi and Kanade [17]. Among the other feature-based method we suggest the technique proposed by Faugeras, Lustman, and Toscani [5]. The epipole's method for reconstruction from optical flow is based on the paper by Rieger and Lawton [16] and on an idea originally suggested by Longuet-Higgins and Prazdny [11]. Part of the algorithm MOTSTRUCT_FROM_FLOW is based on the implementation proposed in the appendix of the paper by Heeger and Jepson [6], the main topic of which is an alternative algorithm to MOTSTRUCT_FROM_FLOW.

For a thorough analysis of change detection see Hsu, Nagel and Reker [9]. As to motion-based segmentation in the general case, you may want to look at the work of Irani, Rousso, and Peleg [10], for example, and references therein.

8.9 Review

Questions

❏ *8.1* What are the properties of the motion field generated by a planar surface? Would an arbitrarily oriented planar surface generate a motion field across the entire image plane?

❏ *8.2* What is the relation between the instantaneous epipole and the focus of expansion (or contraction)?

❏ *8.3* What is the difference between motion field and optical flow?

❏ *8.4* The derivation of (8.17) assumes that the image brightness is continuous and can be differentiated. How plausible is this assumption in general? Why?

❏ *8.5* What are the assumptions behind the algorithm CONSTANT_FLOW?

❏ *8.6* Given our discussion of the Kalman filter for feature tracking, how would you decide the shape and size of the search regions given the uncertainties produced by the filter?

❏ *8.7* How would you decide whether the effective rank of $\tilde{W}$ in algorithm MOTSTRUCT_FROM_FEATS is 3?

❏ *8.8* What happens if you apply MOTSTRUCT_FROM_FLOW to the optical flow of a planar surface?

❏ *8.9* Why change detection methods are not useful (in themselves) for motion-based segmentation in the general case?

Exercises

○ *8.1* Estimate the ratio between the quadratic and linear terms, the quadratic and constant terms, and the linear and constant terms in (8.13). Set the motion and structure parameters of the planar surface to some arbitrary but reasonable values.

○ *8.2* Show that the aperture problem can be solved if a corner is visible through the aperture.

○ *8.3* Extend algorithm CONSTANT_FLOW by assuming that the motion field is locally approximated by a linear vector field.

○ *8.4* Create a simple, synthetic image sequence by shifting a given image by $[v_x, v_y]^T$ pixel per frame (with v_x and v_y not integer), and using bilinear interpolation in the resampling stage. Then, apply the *inverse* transformation; that is, create an image sequence starting from the last frame and applying a motion of $[-v_x, -v_y]^T$ pixel per frame. Compare the last frame of this sequence with the original image. Do you expect them to be *exactly* equal? If not, why?

○ *8.5* Show that the distance between system states computed through the covariance matrix, named *Mahalanobis distance*, is not isotropic. What is the advantage of using Mahalanobis distance instead of the usual Euclidean distance?

○ *8.6* Write the Kalman filter equations for a one-dimensional state vector.

○ *8.7* The celebrated structure-from-motion theorem, due to Shimon Ullman, states that, under orthographic projection, at least N views of at least n noncoplanar points are needed to uniquely recover structure. Can you guess the values of N and n from the factorization method?

○ *8.8* The nonlinear system (8.39) can be solved by means of *Newton's method*, an iterative procedure, the main idea of which is rather simple. Write down the full system of $3 \times N$ quadratic equations in the nine entries of the matrix Q, say $q_1, \ldots, q_9$. Starting from an initial guess for the values of $q_1, \ldots, q_9$ (like the identity matrix for example), take the partial derivatives of each equation with respect to each unknown, and evaluate the expressions obtained at the current value of the entries. If M_{ij} denotes the partial derivatives of the i-th equation with respect to q_j, the matrix M_{ij} can be viewed as the system matrix of the linear system

$$M \Delta \mathbf{q} = \epsilon,$$

where $\Delta \mathbf{q} = [\Delta q_1, \ldots, \Delta q_9]^\top$, the components of which should be used to update the current estimate of $q_1, \ldots, q_9$. The components of the $3N$-D vector ϵ are the residuals of all equations, computed by means of the previous estimates of $q_1, \ldots, q_9$. The procedure is iterated until the components of $\Delta \mathbf{q}$ are sufficiently small. Implement and solve this method for estimating Q.

○ *8.9* Show that Step 3 of MOTSTRUCT_FROM_FLOW works equally well even if the epipole is at infinity.

Projects

● *8.1* Implement a coarse-to-fine version of CONSTANT_FLOW. Build the coarser levels by iteratively averaging over the four neighboring pixels at the immediately finer levels. Start CONSTANT_FLOW at the coarsest level and propagate the estimates to each corresponding group of four pixels at the finer level. Compare the results with a standard implementation of CONSTANT_FLOW. Why should the two methods differ? Which method is expected to perform better in the presence of large image displacement?

● *8.2* Implement FEATURE_POINT_MATCHING and MOTSTRUCT_ FROM_FEATS. Use the set of matched features, output of the former, as input of the latter.

● *8.3* Implement CONSTANT_FLOW and MOTSTRUCT_FROM_FLOW. Use the optical flow estimates, output of the former, as input of the latter.

● *8.4* Implement a tracking system based on the Kalman filter, and use it to track a moving object viewed from a fixed camera. Use the centroid of the largest set of connected pixels in which a change is detected, and measure the object's position through CHANGE_DETECTION. Let the centroid position and velocity be the system state. Devise a simple data association algorithm to deal with the possibility of detecting changes in more than one large set of connected pixels.

References

[1] Y. Bar-Shalom (ed.), *Multi-Target Multi-Sensor Tracking*, Artec House (1990).

[2] J.L. Barron, D.J. Fleet, and S. Beauchemin, Performance of Optical Flow Techniques, *International Journal of Computer Vision*, Vol. 12, pp. 43–77 (1994).

[3] M. Campani and A. Verri, Motion Analysis from First Order Properties of Optical Flow. *CVGIP: Image Understanding*, Vol. 56, pp. 90–107 (1992).

[4] W.S. Cooper, Use of Optimal Estimation Theory, in Particular the Kalman Filter, in Data Analysis and Signal Processing, *Review of Scientific Instrumentation*, Vol. 57, pp. 2862–2869 (1986).

[5] O.D. Faugeras, F. Lustman, and G. Toscani, 3-D Structure from Point and Line Matches, *Proc. 1st International Conference on Computer Vision*, London (UK), pp. 25–34 (1987).

[6] D.J. Heeger and A.D. Jepson, Subspace Methods for Recovering Rigid Motion I: Algorithm and Implementation, *International Journal of Computer Vision*, Vol. 7, pp. 95–117 (1992).

[7] E.C. Hildreth, *The Measurement of Visual Motion*, MIT Press, Cambridge (MA), (1984).

[8] B.K.P. Horn and B.G. Schunck, Determining Optical Flow, *Artificial Intelligence*, Vol. 17, pp. 185–203 (1981).

[9] Y.Z. Hsu, H.-H. Nagel, and G. Rekers, New Likelihood Test Methods for Change Detection in Image Sequences, *Computer Vision, Graphics, and Image Processing*, Vol. 26, pp. 73–106 (1984).

[10] M. Irani, B. Rousso, and S. Peleg, Computing Occluding and Transparent Motions, *International Journal of Computer Vision*, Vol. 12, pp. 5–16 (1994).

[11] H.C. Longuet-Higgins and K. Prazdny, The Interpretation of a Moving Retinal Image, *Proc. Royal Soc. London B*, Vol. 208, pp. 385–397 (1980).

[12] B.D. Lucas and T. Kanade, An Iterative Image Registration Technique with an Application to Stereo Vision, *Proc. 7th International Joint Conference on Artificial Intelligence*, Vancouver (CA), pp. 674–679 (1981).

[13] P.S. Maybeck, *Stochastic Models, Estimation, and Control*, Vol. I, Academic Press, New York (1979).

[14] H.-H. Nagel, Displacement Vectors Derived from 2nd Order Intensity Variations in Image Sequences, *Computer Vision, Graphics, and Image Processing*, Vol. 21, pp. 85–117 (1983).

[15] T. Poggio, J.J. Little, E. Gamble, Parallel Optical Flow, *Nature*, Vol. 301, pp. 375–378 (1986).

[16] J.H. Rieger and D.T. Lawton, Processing Differential Image Motion, *Journal of the Optical Society of America A*, Vol. 2, pp. 354–359 (1985).

[17] C. Tomasi and T. Kanade, Shape and Motion from Image Streams under Orthography: a Factorization Method, *International Journal of Computer Vision*, Vol. 9, pp. 137–154 (1992).

[18] S. Ullman, *The Interpretation of Visual Motion*, MIT Press, Cambridge (MA) (1979).

[19] A. Verri and T. Poggio, Motion Field and Optical Flow: Qualitative Properties, *IEEE Transactions on Pattern Analysis Machine Intelligence*, Vol. 11, pp. 490-498 (1989).

[20] T. Viéville and O.D. Faugeras, Feed-Forward Recovery of Motion and Structure from a Sequence of 2D-lines Matches, *Proc. 3rd International Conference on Computer Vision*, Osaka (Japan), pp. 517–520 (1990).

9

Shape from Single-image Cues

Je voudrais pas crever
Sans savoir si la lune
Sous son faux air de thune
A un côté pointu[1]

Boris Vian, *Je Voudrais pas Crever*

The subject of this chapter, inferring the shape of objects from a single intensity image, is a classic problem of computer vision. We do not consider range images, in which shape is already explicit.

Chapter Overview

Section 9.1 lists the main methods for inferring shape from intensity images.

Section 9.2 introduces the concept of *reflectance map* and the problem of *shape from shading* from a physical and mathematical viewpoint.

Section 9.3 discusses a method for estimating *albedo* and *illuminant direction*.

Section 9.4 describes a method for extracting the shape of an object from a shading pattern.

Section 9.5 shows how shape can be computed from the distortion of 3-D textures caused by the imaging projection, considering deterministic and statistical textures.

What You Need to Know to Understand this Chapter

- Working knowledge of Chapters 2 and 4.
- Working knowledge of Fast Fourier Transform (FFT).

---

[1] I would not want to die without knowing whether the moon, behind her false look of coin, has a pointed side.

9.1 Introduction

A common experience of our everyday life is the perception of solid shape, indeed so common that it is hard to appreciate its full complexity. We perceive without effort the shape of complex objects like faces and cars, irregular surfaces like mountains, and even of changing surfaces like a tree in the wind. In spite of this apparent simplicity, shape reconstruction has proven a very hard problem for computer vision; indeed, one which is solved only partially. Many methods, collectively known as *shape from X*, have been proposed for reconstructing 3-D shape from intensity images, and the algorithms in the previous two chapters can be regarded as shape-from-X methods (i.e., shape from stereo, shape from motion). Shape-from-X methods exploit a large variety of image cues; Table 9.1 mentions the best-known ones, and classifies them according to two important characteristics: the number of images needed, and whether or not the method requires purposive modification of the vision system's parameters; that is, whether the method is *active* or *passive*.

The active-passive distinction is worth commenting. In *active* methods, the vision system's parameters are modified *purposively*; for instance, a shape-from-defocus system controls the focus of the lens to acquire two or more out-of-focus images, and use estimates of the focus level at each pixel to compute local shape. Shape-from-motion systems can be either active (if the sensors are moved purposively to generate a relative motion between sensor and scene) or passive (if relative motion occurs without purposive sensor motion).[2]

This chapter concentrates on *shape reconstruction from a single intensity image*. Notice that we make no assumptions on the shape of the objects in the scene. This situation is similar to trying to make out the shapes of unknown, solid objects from a single photograph. Here is a concise statement of the problem.

Problem Statement

Given a single image of unknown objects, reconstruct the shape of the visible surfaces.

As suggested by the table above, various cues can be exploited to tackle the problem. This chapter discusses two of them in detail, *shading* and *texture*, and gives algorithms for computing *shape from shading* and *shape from texture*. The reasons for choosing these two methods are that they do not require special hardware (unlike for instance shape from zoom or focus/defocus, which require a motorized lens), and do not depend on image preprocessing (unlike for instance shape from contours, which assumes that the contours of objects have been identified).

---

[2] *Active vision* denotes a whole area of computer vision, the scope of which goes far beyond reconstruction: in general, the term refers to *strategies for observation*. In the active vision paradigm, images are analyzed and acquired purposefully, in order to accomplish specific tasks. Thus, sensor and observer interact continuously, and vision is not just about interpreting isolated snapshots or sequences. For more on active vision, see the Further Readings.

Shape from	How many images	Method type	Find more in
Stereo	2 or more	passive	Chapter 7
Motion	a sequence	active/passive	Chapter 8
Focus/defocus	2 or more	active	Further Readings
Zoom	2 or more	active	Further Readings
Contours	single	passive	Further Readings
Texture	single	passive	this chapter
Shading	single	passive	this chapter

Table 9.1   Shape-from-X methods and their classification.

☞ Notice that our problem statement implies the use of intensity images, as range images are themselves a representation of 3-D shape. In fact, the problem of this chapter can be regarded as one of *producing a range image from an intensity one*.

9.2 Shape from Shading

Shape from shading uses the pattern of lights and shades in an image to infer the shape of the surfaces in view. This sounds very useful, also considering that single intensity images are easier to acquire than stereo pairs or temporal sequences. Unfortunately, the problems we face in shape from shading are considerably more complicated that any others encountered so far. There are at least two main reasons for this fact: the first has to do with the *physics* of the problem, the second with the *mathematics*. We will have to look into both in order to arrive at a method for computing shape from shading.

In spite of its difficulties, shape-from-shading has been used successfully in various applications. A typical example is astronomy, where shape from shading is used to reconstruct the surface of a planet from photographs acquired by a spacecraft. An example is given in Figure 9.1, which takes forward the Magellan example of Chapter 7. Here, a shape-from-shading algorithms refines an initial range image obtained by stereopsis.

9.2.1 The Reflectance Map

The fact that shape *can* be reconstructed from shading is due to the existence of a fundamental equation that links image intensity and surface slope. In order to establish this equation we need to introduce the important concept of *reflectance map*. We first discuss the simple case of Lambertian surfaces.

In Chapter 2 we have seen that a uniformly illuminated Lambertian surface appears equally bright from all viewpoints. The radiance L at a 3-D point, $\mathbf{P}$, is proportional to the cosine of the angle between the surface normal and the direction of the illuminant:

$$L(\mathbf{P}) = \rho \mathbf{i}^\top \mathbf{n}, \qquad (9.1)$$

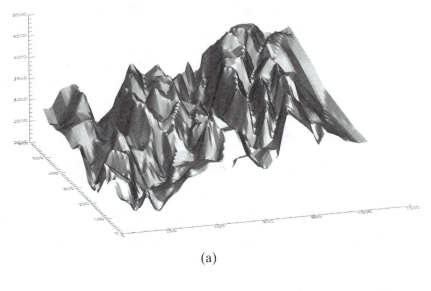

(a)

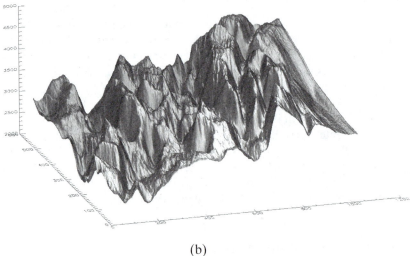

(b)

Figure 9.1 (a): 3-D rendering of the surface reconstructed by stereopsis from the stereo pair on page 142. (b): the refined surface obtained by the shape-from-shading algorithm; notice the increased level of detail. Courtesy of Alois Goller, Institute for Computer Graphics and Vision, Technical University of Graz.

with **i** the vector which gives the illuminant direction (pointing towards the light source), **n** the surface normal at $\mathbf{P} = [X, Y, Z]^\top$, and ρ the *effective albedo*; that is, the real albedo times the intensity of illuminant. To stress the dependence of the radiance on the surface normal, we rewrite (9.1) as

$$R_{\rho, \mathbf{i}} = \rho \mathbf{i}^\top \mathbf{n}. \tag{9.2}$$

We use R instead of L because (9.2) tells you how light is reflected by a Lambertian surface in terms of the surface normal for *any* given albedo and illuminant. This is a particular example of *reflectance map*. In general, the function $R_{\rho,\mathbf{i}}$ is more complicated or known only numerically through experiments.

9.2.2   The Fundamental Equation

In Chapter 2, we also met the *image irradiance equation*, which, denoting with $\mathbf{p} = [x, y]^\top$ the image of $\mathbf{P}$, we wrote as

$$E(\mathbf{p}) = L(\mathbf{P})\frac{\pi}{4}\left(\frac{d}{f}\right)^2 \cos^4 \alpha, \tag{9.3}$$

with $E(\mathbf{p})$ the brightness measured on the image plane at $\mathbf{p}$. We now make two important assumptions:

1. We neglect the constant terms in (9.3) and assume that the optical system has been calibrated to remediate the $\cos^4 \alpha$ effect.
2. We assume that all the visible points of the surface receive direct illumination.

Assumption 1 simplifies the mathematics of the problem, and Assumption 2 avoids dealing with secondary reflections. In these assumptions, combining (9.3) and (9.2) gives

$$E(\mathbf{p}) = R_{\rho,\mathbf{i}}(\mathbf{n}). \tag{9.4}$$

Equation (9.4) constrains the direction of the surface normal at $\mathbf{P}$, and is the *fundamental equation of shape from shading*.

In order to make profitable use of (9.4), we need to express the normal in terms of the surface slopes. To this purpose, we further assume that the visible surface

3. is far away from the viewer
4. can be described as $Z = Z(X, Y)$

Assumption 3 enables us to adopt the *weak-perspective camera model*, and write

$$x = f\frac{X}{Z_0}$$

and

$$y = f\frac{Y}{Z_0},$$

with Z_0 the average distance of the surface from the image plane. Thus, by a suitable rescaling of the horizontal and vertical axis, the surface, Z, can be thought of as a function of the (x, y) coordinates on the image plane,

$$Z = Z(x, y).$$

The surface slopes can now be computed by simply taking the x and y partial derivatives of the vector $[x, y, Z(x, y)]^\top$, which gives

$$[1, 0, \partial Z/\partial x]^\top \quad \text{and} \quad [0, 1, \partial Z/\partial y]^\top.$$

Both vectors lie on the tangent plane to the surface, and hence their normalized vector product is the unit normal $\mathbf{n}$ to the surface at (x, y) (Appendix, section A.5); that is,

$$\mathbf{n} = \frac{1}{\sqrt{1 + p^2 + q^2}}[-p, -q, 1]^\top, \tag{9.5}$$

where we have denoted with p and q the partial derivatives $\partial Z/\partial x$ and $\partial Z/\partial y$ respectively. By means of (9.2) and (9.5), we can rewrite (9.4) as

$$E(x, y) = \frac{\rho}{\sqrt{1 + p^2 + q^2}}\mathbf{i}^\top[-p, -q, 1]. \tag{9.6}$$

Equation (9.6) is the typical starting point of many shape from shading techniques. Unfortunately, it is of a great mathematical complexity: It is a nonlinear partial differential equation through $p = p(x, y)$ and $q = q(x, y)$, the slopes of the unknown surface $Z = Z(x, y)$, and depends on quantities not necessarily known (albedo, illuminant, and boundary conditions).[3]

The best way to grasp the meaning of (9.6) is to produce a synthetic shaded image yourself. To do this, you adopt the Lambertian assumption, write the equation of your favorite surface in the form $Z = Z(x, y)$, choose values for ρ and $\mathbf{i}$, and compute the numerical partial derivatives of the surface Z over a pixel grid. You can then compute the corresponding image brightness according to (9.6). Two examples of this simple procedure to produce a shaded image are shown in Figure 9.2.

☞   Note that, for certain illuminants, some image locations might violate Assumption 2 (Figure 9.2(b)), and the image brightness computed through (9.6) become negative; in this case, the brightness should be set to 0. Therefore, the right hand side of (9.6) should always be regarded as

$$\max\{0, \frac{\rho}{\sqrt{1 + p^2 + q^2}}\mathbf{i}^\top[-p, -q, 1]\}. \tag{9.7}$$

The number of unknowns in (9.6) seems to suggest that this equation does not provide enough constraints to reconstruct p and q at all pixels: in the discrete setting ($N \times N$ pixels), it seems that we end up with N^2 equations, one for each pixel, in $2N^2$ unknowns, the p and the q. But this is not true, because the p and q are not independent: since $Z_{xy} = Z_{yx}$, we have N further equations of the kind

$$p_y = q_x.$$

We will come back to this point in the discussion of a method for solving (9.6).

---

[3] Just like an ordinary differential equation depends on initial conditions, a partial differential equation depends on boundary conditions. In essence, the solution is determined uniquely only if certain *a priori* information on the solution is available. This information is typically given at the boundary of the domain of interest (in our case, the image boundaries) or on some particular curves.

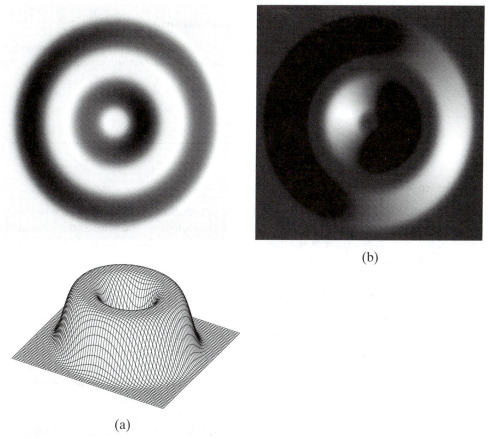

(b)

(a)

Figure 9.2 Two images of the same Lambertian surface seen from above but illuminated from different directions and 3-D rendering of the surface. Practically all the points in the top left image receive direct illumination ($\mathbf{i} = [0.20, 0, 0.98]^\top$); some regions of the top right image are in the dark due to self-shadowing effects ($\mathbf{i} = [0.94, 0.31, 0.16]^\top$).

We can now summarize the various assumptions made, and state the problem of shape from shading in more technical terms.

Assumptions

1. The acquisition system is calibrated so that the image irradiance, $E(\mathbf{p})$, equals the scene radiance, $L(\mathbf{P})$, with $\mathbf{p} = [x, y]^\top$ image of the 3-D point $\mathbf{P} = [X, Y, Z]^\top$.

2. All the visible surface points receive direct illumination.

3. The surface is imaged under weak perspective.

4. The optical axis is the Z axis of the camera, and the surface can be parameterized as

$$Z = Z(x, y).$$

Problem Statement

Given the reflectance map of the viewed surface, $R = R_{\rho,\mathbf{i}}(p, q)$, and full knowledge of the parameters ρ and $\mathbf{i}$ relative to the available image, reconstruct the surface slopes, p and q, for which

$$E(x, y) = R_{\rho,\mathbf{i}}(p, q),$$

and the surface $Z = Z(x, y)$ such that $\partial Z/\partial x = p$ and $\partial Z/\partial y = q$.

9.3 Finding Albedo and Illuminant Direction

We are one last step away from a shape-from-shading method: We still have to deal with the problem that the parameters of the reflectance map (albedo and illuminant) can be unknown. This problem is difficult *per se*; in what follows, we restrict our attention to the relatively simpler Lambertian case.

9.3.1 Some Necessary Assumptions

Similarly to shape from shading itself, the problem of estimating albedo and illuminant direction is only apparently simple; in fact, an effective way to provide good estimates under general circumstances, or even in the simpler Lambertian case, must still be found. The major difficulty is the fact that *the problem is heavily underconstrained*. To overcome this difficulty, we must make assumptions on either the shape of the viewed surface or the distribution of the surface normals; here, we favor the latter view.

Our task is therefore to derive a method which, under appropriate assumptions, estimates albedo and illuminant direction. We begin by stating the problem in more formal terms.

Assumptions and Problem Statement

Under the same assumptions of shape from shading, and with the further assumptions that

1. the surface imaged is Lambertian, and
2. the directions of the surface normals are distributed uniformly in 3-D space,

determine albedo and illuminant direction from a single image of the surface.

Let us fix the notation and select a reasonable distribution of the surface normals in a generic scene. It is customary to describe surface normals by means of the *tilt* and *slant* angles, α and β (see Figure 9.3 with $\alpha = \tau$ and $\beta = \sigma$ respectively). By definition, we have that $\alpha \in [0, 2\pi]$, $\beta \in [0, \pi/2]$ and

$$\mathbf{n} = [\cos \alpha \sin \beta, \sin \alpha \sin \beta, \cos \beta]^\top. \tag{9.8}$$

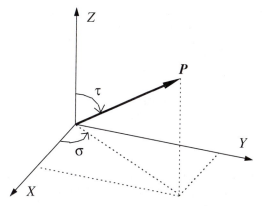

Figure 9.3 Geometric interpretation of tilt and slant.

In our assumptions, the distribution of the normal vectors, $\mathcal{P}$, *as seen from the image plane*, is

$$\mathcal{P}(\alpha, \beta) = \frac{\cos \beta}{2\pi}. \tag{9.9}$$

In agreement with intuition, $\mathcal{P}$ is uniform in α but, due to foreshortening, the number of normal vectors with slant equal to β is proportional to $\cos \beta$.

☞   Like all approximations, especially coarse ones, (9.9) does not describe accurately the distribution of the normals of *every possible* surface as seen from the image plane. However it does a good job on average, and makes it possible to obtain estimates of albedo and illuminant direction in the absence of any guess.

9.3.2   A Simple Method for Lambertian Surfaces

This section describes method for recovering albedo and illuminant direction composed by three steps:

1. precompute the averages of the image brightness and its derivatives, using the hypothesized distribution
2. evaluate the same averages from the image brightness
3. enter the results in equations that can be solved for albedo and illuminant direction

We start by denoting with σ and τ, respectively, the slant and tilt angle of the illuminant, so that we can write

$$\mathbf{i} = [\cos \tau \sin \sigma, \sin \tau \sin \sigma, \cos \sigma]^\top. \tag{9.10}$$

We now compute the average of the image brightness, as given by (9.4), using the distribution in (9.9). By means of (9.8) and (9.10), the image brightness can be written as a function of α and β as

$$E(\alpha, \beta) = \rho(\cos \alpha \sin \beta \cos \tau \sin \sigma + \sin \alpha \sin \beta \sin \tau \sin \sigma + \cos \beta \cos \sigma). \quad (9.11)$$

The average, $< E >$, becomes therefore

$$< E >= \int_0^{2\pi} d\alpha \int_0^{\pi/2} d\beta \mathcal{P}(\alpha, \beta) E(\alpha, \beta),$$

This integral breaks down into three additive terms, of which the first two vanish (because the tilt, α, is integrated over a full period), and the third yields

$$< E >= \frac{\pi}{4} \rho \cos \sigma. \quad (9.12)$$

The interesting fact about this result is that, as we assumed that the surface normals are distributed according to (9.9), we can compute $< E >$ as the average of the brightness values of the given image, and hence look at (9.12) as an equation for albedo and slant.

A similar derivation for the average of the square of the image brightness, $< E^2 >$, (see Exercise 9.1 for some hints) give

$$< E^2 >= \frac{1}{6} \rho^2 \left(1 + 3 \cos^2 \sigma \right). \quad (9.13)$$

From (9.12) and (9.13), it is immediate to recover albedo and slant as

$$\rho = \frac{\gamma}{\pi} \quad (9.14)$$

and

$$\cos \sigma = \frac{4 < E >}{\gamma}, \quad (9.15)$$

where $\gamma = \sqrt{6\pi^2 < E^2 > -48 < E >^2}$.

We are still left with the problem of estimating τ, the tilt of the illuminant. This can be obtained from the spatial derivatives of the image brightness, E_x and E_y. Through some simple but lengthy algebra that we omit (if you are curious, see the Further Readings), one finds

$$\tan \tau = \frac{< \hat{E}_y >}{< \hat{E}_x >} \quad (9.16)$$

with $< \hat{E}_x >$ and $< \hat{E}_y >$ the averages of the horizontal and vertical components of the direction of the image spatial gradient, $\left[\hat{E}_x, \hat{E}_y \right]^\mathsf{T} = (E_x^2 + E_y^2)^{\frac{1}{2}} \left[E_x, E_y \right]^\mathsf{T}$. We now summarize this method:

Algorithm APPRX_ALBEDO_ILLUM_FINDER

The input is an intensity image of a Lambertian surface.

1. Compute the average of the image brightness, $< E >$, and of its square, $< E^2 >$.
2. Compute the spatial image gradient, $[E_x, E_y]^\top$, and let $[\hat{E}_x, \hat{E}_y]^\top$ be the unit vector giving the direction of $[E_x, E_y]^\top$. Compute the average of both components, $< \hat{E}_x >$ and $< \hat{E}_y >$.
3. Estimate ρ, $\cos \sigma$, and $\tan \tau$ through (9.14), (9.15), and (9.16).

The output are estimates of ρ, $\cos \sigma$, and $\tan \tau$.

☞   This method gives reasonably good results, though it fails for very small and very large slants. However, one of the hypothesis underlying the derivation of this method is *inconsistent*. It might have occurred to you that a surface whose normal vectors are uniformly distributed in 3-D space is likely to give rise to self-shadowing, especially for large slant of the illuminant direction. This means that in some of our precomputed integrals the image brightness was negative. To avoid this inconsistency, the integrals should be evaluated numerically using (9.7) as a definition of image brightness. For details on the consistent (but considerably more complicated) version of APPRX_ALBEDO_ILLUM_FINDER refer to the Further Readings. Curiously enough, the "consistent" method does not improve much the final result.

We are now ready to search for a solution to the shape from shading problem.

9.4   A Variational Method for Shape from Shading

We picked one of the many existing methods based on a variational framework, the solution of which gives the slopes of the unknown surface Z.

9.4.1   The Functional to be Minimized

Even under the simplifying Lambertian assumption, the direct inversion of (9.6) is a very difficult task. In essence, one has to solve a nonlinear partial differential equation in the presence of uncertain boundary conditions. For many such equations, even a slight amount of noise can means that the solution (*a*) does not exist, (*b*) is not unique, or (*c*) does not depend continuously on the data.[4] Our equation is no exception. If you are interested to the mathematical aspects of this problem you will find pointers in the Further Readings.

A typical trick to circumvent at least existence and continuity problems (conditions (*a*) and (*c*)) is to recast the problem in the *variational framework*. Instead of looking for an exact solution to (9.6), we allow for some small deviations between the

---

[4] In the mathematical literature, a problem for which at least one of the conditions (*a*), (*b*), or (*c*) holds is said to be *ill posed*.

image brightness and the reflectance map, and enforce a *smoothness constraint* which controls the smoothness of the solution. One possible way to implement this idea is to look for the minimum of a functional $\mathcal{E}$ of the form

$$\mathcal{E} = \int dx dy \left((E(x, y) - R(p, q))^2 + \lambda(p_x^2 + p_y^2 + q_x^2 + q_y^2) \right), \qquad (9.17)$$

in which the smoothness constraint is given by the sum of the spatial derivatives of p and q. The parameter λ is always positive and controls the relative influence of the two terms in the minimization process. Clearly, a large λ encourages a very smooth solution not necessarily close to the data, while a small λ promotes a more irregular solution closer to the data.

Unlike the case of deformable contours, the minimization of this functional cannot be performed effectively by means of a greedy algorithm and we have to make use of the full machinery of the calculus of variations.

9.4.2   The Euler-Lagrange Equations

The calculus of variations gives you a straightforward procedure to derive equations minimizing a generic functional,[5] the *Euler-Lagrange equations*. This section simply tells you how to set up these equations; refers to the Further Readings for more information.[6]

For a functional $\mathcal{E}$ which, like (9.17), depends on two functions p and q of two real variables x and y, and on their first order spatial derivatives, the Euler-Lagrange equations read

$$\frac{\partial \mathcal{E}}{\partial p} - \frac{\partial}{\partial x}\frac{\partial \mathcal{E}}{\partial p_x} - \frac{\partial}{\partial y}\frac{\partial \mathcal{E}}{\partial p_y} = 0,$$

and

$$\frac{\partial \mathcal{E}}{\partial q} - \frac{\partial}{\partial x}\frac{\partial \mathcal{E}}{\partial q_x} - \frac{\partial}{\partial y}\frac{\partial \mathcal{E}}{\partial q_y} = 0.$$

Since R is the only function of p and q in (9.17), and neither E or R depend on p_x, p_y, q_x, and q_y, the Euler-Lagrange equations associated with (9.17) become

$$-2(I - R)\frac{\partial R}{\partial p} - 2\lambda p_{xx} - 2\lambda p_{yy} = 0$$

and

$$-2(I - R)\frac{\partial R}{\partial q} - 2\lambda q_{xx} - 2\lambda q_{yy} = 0,$$

---

[5] It is far easier to write the equations than to solve them, however!

[6] For our purposes, the derivation of the Euler-Lagrange equations associated to a variational method is a rather boring and not much informative exercise of calculus. The real problem is not the derivation of the equations, but finding a good numerical algorithm for solving them.

which can be simplified to give

$$\Delta p = -\frac{1}{\lambda}(I - R)\frac{\partial R}{\partial p} \tag{9.18}$$

and

$$\Delta q = -\frac{1}{\lambda}(I - R)\frac{\partial R}{\partial q}, \tag{9.19}$$

with Δp and Δq denoting the *Laplacian* of p and q (that is, $\Delta p = p_{xx} + p_{yy}$ and $\Delta q = q_{xx} + q_{yy}$). Our next task is to solve (9.18) and (9.19) for p and q.

9.4.3 From the Continuous to the Discrete Case

It turns out that solving (9.18) and (9.19) is easier in the discrete than in the continuous case. Thus we immediately proceed to find the discrete counterpart of (9.18) and (9.19).

We start by denoting with $p_{i,j}$ and $q_{i,j}$ the samples of p and q over the pixel grid at the location (i, j). Through the usual formula for the numerical approximation of the second derivative (Appendix, section A.2), (9.18) and (9.19) become

$$-4p_{i,j} + p_{i+1,j} + p_{i-1,j} + p_{i,j+1} + p_{i,j-1} = -\frac{1}{\lambda}(E(i, j) - R(p_{i,j}, q_{i,j})\frac{\partial R}{\partial p} \tag{9.20}$$

and

$$-4q_{i,j} + q_{i+1,j} + q_{i-1,j} + q_{i,j+1} + q_{i,j-1} = -\frac{1}{\lambda}(E(i, j) - R(p_{i,j}, q_{i,j})\frac{\partial R}{\partial q}. \tag{9.21}$$

☞   The partial derivatives of the reflectance map in (9.20) and (9.21) are either computed analytically (in the Lambertian case, for example) and then evaluated at $p_{i,j}$ and $q_{i,j}$, or evaluated numerically from the reflectance map itself.

The problem is now reduced to finding the slopes $p_{i,j}$ and $q_{i,j}$, solutions of (9.20) and (9.21), and determining the unknown surface $Z = Z(x, y)$ from them.

9.4.4 The Algorithm

We observe that (9.20) and (9.21) can be rewritten as

$$p_{i,j} = \bar{p}_{i,j} + \frac{1}{4\lambda}(E - R)\frac{\partial R}{\partial p} \tag{9.22}$$

and

$$q_{i,j} = \bar{q}_{i,j} + \frac{1}{4\lambda}(E - R)\frac{\partial R}{\partial q}, \tag{9.23}$$

with (Appendix, section A.2)

$$\bar{p}_{i,j} = \frac{p_{i+1,j} + p_{i-1,j} + p_{i,j+1} + p_{i,j-1}}{4}$$

and

$$\bar{q}_{i,j} = \frac{q_{i+1,j} + q_{i-1,j} + q_{i,j+1} + q_{i,j-1}}{4}.$$

As $\bar{p}_{i,j}$ and $\bar{q}_{i,j}$ are the averages of $p_{i,j}$ and $q_{i,j}$ over the four nearest neighbors, (9.22) and (9.23) can be turned into an *iterative scheme* that, starting from some initial configurations for the $p_{i,j}$ and $q_{i,j}$, advances from the step k to the step $k + 1$ according to the updating rule

$$p_{i,j}^{k+1} = \bar{p}_{i,j}^{k} + \frac{1}{4\lambda}(E - R)\frac{\partial R}{\partial p}\bigg|^{k} \tag{9.24}$$

and

$$q_{i,j}^{k+1} = \bar{q}_{i,j}^{k} + \frac{1}{4\lambda}(E - R)\frac{\partial R}{\partial q}\bigg|^{k}. \tag{9.25}$$

However, where have all those boundary conditions gone? The answer is not easy, because in many cases of interest the boundary conditions are practically unknown. In this case, one attempts to impose "neutral" boundary conditions, like the so-called *natural* boundary conditions (p and q constant over the image boundary), or the *cyclic* boundary conditions (p and q wrapped around at the image boundary). Since we are going to compute the Fourier transform of p and q, in what follows we adopt the *cyclic* boundary conditions. If the boundary conditions are known, they should be enforced in the iterative scheme at every step.

9.4.5 Enforcing Integrability

As most pioneers of shape from shading, we too have left out what turns out to be a very important detail. If you actually attempt to reconstruct the viewed surface from the normals obtained by iterating (9.24) and (9.25), you will soon find out that the solution is *inconsistent*! This is hardly surprising: since the functional was not told that p and q were the partial derivatives of the same function Z, there is no Z such that $Z_x = p$ and $Z_y = q$. To circumvent this inconvenience, a good idea is to insert a step enforcing integrability after each iteration. It is more complicated to explain the reason why it works than doing it, so we first tell you how it works and then discuss why.

At each iteration, we compute the Fast Fourier Transform (FFT) of p and q. In complex notation,[7] with i as the imaginary unit, we can write

---

[7] If you are not familiar with the complex notation for the FFT, you may just skip the derivation and go all the way to the description of the algorithm. Of course, you need at least to be able to use an FFT routine!

$$p = \sum c_p(\omega_x, \omega_y) e^{i(\omega_x x + \omega_y y)}$$

and

$$q = \sum c_q(\omega_x, \omega_y) e^{i(\omega_x x + \omega_y y)},$$

where the sums range over all possible values of ω_x and ω_y (multiple of a fundamental frequency) and the c_p and c_q are the Fourier coefficients. Then let

$$Z = \sum c(\omega_x, \omega_y) e^{i(\omega_x x + \omega_y y)}, \tag{9.26}$$

with

$$c(\omega_x, \omega_y) = \frac{-i\omega_x c_p(\omega_x, \omega_y) - i\omega_y c_q(\omega_x, \omega_y)}{\omega_x^2 + \omega_y^2}. \tag{9.27}$$

The function Z in (9.26) has three important properties.

- It provides a solution to the problem of reconstructing a surface from a set of nonintegrable p and q.
- Since the coefficients $c(\omega_x, \omega_y)$ do not depend on x and y, (9.26) can be easily differentiated with respect to x and y to give a new set of *integrable* p and q, say p' and q', such that

$$p' = \frac{\partial Z}{\partial x} = \sum (i\omega_x c(\omega_x, \omega_y)) e^{i(\omega_x x + \omega_y y)} = \sum c'_p(\omega_x, \omega_y) e^{i(\omega_x x + \omega_y y)} \tag{9.28}$$

and

$$q' = \frac{\partial Z}{\partial y} = \sum (i\omega_y c(\omega_x, \omega_y)) e^{i(\omega_x x + \omega_y y)} = \sum c'_q(\omega_x, \omega_y) e^{i(\omega_x x + \omega_y y)}. \tag{9.29}$$

- Most importantly, p' and q' are the integrable pair *closest* to the old pair of p and q.

In more technical terms, we have projected the old p and q onto a set which contains only integrable pairs. It is a projection in a mathematical as well as intuitive sense, because if you do it twice (or more times) you invariably get the result of the first time. This can easily be seen by plugging the coefficients c'_p and c'_q of (9.28) and (9.29) in (9.27).

We now summarize the procedure we have discussed:

Algorithm SHAPE_FROM_SHADING

The input is formed by an image of an unknown surface, Z, the reflectance map of the surface, the surface's albedo, and the direction and intensity of the illuminant. The same assumptions of section 9.2 hold. Moreover, the surface slopes, p and q, (initialized to 0) are assumed to wrap around the image boundaries (cyclic boundary conditions).

Let ρ be the effective albedo and **i** the illuminant direction.

Until a suitable stopping criterion is met, iterate the following two steps.

1. Update p and q through (9.22) and (9.23).
2. Compute the FFT of the updated p and q, estimate Z according to (9.26), and p' and q' according to (9.28) and (9.29). Set $p = p'$ and $q = q'$.

The output is formed by the estimate of Z, p and q.

☞ At each iteration, make sure that the reflectance map is positive at each pixel; set negative values to 0. The parameter λ is usually set to a rather large value, for instance 1000.

☞ Notice that the step enforcing integrability can also be employed as a stand-alone procedure for reconstructing a surface Z from nonintegrable pairs of slopes p and q over a pixel grid.

9.4.6 Some Necessary Details

What can be said about the convergence properties of SHAPE_FROM_SHADING, the optimal λ and the stopping conditions?

Convergence. Unfortunately, not much can be said in general about convergence, except that it depends on how far the initial p and q are from the true values, and on the precision with which the reflectance map and the illuminant are known.

The optimal λ. Something more can be said on λ. The algorithm tends to converge faster for smaller λ (as it can be realized by looking at the right-hand-side of (9.24) and (9.25)). Not surprisingly, though, if λ becomes too small (typically below a few hundreds), the algorithm becomes unstable. For better (and more sophisticated) ways to speed up the iterative step of SHAPE_FROM_SHADING, look into the Further Readings. Instead, if λ is too large, the algorithm tends to promote very regular solutions and can even walk away from the "correct" solution . . . having started from it!

Stopping Condition. It is not easy to give a stopping condition valid in all cases. Looking at the residual,

$$\int dx dy (E - R)^2,$$

often helps but is not always appropriate, and the same can be said for the functional $\mathcal{E}$ itself. The problem is, the history of values of a residual function is not always a faithful indication of what is going on in the minimization process. It may happen that the residual appears stuck while the solution is actually getting closer to the desired minimum. It is also not unusual that the residual does not change appreciably for many iterations and then starts a relatively rapid descent toward the minimum value.

We leave you with the somewhat uncomfortable feeling that implementing a shape-from-shading algorithm means dealing with a number of open issues. Ultimately, you must get acquainted with the particular problem at hand, and develop your own ideas about it.

We conclude this part of the chapter with an example of SHAPE_FROM_SHADING when run on the data of Figure 9.2 (a). Figure 9.4 displays (from left to right) the surface reconstructed after 100 iterations, 1000 iterations, and 2000 iterations.

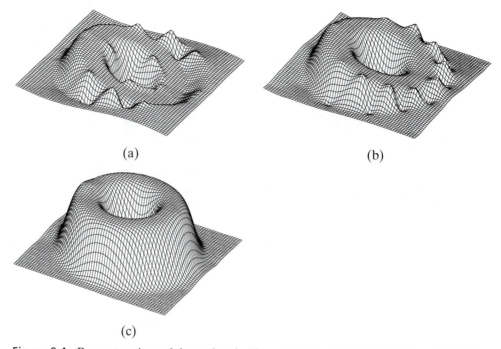

(a)   (b)

(c)

Figure 9.4   Reconstructions of the surface in Figure 9.2 after 100 (a), 1000 (b) and 2000 (c) iterations. The initial surface was a plane of constant height. The asymmetry of the first two reconstruction is due to the illuminant direction.

9.5   Shape from Texture

We now move on to the second theme of this chapter, *shape from texture*. First of all, we must specify what we mean by texture.

9.5.1   What is Texture?

Definition: Texture, Texels

A *surface texture* is created by the regular repetition of an element or pattern, called *surface texel*, on a surface.

An *image texture* is the image of a surface texture, itself a repetition of *image texels*, the shape of which is distorted by the projection across the image.

Figures 9.5 and 9.6 illustrate our definition. Figure 9.5 shows images of a regular grid of circles covering a plane (a) and a cylinder (b). Notice the distortion of the ellipses (image texels), projections of the circles in the scene (surface texels), across the image; we shall comment on this important feature soon. Figure 9.6 shows textures of natural surfaces: from left to right, wood sticks and small leaves, rock, sand. The texture in

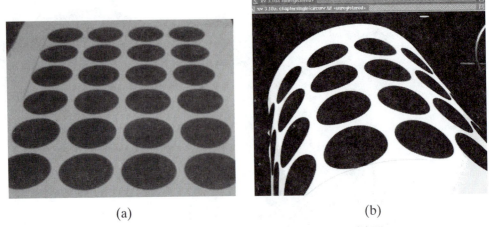

(a)   (b)

Figure 9.5   (a) Image of a plane covered by a deterministic texture. (b) The same texture on a curved surface.

Figure 9.5 is called *deterministic*, the one in Figure 9.6 *statistic*. As we shall see soon, there is an important difference between deterministic and statistic textures in practice.

Definition: Deterministic and Statistic Textures

Deterministic textures are created by the repetition of a fixed geometric shape such as a circle, a square, a decorative motif.
Statistic textures are created by changing patterns with fixed statistical properties.

(a)   (b)   (c)

Figure 9.6   Three natural textures. (a) Wood sticks and small leaves. (b) The surface of a rock. (c) Linear patterns on sand.

Examples of deterministic textures are images of patterned wallpaper, bricks walls, and decorative tiles. Most natural textures are statistic: think for example of pebbles, gravel, wood, or lawns. To recover shape from both types of texture we need to learn a few, basic facts and make some choices.

9.5.2 Using Texture to Infer Shape: Fundamentals

Why Does it Work?   Notice the strong impression of shape you get from Figures 9.5, in which texture is the only cue present.[8] How does this happen? In the image of a textured 3-D surface, *the texels appear distorted*. This is the key fact behind shape from texture: the distortion of the individual texels and its variation across the image create the 3-D impression.[9] We can therefore formulate the computational problem of shape-from-texture as follows.

Problem Statement: Shape from Texture

Given a single image of a textured surface, estimate the shape of the observed surface from the distortion of the texture created by the imaging process.

Representing Image Texture.   How do we represent image texels? Deterministic and statistic textures are represented by qualitatively different techniques, and methods vary accordingly.

Deterministic texels are represented naturally by the *shape parameters* of the specific shape at hand. For instance, the ellipses in Figure 9.5 can be represented by the parameters of the ellipse equation (Chapter 5). Statistic textures are represented typically in terms of *spatial frequency properties*; for instance, by the power spectrum computed over image regions.

Representing Texture Distortion.   What kind of texture distortion does the imaging process introduce? Considering Figure 9.5, we notice two distortions:

perspective distortion: due to the perspective projection, which makes circles increasingly far from the camera project to smaller and smaller ellipses;

foreshortening: which makes circles not parallel to the image plane appear as ellipses.

Under suitable assumptions, the amount of both distortions can be measured from an image. For example, in Figure 9.5 (a) perspective distortion can be quantified by the area variation across the ellipses, and foreshortening by the ratio of the ellipse's semiaxes. In fact, we can use two different classes of texture-based measures to estimate shape:

---

[8] Texture can be also a powerful mean to segment images: In Figure 9.6, although intensity varies with no apparent regularity, we have a strong impression of a uniform texture in each image, and it is perfectly obvious that the three images represent different surface types.

[9] Indeed, both are important cues for shape perception in human vision.

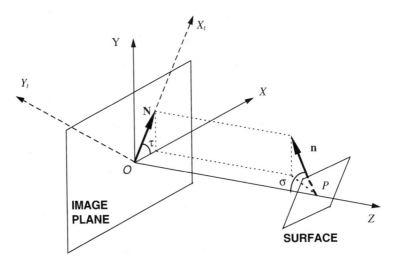

Figure 9.7 The angles tilt, τ, and slant, σ, defining a unit normal **n**. Notice that this figure does *not* illustrate a projection, which is why the origin of the reference frames is on the image plane, and that these angles are *different* from those used in shape from shading (Figure 9.3).

1. a measure of shape distortion (applicable to individual image texels)
2. the rate of change of a measure of shape distortion, called *texture gradient* or *distortion gradient* for that measure (applicable to regions containing several image texels)

Representing Surface Shape. In general, the shape of a surface at any point is completely identified by the surface's orientation (i.e., the normal) and curvatures (Appendix, section A.5). However, it turns out that estimating curvatures from texture is far from trivial. We shall therefore concentrate on recovering just surface normals, as we did in section 9.4. As we know from our discussion of shape from shading, a map of normals specifies the surface's orientation only at the points at which the normals are computed (e.g., the center of deterministic texels), but, assuming that the normals are dense enough and the surface is smooth, the map can be integrated to recover surface shape (for example, as explained in section 9.4.5).

Representing Normals. In the literature of shape from texture, it is common to represent unit normals by the tilt and slant angles illustrated in Figure 9.7.

☞   Notice that these angles are defined in a different way from their shape-from-shading namesakes (Figure 9.3).

Let **n** be the normal to the surface at P. The surface is locally approximated by its tangent plane, perpendicular to **n**. The tilt, τ, is the angle taking the X axis of the camera frame, XYZ, on the projection of the normal on the image plane, **N**. The slant, σ, is the angle taking **n** onto the $-Z$ axis of the camera frame. Rotating the frame $OXYZ$ around Z by τ generates the new reference frame OX_tY_tZ; in this frame, the normal **n** lies in the X_tZ plane, so that the tangent plane is parallel to Y_t.

The General Structure of Shape-from-Texture Algorithms.   We have now all the elements necessary to state the general structure of shape-from-texture methods.

1. Select a representation adequate for the image texture at hand.
2. Compute the chosen distortion measures (if required, their gradients) from the image, in terms of the representation selected.
3. Use local distortion (if required, texture gradients) to estimate the local orientation of the surface.

The next section gives a simple shape-from-texture algorithm which estimates the orientation of a plane from a statistic texture.[10]

9.5.3   Surface Orientation from Statistic Texture

As usual, we begin by stating a set of assumptions.

Assumptions

1. The 3-D texels are small line segments, called *needles*.
2. The needles are distributed uniformly on the 3-D surface, and their directions are all independent.
3. The surface is approximately planar.
4. The image projection is orthographic.

We characterize the needles by their orientation only: their positions and lengths are irrelevant. The reason for choosing such apparently odd texels is that they allow us to establish a deterministic, geometric relation between the orientation of a 3-D texel and the one of its corresponding image texel. The idea is to use this relation to write the probability, say p, of the observed image given an orientation (σ, τ) of the 3-D plane, which allows us to estimate the orientation of the plane as the pair $(\hat{\sigma}, \hat{\tau})$ maximizing p (a *maximum likelihood* approach, see Appendix, section A.7). In actual fact, an approximate solution $(\hat{\sigma}, \hat{\tau})$ can be derived in closed form, and this is what we shall use in STAT_SHAPE_FROM_TEXT.

☞   Image needles can be extracted from images of 3-D textures not necessarily made up by small line segments. To do this, you run an edge detector (Chapter 4) with an adequately small kernel to extract short contours, followed by a module detecting and describing

---

[10] An algorithm recovering the orientation of circles in space is given in Chapter 10.

small, rectilinear segments, e.g., a variation of the Hough line detector of Chapter 5. Remember though, if the 3-D texture is not composed of small line segments, the conditions of assumptions 1 and 2 are only approximated, and this is likely to worsen results.

We now sketch the derivation behind STAT_SHAPE_FROM_TEXT, omitting much detail. Assume there are N needles in the image, and let α_i be the angle formed by the i-th image needle with the image's x axis. In our assumptions, α_i is a uniformly distributed random variable in $[0, \pi]$. We now introduce an auxiliary vector, $[\cos 2\alpha_i, \sin 2\alpha_i]^T$, itself a random quantity. This vector is characterized by a probability distribution called *distribution on the unit circle*, which is a function of the distribution of the α_i. Its *center of mass* is defined by

$$C = \frac{1}{N} \sum_{i=1}^{N} \cos 2\alpha_i$$

$$S = \frac{1}{N} \sum_{i=1}^{N} \sin 2\alpha_i. \tag{9.30}$$

It can be proven that, in orthographic projections (Assumption 4), the center of mass is

$$C = \cos 2\tau \frac{1 - \cos \sigma}{1 + \cos \sigma}$$

$$S = \sin 2\tau \frac{1 - \cos \sigma}{1 + \cos \sigma}. \tag{9.31}$$

Solving for σ and τ, we find

$$\sigma = \arccos \frac{1 - Q}{1 + Q}$$

$$\tau = \psi \pm \frac{\pi}{2} \ (\text{mod } 2\pi), \tag{9.32}$$

where Q and ψ are the polar coordinates of the center of mass,

$$Q = \sqrt{C^2 + S^2}, \qquad \psi = \frac{1}{2} \arctan \frac{S}{C}. \tag{9.33}$$

☞ Notice the ambiguity in the estimate of tilt.

The complete algorithm is stated below:

Algorithm STAT_SHAPE_FROM_TEXTURE

The input is an image containing N needles, each forming an angle α_i with the x axis of the image. The assumptions stated above hold.

1. Compute C, S using (9.30).

2. Compute the polar coordinates Q, ψ of the center of mass of the distribution on the unit circle using (9.33).

3. Estimate the orientation of the 3-D plane, $(\hat{\sigma}, \hat{\tau})$, using (9.32).

The output is $(\hat{\sigma}, \hat{\tau})$, the estimate of the orientation of the 3-D plane.

9.5.4   Concluding Remarks

Shape from Texture and Texture Segmentation.   In our discussion of shape from texture, we assumed a uniform texture throughout the image. In reality, this is something of a special case: Images are likely to contain different textures, or textured areas surrounded by non-textured ones. In general, differently textured regions need separating before shape-from-texture algorithms can be applied. This problem is called *texture segmentation*, and it is a classic problem of image processing. As texture is an ubiquitous feature of surfaces, texture segmentation is frequently used to identify objects of interest in the image; for instance, it proves very useful in many defect detection systems.

In texture segmentation, image pixels are classified on the basis of several textural measures, called *texture features*, computed in local neighborhoods. These measures are usually statistical properties of the intensity values, or spatial frequency measures like the power spectrum. Unfortunately, segmentation relies on the assumption that texture features are constant within regions of uniform texture, whereas texture (and feature) distortions are exactly what shape-from-texture methods rely on! For this reason, performing texture segmentation and shape from texture at the same time is not trivial at all, which is why we have treated shape from texture as independent of texture segmentation. The Further Readings point you to an introduction to the literature of texture segmentation.[11]

Texture Depends on Spatial Scale.   Textures appear and disappears at different spatial scales. For example, imagine to zoom in on a floor made of wooden planks: When the image contains many planks, the main image texture is given by the planks' contours; as you close in on one plank, the main texture is given by the wood's fibers. These textures look different: The wooden planks create a deterministic pattern, the fibers a statistical one. Therefore, "the texture of a surface" actually refers to the texture of a surface *at a given spatial scale*.

9.6   Summary

After working through this chapter you should be able to:

❑ explain the nature and objectives of shape-from-X methods

❑ explain the purpose and nature of shape from shading

---

[11] The reason why texture segmentation has not been included in our discussion of feature detection (Chapters 4 and 5) is exactly that it was not needed to support shape from texture.

❑ design an algorithm for shape from shading, and recover a surface from a map of normals

❑ explain the purpose and nature of shape from texture

❑ recover the orientation of a plane covered by a statistical texture

9.7 Further Readings

Table 9.1 mentions several shape-from-X methods; here are some introductory references, chosen from a vast literature. You can begin an investigation into *shape from focus and defocus* from [17, 21]. Ma and Olsen's work [19] is a good starting point for *shape from zoom*. A good illustration of *shape from contours* is given by [16], and a modern shape-from-contour approach is discussed by Cipolla and Blake [5]. *Active vision* is a recent, influential paradigm of computer vision; for an introduction, see [3], or the report of the recent panel discussion in [24], or again the seminal paper by Aloimonos *et al.* [1].

The approximate method for determining albedo and illuminant has been adapted from [26], which also explains in detail how to obtain (9.16). The field of shape from shading has been strongly influenced by a number of pioneering works due to Horn and coworkers [11, 14, 12]. More on the reflectance map can be found in [13]. The original method proposed by Horn more than twenty years ago in [11] is still a classic, though not easy to explain and implement. The algorithm described in this chapter has been proposed by Frankot and Chellappa [6] as an improvement of the method described in [12]. A rigorous account on ill-posed problem of computer vision and methods for their solution can be found in [2].

Much of our discussion of shape from texture is based on Gårding's work [7, 8] and references therein, in which you can also find a detailed treatment of texture-based curvature estimation. Methods for recovering shape from deterministic textures under perspective projections are discussed in [4, 15, 9]. The closed-form solution in STAT_SHAPE_FROM_TEXTURE is due to Gårding [8], and is a variation of Witkin's maximum-likelihood estimator [25] which involves a nontrivial minimization. Recent examples of shape-from-texture algorithms reconstructing curved surfaces covered by statistical textures, using spatial-frequency descriptors, are given in [23] and [20]. Approaches to the problem of segmenting texture and computing shape from texture simultaneously exists, but the methods are not trivial; an example is reported in [18]. Texture segmentation in itself is a classic topic of image processing, and its literature is vast. To begin its exploration, see [10] for a collection of texture-based segmentation algorithms, and [22] for a recent survey of the field.

9.8 Review

Questions

❒ *9.1* Why the methods in this chapter recover shape but not distance?

❒ *9.2* Why the methods in this chapter are applied to intensity images, and not to range images?

❐ 9.3 Explain the difference between the tilt and slant used in shape from shading and shape from texture.

❐ 9.4 Are the tilt and slant angles used in shape from shading and shape from texture the same as spherical coordinates? If not, what are the differences?

❐ 9.5 Explain why, in algorithm APPRX_ALBEDO_ILLUM_FINDER, it could happen that $\cos \sigma > 1$. What would you do to avoid this inconsistency?

❐ 9.6 Identify the ambiguity intrinsic to the reflectance map of a Lambertian surface. Discuss the consequences on shape from shading.

❐ 9.7 Why are the boundary conditions necessary in SHAPE_FROM_SHADING? Could you run the algorithm ignoring them?

❐ 9.8 How would you set a value for λ in SHAPE_FROM_SHADING? (*Hint:* Look at the right-hand side of (9.24) and (9.25).)

❐ 9.9 Explain the difference between foreshortening and perspective distortion.

❐ 9.10 Can you assume orthographic projections when using the area gradient? Why?

❐ 9.11 Identify the elements given in the general structure of shape-from-texture algorithms in STAT_SHAPE_FROM_TEXT.

❐ 9.12 How would you choose the parameters of an edge detector used to extract needles for STAT_SHAPE_FROM_TEXT?

Exercises

○ 9.1 Compute the integral

$$< E^2 >= \int_0^{2\pi} d\alpha \int_0^{\pi/2} d\beta \mathcal{P}(\alpha, \beta) E^2(\alpha, \beta)$$

with $E(\alpha, \beta)$ as in (9.11). *Hint:* Recall that

$$\int_{-\pi}^{\pi} \sin^2 x dx = \int_{-\pi}^{\pi} \cos^2 x dx = \pi.$$

○ 9.2 Show that the Fourier coefficients c_p' and c_q' of (9.28) and (9.29) leave the Fourier coefficients c of (9.26) unchanged.

○ 9.3 Explain why replacing the term multiplying λ in (9.17) with $(p_y - q_x)^2$ is an alternative way of enforcing integrability in shape from shading. Do you think this way of enforcing integrability is as effective as Step 2 in SHAPE_FROM_SHADING? Why does the new functional also enforce smoothness?

○ 9.4 Derive and discretize the Euler-Lagrange equations for the new functional of Exercise 9.3. Find an iterative scheme similar to the one in SHAPE_FROM_SHADING.

○ 9.5 Consider the textured plane shown in Figure 9.5 (a). Write an expression linking slant, the (known) size of an image texel, and the distance of the corresponding 3-D texel from the camera. Assume there is no tilt.

○ *9.6* Extend your solution to Exercise 9.5 by removing the assumption of zero tilt. You have now to write an expression for the tilt and one for the slant.

Projects

● *9.1* Implement SHAPE_FROM_SHADING. Test the algorithm on synthetically generated images of a Lambertian surface following the suggestions given in section 9.2. Study the sensitivity of the algorithm to noise, direction of illuminant and uncertainty in the knolwdge of albedo and illuminant.

● *9.2* Implement STAT_SHAPE_FROM_TEXT and test your implementation with synthetic images of planar textures corrupted by additive noise. Study the variation of the error as a function of the amount of additive noise and of the orientation of the plane. For which orientations do you expect higher errors? Why? Is your expectation confirmed by experiments?

References

[1]   Y. Aloimonos, I. Weiss and A. Bandopadhay, Active Vision, *International Journal of Computer Vision*, Vol. 7, pp. 333 – 356 (1988).

[2]   M. Bertero, T. Poggio and V. Torre, Ill-posed Problems in Early Vision, *Proc. IEEE*, Vol. 76, pp. 869–889 (1988).

[3]   A. Blake and A. Yuill, *Active Vision*, MIT Press, Cambridge (MA) (1992).

[4]   D. Blostein and N. Ahuja, Shape from Texture: Integrating Texture-Element Extraction and Surface Estimation, *IEEE Transactions on Pattern Analysis and Machine Intelligence*, Vol. PAMI-11, no. 12, pp. 1233–1251 (1989).

[5]   R. Cipolla and A. Blake, Surface Shape from the Deformation of Apparent Contours, *International Journal of Computer Vision*, Vol. 9, pp. 83 – 112 (1992).

[6]   R.T. Frankot and R. Chellappa, A Method for Enforcing Integrability in Shape from Shading Algorithms, *IEEE Transactions on Pattern Analysis and Machine Intelligence*, Vol. 10, no. 4, pp. 439 – 451 (1988).

[7]   J. Gårding, Shape from Texture for Smooth Curved Surfaces, *Proc. European Conf. on Computer Vision*, S. Margherita (Italy), pp. 630–638 (1992).

[8]   J. Gårding, Direct Estimation of Shape from Texture, *IEEE Transactions on Pattern Analysis and Machine Intelligence*, Vol. PAMI-15, no. 11, pp. 1202–1207 (1993).

[9]   J. Gårding, Shape from Texture for Smooth Curved Surfaces in Perspective Projection, *Int. Journ. of Mathematical Imaging*, Vol. 2, no. 4, pp. 329 – 352 (1992).

[10] R.M. Haralick and L.G. Shapiro, *Computer and Robot Vision*, Vol. I, Addison-Wesley (1992).

[11] B.K.P. Horn, Obtaining Shape from Shading Information, in P.H. Winston (ed.), *The Psychology of Computer Vision*, McGraw-Hill, New York, pp. 115–155 (1975).

[12] B.K.P. Horn and M.J. Brooks, The Variational Approach to Shape from Shading, *Computer Vision Graphics and Image Processing*, Vol. 33, no. 2, pp. 174 – 208 (1986).

[13] B.K.P. Horn and B.G. Sjoberg, Calculating the Reflectance Map, *Applied Optics*, Vol. 18, no. 11, pp. 1770 – 1779 (1979).

[14]   K. Ikeuchi and B.K.P. Horn, Numerical Shape form Shading and Occluding Boundaries, *Artificial Intelligence*, Vol. 17, no. 3, pp. 141 – 184 (1981).

[15]   K. Kanatani and T.-C. Chou, Shape from Texture: General Principle, *Artificial Intelligence*, Vol. 38, pp. 1 – 48 (1989).

[16]   J.J. Koenderink, What Does the Occluding Contour Tell Us About Solid Shape? *Perception*, Vol. 13 (1984).

[17]   E. Krotkow, Focusing, *International Journal of Computer Vision*, Vol. 1, pp. 223 – 237 (1987).

[18]   J. Krumm and S. A. Schafer, Texture Segmentation and Shape in the Same Image, Proc. IEEE Int. Conf. on Comp. Vision, Cambridge (MA), pp. 121–127 (1995).

[19]   J. Ma and S. I. Olsen, Depth from Zooming, *Journ. of the Optical Society of America A*, Vol. 7, no. 10, pp. 1883 – 1890 (1990).

[20]   J. Malik and R. Rosenholtz, Recovering Surface Texture and Orientation from Texture Distortion, *Proc. European Conf. on Computer Vision*, Stockholm, pp. 353–364 (1994).

[21]   A. P. Pentland, A New Sense for Depth of Field, *IEEE Transactions on Pattern Analysis and Machine Intelligence*, Vol. PAMI-9, no. 4, pp. 523–531 (1987).

[22]   T.R. Reed and J.M. Hand du Buf, A Review of Recent Texture Segmentation and Feature Extraction Techniques, *Computer Vision Graphics and Image Processing: Image Understanding*, Vol. 57, no. 3, pp. 359–372 (1993).

[23]   B.J. Super and A.C. Bovik, Shape from Texture Using Local Spectral Moments, *IEEE Transactions on Pattern Analysis and Machine Intelligence*, Vol. PAMI-17, no. 4, pp. 333–343 (1995).

[24]   M.J. Swain and M.A. Stricker, Promising Directions in Active Vision, *International Journal of Computer Vision*, Vol. 11, no. 2, pp. 109 – 126 (1993).

[25]   A.P. Witkin, Recovering Shape and Orientation from Texture, *Artificial Intelligence*, Vol. 17, pp. 17–45 (1981).

[26]   Q. Zheng and R. Chellappa, Estimation of Illuminant Direction, Albedo, and Shape from Shading, *IEEE Transactions on Pattern Analysis and Machine Intelligence*, Vol. PAMI-13, no. 7, pp. 680 – 702 (1991).

10

Recognition

Ceci n'est pas une pipe.[1]

Rene Magritte

This chapter introduces object recognition and its two subproblems, *object identification* and *object location*, and presents algorithms for identifying 3-D objects from intensity and range images. Location is the subject of the next chapter.

Chapter Overview

Section 10.1 lays out the basic ideas behind object identification.

Section 10.2 introduces *interpretation trees*, a class of feature-based algorithms rooted in the state-space search methods of classic artificial intelligence.

Section 10.3 introduces *invariants* and how they can lead to object characterizations independent of the imaging transformation.

Section 10.4 describes *parametric eigenspaces*, a method for recognizing 3-D objects from their appearance; that is, using images instead of features to build models.

Section 10.6 brings together various issues from previous sections in a discussion of *3-D object modelling*; that is, how objects can be represented for computer vision purposes.

What You Need to Know to Understand this Chapter

- Basic concepts of projective geometry (Appendix, section A.4).
- The essentials of state-space search.
- Eigenvalues and eigenvectors.

---

[1] This is not a pipe.

10.1 What Does it Mean to Recognize?

In the previous chapters, we learned how to reconstruct the shape of unknown objects from different information contained in images. We now come to the important problem of *object recognition*. By definition, recognition implies that object descriptions, or *models*, are already available; you cannot recognize what you do not know yet. We speak therefore of *model-based recognition*; the key idea is *the comparison of image data with a database of models*. When a model is found to correspond to a subset of the data (for example, particular configurations of contours, or range surfaces of specific shapes), we say that *a match has been found*, or that *the model matches the data*. The matched model is the *identity* of the object imaged, in the sense that both data and model represent the same object in the scene.

For our purposes, object recognition entails two basic operations, identification and location. *Identification* determines the nature of the objects imaged. For instance, you may want to know whether there is a car among the many objects in an image, or whether the only object you are looking at is indeed a car. *Location* determines the position in 3-D space of the objects in view. This chapter deals with identification, and the next one with location.

Problem Statement: Model-Based Object Recognition

Given a database of object models, and an input image, model-based object recognition addresses two problems.

Identification: Which models in the database match the data in the image?

Location: Given that an object in the image matches a given model, what is the location in space (rotation and translation) of the 3-D object imaged?

We hasten to add that the separation between identification and location is rather blurred in many computer vision systems, some of which perform both at the same time. Notice also that our statement of the location problem implies that *identification must be solved first*: We first assess that we are looking at an instance of model X, then we use model X to find where the object is in 3-D space. This "first-identify-then-locate" sequence is by no means the only one possible, but it has been adopted here as it seems an easy and intuitive way to introduce object recognition.

This short discussion makes us realize that identification may involve a few different scenarios, depending on two main points:

1. Do we need to search the database of models, or do we use only a single, specific model?

2. Do we need to search the image, or do we know already which image data correspond plausibly to a single (yet unknown) object?

These questions influence the nature and amount of the search to be performed, and consequently the structure and complexity of identification algorithms. A useful thing to do is therefore to write a tidy list of all the possible scenarios.

Problem 1: What objects are we looking at? (*Model search needed, image region search needed*). Are they bottles, cars, cabbages, kings? And where are they in the image? This is the most general and complex question: We want to identify all objects in view. Answering involves finding which parts of the image, or groups of image features, correspond plausibly to individual objects, as well as searching the database of models to find those which match the various parts of the image.

Problem 2: Is this part of the image an instance of X? (*Given model, given image region*). This is the identity test introduced above, and the most specific question we can ask. It implies that we have already picked one object model ("X"), and we are trying to match it to a precise part of the image.

Problem 3: What is this part of the image? (*Model search needed, given image region*). We want to determine the identity of a part of the image, say a particular group of image features. We require an algorithm indexing a possibly large database of models efficiently.

Problem 4: Are there any instances of X in the image? (*Given model, image region search needed*). For instance, is there my aunt Amelia in this family photograph? Again this implies that we have *already* picked one object model but are trying to find instances of it in the whole image. Of course, we need a way to avoid running identity tests on all possible subsets of features, which could be very expensive.

☞   Notice that asking "what is the object in this image?" actually means "does it correspond to any model I know?" If not, the object is unknown *to my database*. If yes, the object is known, and we might want to estimate how well data and model correspond.

The algorithms in this chapter address a subset of these problems. *Interpretation trees* (section 10.2) identify all subsets of image data which plausibly fit a given model (Problem 4); *invariants* (section 10.3) support direct, feature-based model indexing, and are therefore well-suited to identify specific subsets of image data (Problems 1 and 3); *appearance-based methods* (section 10.4) are an example of methods using images (*appearance*), not features, as the basic model elements (Problem 2).

The crucial ingredient of identification is the comparison of image data and models, which establishes whether data and model match. Comparisons make sense only between similar objects. We know that an image is a matrix of numbers, and that we can extract some features from it; but *what* is exactly an object model? We shall hint to various answers while explaining identification algorithms, and bring all issues on modelling together in section 10.6.

10.2   Interpretation Trees

Interpretation trees (henceforth called ITs) are a class of feature-based identification algorithms rooted in artificial intelligence, especially graph search and constraint satisfaction. We begin by stating the assumptions of the method.

Assumptions

- All image features of a given class (e.g., surface patches) have been detected, and a list of symbolic descriptors formed (as described in Chapter 4 and 5).
- A list of symbolic descriptors covering all the features of a given object model has been formed.
- The symbolic descriptors of image and model features are expressed in the same format, and therefore comparable directly.
- All properties in the feature descriptors are geometric.

Recall that symbolic feature descriptors (introduced in Chapters 4 and 5) list the numerical values of the feature's properties, like area and centroid of a range patch. Obviously, a model feature can correspond to an image feature only if their properties are the same (in practice, similar enough).

In the example we give below, we focus on range images, and assume that the features are surface patches (Chapter 4), characterized by a set of geometric properties. Notice, however, that ITs can be used with intensity images and nongeometric properties as well. Here is what we want to do.

Problem Statement

Given:

- the list of feature descriptors from a given object model
- the list of feature descriptors detected in a range image
- a list of geometric constraints that model features must satisfy

Find a mapping between model features and image features such that the constraints satisfied by the model features are satisfied by the corresponding image features.

The mapping sought is called an *interpretation*. It identifies a set of image features corresponding to the given model; that is, one instance of the model in the image. If no interpretation can be found, we conclude that no instance of the model appears in the image. The geometric constraints we consider can be *unary constraints*, like area or centroid, representing properties of individual features; or *binary constraints*, involving a pair of features, like adjacency between two patches. Constraints characterize features and their relations, and allow us to discard match hypotheses that are locally inconsistent.

The problem is *to find consistent interpretations without exploring all possible ways of matching image and model features*, which would be normally prohibitive combinatorially. ITs use *constraint satisfaction* to reduce the complexity of the search. An interpretation is modelled by a path in the space of all possible image-model correspondences, which, in turn, is represented by a tree (the interpretation tree).

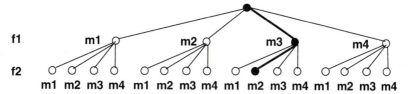

Figure 10.1   An interpretation tree representing the complete search space for a problem involving two image features, $\{f_1, f_2\}$, and four model features, $\{m_1, \ldots m_4\}$. Each level represents all possible matches between an image features f_i and a model feature m_j. Therefore, the node m_j on level i represents the match (f_i, m_j).

Finding interpretations means finding paths through the tree. Each tree node represents a correspondence hypothesis, as illustrated by Figure 10.1.

10.2.1   An Example

Let us illustrate the above with a simple example. Figure 10.2(b) shows an ideal, surface-based object model, and the labels attached to the model patches, $m_1, \ldots m_8$. Assume that Figure 10.2(a) shows the surface patches extracted from an imaginary range view of the object, and the labels attached to the image features. Five patches are visible, $f_1, \ldots f_5$. Assume that the only constraint in the symbolic patch descriptor is the shape type (rectangular, square, or L-shaped). Figure 10.3 shows a portion of the corresponding IT. The nodes at level d represent all possible correspondence hypotheses for feature f_d. The constraints on shape discard several inconsistent matches; for instance, f_1 can

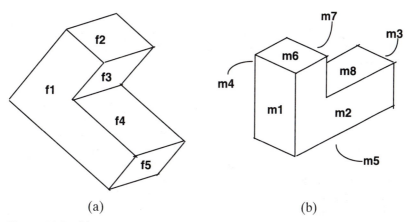

(a)   (b)

Figure 10.2   (a) Ideal range view of a simple object, showing the labels (f_i) of image features (surface patches). m_1, m_5, m_8 are rectangular; m_3, m_6, m_7 are square; m_2, m_4 are L-shaped. (b) Visualization of a surface-based model of the same object, showing the labels (m_j) of the object features.

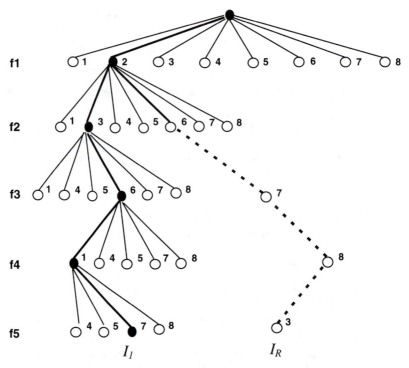

Figure 10.3 Portion of the interpretation tree for the problem of Figure 10.2 expanded when reaching the first interpretation, I_1 (solid path). The figure shows also the path (dashed) corresponding to the correct interpretation, I_R (but not the nodes generated between I_1 and I_R).

only be matched to m_2 or m_4. The tree is expanded depth-first, until all image features are matched. The next level down does not include the model features already matched, as each model patch can correspond to only one image patch. The first consistent interpretation we find,

$$I_1 = \{(f_1, m_2), (f_2, m_3), (f_3, m_6), (f_4, m_1), (f_5, m_7)\},$$

is *wrong*. Why? Because the constraints enforced are associated to local features, and therefore *local* in nature. In general, even if we introduce more constraints, even involving groups of some features, there is no guarantee that all interpretations found make *global* sense.

One way to check global consistency is to compute the transformation which brings the matched model patches onto the image patches, project the located model onto the image, and check that all model patches end up in the expected positions. This process is called *verification*; it uses the image and 3-D positions of the features matched to estimate the imaging transformation, *backprojects* the mode features onto the image plane, and checks whether model features are imaged close enough to their corresponding image features in the original image. Since this amounts to solving the location problem, which is the subject of the next chapter, we leave out the details for now.

Going back to our example, we verify that I_1 is inconsistent, and start backtracking. We move to the first alternative possible, (f_3, m_5), expand a new subtree depth-first, and look for new interpretations. We come across a few more inconsistent ones, and eventually reach the correct one.

☞   Notice that depth-first search with backtracking is a reasonable choice, although heuristic search and more sophisticated algorithms are also possible (see Further Readings).

10.2.2 Wild Cards and Spurious Features

Our example above is unrealistic in at least two ways. First, real images often contain several instances of the target object; we would like a recognition algorithm to find them all. In this sense, we should make ITs address the question "are there any instances of X in the image?" (section 10.1). Second, feature correspondence is not generally one-to-one: Some object features are certainly hidden to the viewer (m_1, m_4, m_5 in Figure 10.2(a)), some maybe occluded by other objects; the feature detector might miss some image features, and introduce some spurious ones. We would like to recognize objects in spite of reasonable amounts of missing and spurious features.

Spurious features are accommodated by a *wild card*; that is, a fictitious feature which matches any image feature for which no real match is found. In practice, an additional "wild card node" is simply appended to the list of model features. Without the wild card, the search would backtrack as soon as encountering a spurious image feature, failing potentially acceptable interpretations. Unfortunately, the wild card increases the search complexity seriously. Assume that R image features are real object features, and S are spurious. The wild-card search finds the true interpretation, in which all the R real image features have been matched, but also the interpretations containing $R - k$ real features and $S + k$ wild card matches, of which there are many.

A method reducing unnecessary efforts is *branch-and-bound*. Suppose we are looking for all the instances of a model in the image. When we find the first interpretation accepted by verification, we record the number, r, of its nonwildcard matches. We then backtrack to look for other interpretations, but terminate any path if we realize that any interpretation below the current node cannot possibly include more than r real matches. For instance, if $r = 6$, the current interpretation has two nonwildcard matches, and only 3 image features have not been explored, we can safely quit this path and backtrack. Of course, we must also update r when discovering a verified interpretation with $r' > r$ real matches.

10.2.3 A Feasible Algorithm

We are now ready to give an algorithm which takes into account all the points touched on in our discussion.

Algorithm INT_TREE

Let

    Open the ordered list of nodes to expand, left to right;

    Interp the list of consistent matches forming an interpretation;

    Maxsize the maximum number of nonwildcard matches in any interpretation found;

    W the wild card;

    size(*Interp*) the number of nonwildcard matches in Interp;

    root a label for the tree root;

consistent(X), X a non-wildcard match, is true if X is consistent with constraints or X=root, and false otherwise. Let [ ] indicate an empty list, and $[a, b, \ldots n]$ a list of elements. Notice that the wildcard is, by definition, consistent with any constraints.

```
Open = [root], Interp = [ ]; Maxsize = 0;
WHILE (open ≠ [ ])
   BEGIN
      remove leftmost match, X = (fⱼ, mₖ), from open;
      IF (consistent(X)) AND
      max possible size of interpretation on this path ≥ Maxsize)
         BEGIN
         add X to Interp;
         IF (leaf node reached on this path)
            BEGIN
            verify Interp;
            IF (verification succeeds)
               BEGIN
               save Interp;
               Interp = [ ];
               Maxsize = size(Interp);
               END
            return(failure);
            END
         ELSE (* not leaf, but consistent *)
            let L = {(fⱼ₊₁, m₁), . . . (fⱼ₊₁, m_M), (fⱼ₊₁, W)}
               be the expansion of the current node;
            drop model features already matched from L;
            add L to the left of Open;
         END
      ELSE (* inconsistent match *)
         continue;
   END
END
```

10.3 Invariants

We now move on to *invariants*, the second identification method of this chapter. Invariants are functions of feature groups that allow us to index into the database of models. Therefore, with reference to our classification of recognition tasks (section 10.1), invariants address the question "what is this part of the image?" (Problem 3). This section suggests also how to use invariants to address the more general question "what objects are we looking at" (Problem 1)? For reasons which will be clear soon, we focus on *intensity images*.

10.3.1 Introduction

In general, *invariants* can be defined as *properties (functions) of geometric configurations which do not change under a certain class of transformations*. For instance, the length of a line segment does not change if the segment is rotated and translated: Length is invariant under rigid motion. How can invariants help in object recognition?

One of the key problems of recognition from intensity images is that the appearance of an object depends on imaging conditions like viewpoint and intrinsic parameters. If we could identify shape properties which do not change with imaging conditions, the problem would be solved. This is exactly what invariants are, and the basic idea behind invariant-based recognition is very simple. You define invariant functions of some image measurables (for instance, image contours) which yield sufficiently different values for all the shapes of interest, and then use vectors of invariant functions to index a database of models. The invariance of the *model index* (the vectors of invariants) to the imaging conditions means that the *same* index is computed from *any* image of the same object. So here is what we want to do.

Problem Statement

Given a set of image features, define a vector of invariant functions of the image features and use it to index a library of models.

Unfortunately, but not surprisingly, there is a basic limitation: Invariants are known only for some classes of shapes, and defining useful invariants for general objects is not easy at all. For the purposes of our introductory discussion, we make the following assumptions.

Assumptions

1. We consider only *scalar, algebraic invariants* of the *geometric* imaging transformation.
2. The objects to be recognized are planar (but their position and pose in space is unconstrained).
3. The invariants are functions of *groups of image contours*.
4. Image contours are formed by line segments and arcs of conics.

Scalar, algebraic invariants are obviously scalar, algebraic functions, in our case of the parameters of the equations representing image contours. Two examples of such invariants are introduced in the next section; the Further Readings point to discussions of non-algebraic invariants. We concentrate on *geometric* invariance.

There are various reasons behind the apparently very restrictive Assumptions 2, 3, and 4. First, all the fundamental concepts of invariants can still be illustrated. Second, several invariants for lines and conics are known. Third, many man-made objects, although not planar, contain coplanar lines or conics (for instance, windows of cars and buildings, or circular holes on a same planar face of a mechanical component). Fourth, it is difficult to determine invariants for general, 3-D objects.

Assumption 4 implies that line and conics detectors, such as those described in Chapter 5, are available to locate and describe lines and conics in the image.

The use of invariants brings about two main advantages.

Easy model acquisition. A library of models suitable for recognition is built easily by acquiring a single, real image of each object in any pose, computing a vector of invariants, and storing the vector as the object index.[2]

Recognition without model search. There is no need to search the model base to recognize an object, as a vector of invariants indexes directly into the library of models. This implies that a suitable indexing strategy (for instance, a hash table) makes the recognition time independent of the size of the model library.[3]

10.3.2   Definitions

We are now ready to give more precise definitions. In particular, we want to formalize the transformation for which we need invariants, as well as the concept of invariant property.

We are after invariants of the geometric imaging transformation introduced in Chapter 2. As we assume all objects planar, *the imaging transformation is a transformation between planes*: it maps an object plane onto the image plane. Therefore, we can consider directly *the mapping between corresponding points on the object plane and the image plane*. The situation is illustrated by the example in Figure 10.4. As the general imaging transformation of Chapter 2, our plane-to-plane mapping is modelled by a *projective transformation* (Appendix, section A.4); hence the name *projective invariants*.

Let $[x, y, 1]^\top$ be the homogeneous coordinates of an image point, $\mathbf{p}$, and $[X, Y, 1]^\top$ the homogeneous coordinates of the corresponding scene point, $\mathbf{P}$. The projective transformation that maps $\mathbf{P}$ into $\mathbf{p}$ (Appendix, section A.4) can be written as

$$k\mathbf{p} = T\mathbf{P},$$

---

[2] Though it is a good idea to acquire and process more than one image per model.

[3] In practice, recognition time will grow with the size of the model library; see the Further Readings for more.

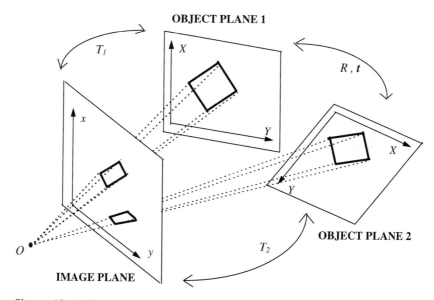

Figure 10.4   The projective transformations T_1 and T_2 model the mapping from object to image plane for two different viewpoints (before and after a rotation R and a translation $\mathbf{t}$ of the object plane).

where T is the 3×3 projection matrix, defined up to a scale factor, and k accounts for the fact that homogeneous coordinates are defined up to a scale factor too. And here is a formal definition of the invariants we need for our purposes.

Definition: Projective Invariant

Let $\mathbf{c}_m$ be a set of contour descriptors of a planar object, expressed in the homogeneous coordinates of the object plane. Let T be the projective transformation modelling the mapping between object plane and image plane points. The function $I = I(\mathbf{c}_m)$ is a *projective invariant* if

$$I(\mathbf{c}_m) = I(T(\mathbf{c}_m))$$

for any projective transformation T.

It is easy to guess how $\mathbf{c}_m$ looks like in our assumptions. A line, $ax + by + cz = 0$, is simply represented by the vector $\mathbf{l} = [a, b, c]^\top$; a conic, $ax^2 + bxy + cy^2 + dx + ey + f = 0$, by the matrix, C, defined by

$$
\begin{bmatrix} x & y & 1 \end{bmatrix}
\begin{bmatrix} a & b/2 & d/2 \\ b/2 & c & e/2 \\ d/2 & e/2 & f \end{bmatrix}
\begin{bmatrix} x \\ y \\ 1 \end{bmatrix}
= \mathbf{p}^\top C \mathbf{p} = 0,
$$

and used in Chapter 5 in the context of ellipse fitting. In order to move on to algorithms, we still need to define invariants suitable for recognizing planar objects delimited by segments of lines and conics. We begin with the *cross-ratio*, defined on the projective line, on which our other invariants are based.

Definition: Cross-ratio

Given four distinct collinear points, described in homogeneous coordinates of the projective line as $\mathbf{p}_i = [x_i, 1]^\top$ ($i = 1, \ldots, 4$), the *cross-ratio* c is defined as

$$c(1, 2, 3, 4) = \frac{d(1, 2)d(3, 4)}{d(1, 3)d(2, 4)}$$

with

$$d(i, j) = x_i - x_j$$

the *determinant* between $\mathbf{p}_i$ and $\mathbf{p}_j$.

Cross-ratio Invariance

The cross-ratio is invariant to projective transformations of the projective line onto itself.

The proof of the invariance and the main mathematical properties of the cross-ratio can be found in the Appendix, section A.4. The importance of the cross-ratio is in the fact that, although four collinear points may seem a rather restrictive configuration, it is possible to obtain them by projective constructions of some sets of lines and conics. We report two of these in the following.

Invariants of Five Coplanar Lines

Given five coplanar lines, labelled $\mathbf{l}_1, \ldots, \mathbf{l}_5$, two independent projective invariants are

$$I_1 = \frac{|M_{431}|\,|M_{521}|}{|M_{421}|\,|M_{531}|} \qquad I_2 = \frac{|M_{421}|\,|M_{532}|}{|M_{432}|\,|M_{521}|}, \tag{10.1}$$

where $M_{ijk} = [\mathbf{l}_i, \mathbf{l}_j, \mathbf{l}_k]$, and $|M_{ijk}|$ is the determinant of M_{ijk}.

☞   Be warned that, for certain configurations, the labelling of lines in each M_{ijk} may make some determinants in the denominators vanish (Exercise 10.5). See the Appendix, section A.4 for an approach to this problem.

The proof of the invariance of I_1 and I_2 is left as an exercise (Exercise 10.6). Notice that each M_{ijk} can be viewed as the area of the triangle the vertices of which are the intersections of the lines $\mathbf{l}_i, \mathbf{l}_j,$ and $\mathbf{l}_k$. Our next invariant involves two coplanar conics.

Invariants of Two Coplanar Conics

Given two matrices describing coplanar conics, C_1 and C_2, such that $|C_1| = |C_2| = 1$, two indepen-
dent projective invariants are

$$I_3 = \text{tr}\left[C_1^{-1}C_2\right] \quad I_4 = \text{tr}\left[C_2^{-1}C_1\right], \tag{10.2}$$

where $\text{tr}[A]$ is the trace of matrix A.

The proof of the invariance of, say, I_3, is instructive. We need to show that

$$\text{tr}\left[C_1^{-1}C_2\right] = \text{tr}\left[c_1^{-1}c_2\right],$$

where c_1 and c_2 are the matrices describing the conics C_1 and C_2 respectively transformed by an arbitrary projective transformation T such that

$$\mathbf{p} = T\mathbf{P}. \tag{10.3}$$

The four conics are defined by

$$\mathbf{P}^\top C_1 \mathbf{P} = 0 \tag{10.4}$$
$$\mathbf{P}^\top C_2 \mathbf{P} = 0 \tag{10.5}$$
$$\mathbf{p}^\top c_1 \mathbf{p} = 0 \tag{10.6}$$
$$\mathbf{p}^\top c_2 \mathbf{p} = 0 \tag{10.7}$$

If we recover $\mathbf{P}$ from (10.3) and plug the result in (10.4) and (10.5), by comparison with
(10.6) and (10.7) we obtain

$$\begin{aligned} c_1 &= T^{-\top}C_1 T^{-1} \\ c_2 &= T^{-\top}C_2 T^{-1}. \end{aligned} \tag{10.8}$$

Equations (10.8) tell you how conics are affected by a projective transformation. By
means of (10.8), it is easy to derive the thesis as

$$\text{tr}\left[c_1^{-1}c_2\right] = \text{tr}\left[TC_1^{-1}T^\top T^{-\top}C_2 T^{-1}\right] = \text{tr}\left[C_1^{-1}C_2\right],$$

where the last passage relies on the *circular property* of the trace of the matrix product,

$$\text{tr}\left[A_1 \dots A_n\right] = \text{tr}\left[A_n, A_1, \dots A_{n-1}\right].$$

10.3.3 Invariant-Based Recognition Algorithms

This section gives two essential, invariant-based algorithms for recognition and model
acquisition. In both cases, the underlying mechanism is rather simple. We identify a set of
four invariants, $I_1, \dots, I_4$, such that one or more can be computed for every interesting
object, and define a vector collecting all the invariants in a fixed order, $\mathbf{g} = [I_1, \dots, I_4]^\top$.

For model acquisition, we take a generic image of each object, O_k, extract lines and conics, evaluate the vector of invariants for each relevant group of contours, and store the resulting vector, $\mathbf{g}_k$, as the index for the model of object O_k.

☞   We are going to use *two* views of each object, not just one, to filter out invariants which may vary too much numerically between different views. The two views should be representative of viewpoints adopted during recognition.

For recognition, we extract lines and conics from the input image, select the groups of contours for which we can compute the invariants, compute the invariant vectors for all such groups, and use the vectors to index the model library. The object models indexed by the larger numbers of invariants are assumed to be present in the image.

☞   Contour grouping is a serious practical problem (recall our discussion of grouping in Chapter 5). How do we know that the features we use to compute an invariant really belong to the *same* object? Simply trying all possible groups of features leads to impractically large numbers of groups, including many spurious ones. The algorithms in this section adopt a simple solution: for lines, they consider only groups of consecutive image lines (grouping by *proximity*); that is, lines with endpoints sufficiently close to each other; for conics, they just assume that the number of conics in the image is low, and the number of all possible groups of conics small (no grouping at all). Notice that even simple proximity requires care, as line detectors are likely to split at least some lines into various segments.

This is the box for the algorithm acquiring models.

Algorithm INV_ACQ

The input is the set of descriptors of lines and conics (defined in Chapter 5) detected in two views of each object.

1. For each object, O_i, $i = 1, \ldots, N$:

    (a) for each of the two views:

        (i) form all possible feature groups, say M in all, for which an invariant $I_1 \ldots I_4$ is defined (either five consecutive lines or pairs of conics);

        (ii) compute the M vectors $\mathbf{g}_1, \ldots, \mathbf{g}_M$, one per group, where $\mathbf{g}_k = [I_1, I_2, I_3, I_4]^\top$ contains the values of the four invariants defined by equations (10.1) and (10.2); mark invariants inapplicable to each group (some groups contain only conics, the others only lines);

    (b) store an object model for O_i, formed by a label (the object's name) and a vector index, $[\mathbf{g}_1, \ldots, \mathbf{g}_M]$, where $\mathbf{g}_k = [I_{1k}, I_{2k}, I_{3k}, I_{4k}]^\top$, and I_{jk} is:

        (i) the average values of the invariant over the two views, if the values are sufficiently close to each other;

        (ii) a symbolic label indicating that the invariant is not reliable, otherwise.

The output is a library of models suitable for invariant-based recognition.

☞ The inevitable fluctuations of the numerical values computed, caused by inaccuracies of feature detection, image noise and so on, mean that a *range* of values, not just one, of each invariant must be associated to each object.

Next we give the skeleton of an invariant-based recognition algorithm. As we adopt only a very limited mechanism for feature grouping, recognition tries to filter out spurious groups by *verification*; that is, backprojecting each model indexed by a set of feature groups, and checking that the model's features actually project onto the corresponding image features in the groups.

Algorithm INV_REC

The input is the set of descriptors of lines and conics (defined in section 10.3.2) detected in an input image, and the library of models generated by INV_ACQ, in which each model, O_j, is associated to a set of numerical vector indices, $\mathbf{g}_1, \ldots, \mathbf{g}_M$, with $\mathbf{g}_k = [I_{k1}, \ldots, I_{k4}]^\top$.

1. Form all possible groups (say there are R) of five consecutive image lines and image conics, as done for INV_ACQ.

2. Compute the R invariant vectors $\mathbf{g}_1, \ldots, \mathbf{g}_R$, one per feature group, where $\mathbf{g}_k = [I_1, I_2, I_3, I_4]^\top$ contains the values of the four invariants defined by equations (10.1) and (10.2); mark invariants inapplicable to the group (some groups contain only conics, the others only lines).

3. Form a list of all object models indexed by at least one invariant vector index of $\mathbf{g}_1, \ldots, \mathbf{g}_R$. Let $\mathbf{g}_{f,1}, \ldots, \mathbf{g}_{f,H}$ the H index vectors pointing to a generic model hypothesis O_f.

4. *(Verification 1)* Discard from the list all O_f for which it is not possible to determine a unique projective transformation, T, compatible with all the features associated to $\mathbf{g}_{f,1}, \ldots, \mathbf{g}_{f,H}$.

5. *(Verification 2)* Discard all O_f, now associated to a unique projective transformation, for which some backprojected features are not sufficiently close to the corresponding image features.

The output is a list of objects detected in the image.

Some comments are in order on steps 3, 4 and 5. For step 3, all invariant values are real, and therefore best represented by ranges of values to account for uncertainty; care must be taken when writing indexing procedures. See Further Readings for examples of indexing procedures.

As to step 4, you can estimate the projective transformation T as follows. For each image *line* $\mathbf{l}_i$ featuring in the definition of $\mathbf{g}_{f,1}, \ldots, \mathbf{g}_{f,H}$ (say $i = 1, \ldots, n$) you write

$$T\mathbf{L}_i = k_i \mathbf{l}_i \tag{10.9}$$

where $\mathbf{L}_i$ is the model line corresponding to $\mathbf{l}_i$, and k_i, as usual in projective geometry, a nonzero, arbitrary real number. By linearly combining the three scalar equations in (10.9), each line $\mathbf{l}_i$ generates two homogeneous equations in the nine entries of T, in which the coefficients are functions of the components of $\mathbf{L}_i$ and $\mathbf{l}_i$. The set of all equations (10.9) forms a linear, overconstrained, homogeneous system in the entries of T that we write

$$A\mathbf{t} = 0, \tag{10.10}$$

with A the $2n \times 9$ matrix of coefficients and $\mathbf{t} = [T_{11}, T_{12}, \ldots, T_{33}]^\top$. The *compatibility* of the lines $\mathbf{l}_i$ and $\mathbf{L}_i$ can be checked by looking at the SVD of the matrix A, $A = UDV^\top$ (Appendix, section A.6). If the effective rank of A is 8 (that is, the least singular value of A is very small) all the lines are compatible and $\mathbf{t}$, as usual, is given by the column of V corresponding to the least singular value. Otherwise, the lines are not compatible.

In the assumption that the matrix T passed the compatibility test for lines, you are left with the problem of testing the compatibility for the conics c_i featuring in the definition of $\mathbf{g}_{f,1}, \ldots, \mathbf{g}_{f,H}$ (say $i = 1, \ldots, m$) with the corresponding model conics C_i. This can be done by looking at the ratio between each of the nine coefficients of c_i and $T C_i T^\top$. If the ratio k_i is approximately the same for all coefficients, the matrix T passed the global compatibility test and you may proceed to step 5. Otherwise, the model hypothesis O_f discarded.

In step 5, there are two points to make. First, how do we backproject lines and conics? For lines, we backproject each $\mathbf{l}_i$ by applying the transformation T to the model line $\mathbf{L}_i$ like in (10.9). For conics, we obtain the backprojection of each c_i by computing $T C_i T^\top$. Notice that the formal difference between this expression and the right hand side of (10.8) is due to the fact that T denotes a projective transformation of lines in the former and of points in the latter. In mathematical terms, T in the former is the inverse transpose of T in the latter. Second, "sufficiently close" (verification 2) means that the distance between backprojected features should never be larger than a few pixels; exact figures depend on resolution.

10.4   Appearance-Based Identification

Finally, we address Problem 2 of section 10.1: "Is this part of the image an instance of X?" And we address the question by *using images instead of features as basic components of object models.*

10.4.1   Images or Features?

The key idea behind appearance-based identification is simple: To store images of 3-D objects as their representation. Instead of representing object O through its geometric features and their spatial relations, *we represent O with the set of its possible appearances*; that is, the set of images taken, ideally, from all possible viewpoints and with all possible illumination directions.[4] In practice, we use a sufficiently large number of viewpoints and illuminations directions. As an example, Figure 10.5 shows a 12-image representation of a toy car. We can create a database of models for identification by building such a set

---

[4] We consider only direction for simplicity, but, as we know from Chapter 2, other illumination parameters play a role in determining pixel values.

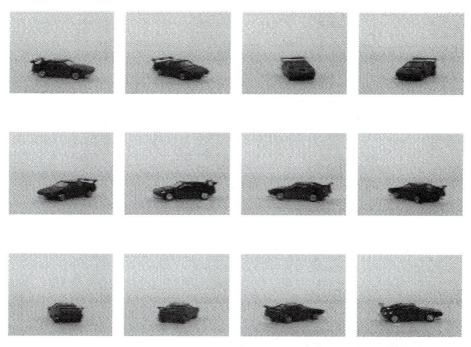

Figure 10.5 A simple database exemplifying appearance-based object representation. Only the viewpoint, not the illumination, was changed to obtain the views shown.

for all objects of interest. Identifying an object, then, means to find the set containing the image which is most similar to the one to be recognized.

Problem Statement

Given an image, I, containing an object to identify, and a database of object models, each one formed by a set of images showing the object under a large number of viewpoints and illumination conditions, find the set containing the image which is most similar to I.

A desirable characteristic of appearance-based identification, as presented, is that *object models can be compared directly with input data*, as both are images. Feature-based models (like the ones used with invariants and interpretation trees), instead, require that features be detected and described before data and model can be compared. Unfortunately there is a price to pay: the database may become extremely large even for limited numbers of objects and illumination conditions. For example, assuming 128×128 images, at one byte per pixel, 100 viewpoints per object, and 10 illumination directions, the representation of a single object would occupy about 64 megabytes of memory! Therefore the practical problem is, *can we devise a way to keep memory occupation within manageable limits while performing appearance-based recognition?*

10.4.2 Image Eigenspaces

We shall arrive at an appearance-based algorithm, the *parametric eigenspace* method in three steps:

1. We define a quantitative method to compare images and introduce some necessary assumptions.
2. We introduce an efficient, appearance-based object representation, which makes it feasible to search a large database of images.
3. We give algorithms to build the representation and to perform identification.

Comparing Images. A simple, quantitative way to compare two images, say I_1 and I_2, both $N \times N$ for simplicity, is to compute their *correlation*, c:

$$c = I_1 \circ I_2 = \frac{1}{K} \sum_{i=1}^{N} \sum_{j=1}^{N} I_1(i, j) I_2(i, j),$$

where K is a normalizing constant, and $\circ$ denotes image correlation. The larger c, the more similar I_1 and I_2.

This is simple enough, but we must take some precautions. As we are really interested in comparing 3-D objects, not just images, we need assumptions to guarantee that correlation is meaningful for our purposes.

Assumptions

1. Each image contains one object only.
2. The objects are imaged by a fixed camera under weak perspective.
3. The images are *normalized in size*; that is, the image frame is the minimum rectangle enclosing the largest appearance of the object.
4. The energy of the pixel values of each image is normalized to 1; that is, $\sum_{i=1}^{N} \sum_{i=1}^{N} I(i, j)^2 = 1$.
5. The object is completely visible and unoccluded in all images.

All these assumptions are consequences of the fact that we want to compare 3-D objects by comparing their images. You should be able to explain the reasons behind each assumption (see Review Questions for hints).

Efficient Image Comparison with Eigenspaces. Our next goal is to devise an efficient method to search a large image database in order to find the image most similar to a new one, in the correlation sense. The database contains the appearance-based representations of several objects; each object is represented by a set of images, taken by different viewpoints and with light coming from different directions. Such a database is suggested by Figure 10.6, which shows only one image per object for reasons of space. Clearly, if we store a full image for each view, and many views for

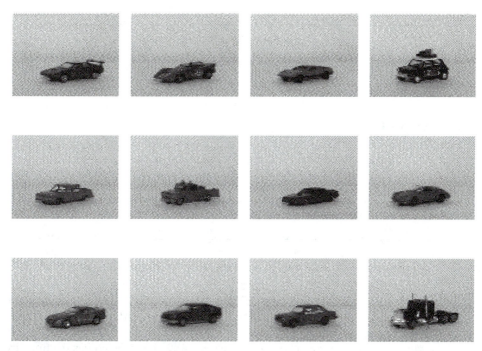

Figure 10.6   Images from a small appearance-based database composed of twelve toy cars. Only one image per object is shown.

each object, the size of such a database becomes prohibitively large, and search based on brute-force correlation unfeasible. Instead, we represent objects in *eigenspace*. To introduce eigenspaces and their advantages, we need to regard images as vector, and state a fundamental theorem.

To transform a 2-D image into a 1-D vector, we just scan the image top to bottom and left to right. In this way, a $N \times N$ image, X, is represented by a N^2-dimensional vector

$$\mathbf{x} = [X_{11}, X_{12} \ldots X_{1N}, X_{21}, \ldots X_{NN}]^\top .$$

Notice that *this representation allows us to write image correlation as the dot product of two vectors.*[5] For instance, the correlation of images X_1 and X_2, represented by vectors $\mathbf{x}_1, \mathbf{x}_2$ respectively, becomes

$$c = X_1 \circ X_2 = \mathbf{x}_1^\top \mathbf{x}_2.$$

From now on, we shall use vectors for images. And here is the fundamental theorem.

---

[5] We assume that the constant K in the correlation definition is 1.

Theorem: Eigenspace Representation

Let $\mathbf{x}_1, \ldots \mathbf{x}_n$ be N^2-dimensional vectors, and $\bar{\mathbf{x}} = \frac{1}{n} \sum_{j=1}^{n} \mathbf{x}_j$ their average. Given the $N^2 \times n$ matrix

$$X = [(\mathbf{x}_1 - \bar{\mathbf{x}})| \ldots (\mathbf{x}_n - \bar{\mathbf{x}})] ,$$

we can write each $\mathbf{x}_j$ as

$$\mathbf{x}_j = \bar{\mathbf{x}} + \sum_{i=1}^{n} g_{ji} \mathbf{e}_i,$$

where $\mathbf{e}_1, \ldots \mathbf{e}_n$ are the eigenvectors of the *covariance matrix*, $Q = XX^\top$, corresponding to the n (nonzero) eigenvalues of Q, and $\mathbf{g}_j = [g_1, g_2, \ldots g_N]^\top$ is the vector of the *components* of $\mathbf{x}_j$ in *eigenspace*.

Now let us go back to our database of all images (all objects, all viewpoints, all illumination directions). Assuming O objects, with P viewpoints and L illumination directions for each object, the database contains OPL images. Using the procedure suggested by the theorem, we can build the covariance matrix, Q, of the whole database, and represent each image, $\mathbf{x}_{pl}^o$, with its vector of eigenspace coordinates, $\mathbf{g}_{pl}^o$. Q is clearly a very large matrix, and $\mathbf{g}_{pl}^o$ is the same size as $\mathbf{x}_{pl}^o$, but here comes the first advantage of eigenspaces: *only the components associated to the largest eigenvalues of Q are significant to represent the images*. In other words, assuming that the nonzero eigenvalues $\lambda_1 \ldots \lambda_n$ of Q are such that $\lambda_1 \geq \lambda_2 \geq \ldots \lambda_n$ and $\lambda_i \approx 0$ for $i > k$, we can write

$$\mathbf{x}_j \approx \sum_{i=1}^{k} g_{ji} \mathbf{e}_i + \bar{\mathbf{x}},$$

and ignore all the remaining $n - k$ components. If $k \ll n$ Each image, $\mathbf{x}_j$, is therefore represented by a point of coordinates $\mathbf{g}_j^\top$ in a k-dimensional eigenspace, a substantially smaller subspace of the original, n-dimensional eigenspace.

So far for one image, but how do we represent a *set of images*; that is, all the views in the representation of the o-th object? Imagine to move through the views of the representation; as pose and illumination change continuously, the point $\mathbf{g}_{pl}^o$ moves continuously in eigenspace, sweeping a so-called *manifold*,[6] $\mathbf{g}^o = \mathbf{g}^o(\mathbf{p}, \mathbf{l})$, where $\mathbf{p}$ and $\mathbf{l}$ are vectors defining the object pose and the illumination direction, respectively. The set of eigenspace points associated to the images of the o-th object is a sampling of the associated manifold.[7]

---

[6] Do not reel at the word "manifold"! You do not need any knowledge of manifolds to understand this section, and manifolds are replaced by curves in the algorithm boxes below. Notice that, if only one parameter is allowed to change between images, i.e., $\mathbf{g}^o(v)$, v a real number, the manifold can be viewed as a curve in eigenspace; if two parameters can change, a surface.

[7] If necessary, we can estimate the continuous manifold by interpolating between adjacent samples.

How do eigenspaces allow us to perform image correlation, and why do they make correlation more efficient? The key is that *the Euclidean distance in eigenspace is equivalent to image correlation*, and the advantage is in the fact that eigenspace points have only k coordinates. To show this, we first notice that the correlation of two image vectors, $\mathbf{x}_1$ and $\mathbf{x}_2$, with grey levels normalized as in our assumptions ($\|\mathbf{x}_1\|^2 = \|\mathbf{x}_2\|^2 = 1$), can be written through their Euclidean distance:

$$\|\mathbf{x}_1 - \mathbf{x}_2\|^2 = 2(1 - \|\mathbf{x}_1^\top \mathbf{x}_2\|),$$

so that maximizing correlation is equivalent to minimizing distance. The distance between image vectors, in turn, can be approximated by the distance in the k-dimensional eigenspace, $\|\mathbf{g}_1 - \mathbf{g}_2\|$:

$$\|\mathbf{x}_1 - \mathbf{x}_2\|^2 = \| \sum_{i=1}^{n} g_{1i}\mathbf{e}_i - \sum_{i=1}^{n} g_{2i}\mathbf{e}_i \|^2$$

$$\approx \| \sum_{i=1}^{k} g_{1i}\mathbf{e}_i - \sum_{i=1}^{k} g_{2i}\mathbf{e}_i \|^2$$

$$= \| \sum_{i=1}^{k} (g_{1i} - g_{2i})\mathbf{e}_i \|^2$$

$$= \sum_{i=1}^{k} (g_{1i} - g_{2i})^2$$

$$= \|\mathbf{g}_1 - \mathbf{g}_2\|^2.$$

And now we realize why correlation is computed more efficiently in eigenspace: Instead of $O(n)$ products needed by $\|\mathbf{x}_1 - \mathbf{x}_2\|^2$, we perform only $O(k)$ products for $\|\mathbf{g}_1 - \mathbf{g}_2\|^2$. As, in practice, n is usually larger than 100 and k smaller than, say, 20, we are reducing a minimum of $n^2 = 10000$ products to a maximum of $k^2 = 400$, saving two orders of magnitude!

Let us summarize it all. First, eigenspace points represent images with fewer numbers. Second, images are correlated efficiently by computing distances in eigenspace. Third, and most importantly, *eigenspaces suggest a way of learning object models automatically*: we acquire the complete set of all possible views for each object o,

$$\left\{ \mathbf{x}_{11}^o, \mathbf{x}_{12}^o, \ldots \mathbf{x}_{1L}^o, \mathbf{x}_{21}^o \ldots \mathbf{x}_{PL}^o \right\},$$

reduce the dimensionality as described above, and compute the corresponding, discrete manifold in eigenspace, $\left\{ \mathbf{g}_{11}^o, \mathbf{g}_{12}^o, \ldots \mathbf{g}_{1L}^o, \mathbf{g}_{21}^o \ldots \mathbf{g}_{PL}^o \right\}$. To identify an object from a new image $\mathbf{y}$, we project $\mathbf{y}$ in eigenspace (using the eigenvectors of the covariance matrix of *all* the OPL images in the database), obtaining a point $\mathbf{g}_y$, then look for the object manifold $\mathbf{g}^o(\mathbf{p}, \mathbf{l})$ closest to $\mathbf{g}_y$. So 3-D appearance-based identification is solved as a minimum-distance problem in eigenspace.

The Parametric Eigenspace Method.   This section contains two algorithm boxes: one to learn appearance-based models, one to identify objects from new images. As the learning stage can be very expensive in terms of memory occupation, we suggest a small-scale version of the algorithm, which uses small images, assumes illumination fixed, and constrains pose changes to rotations around a fixed axis (which means the manifold becomes a curve in eigenspace). You can easily extend our version to consider illumination direction, full pose parameters, and larger images.[8]

Algorithm EIGENSPACE_LEARN

We consider the assumptions stated at the beginning of this section valid; moreover, we assume a fixed camera, fixed illumination conditions, and images of $N \times N$ pixels.

1. For each object o to be represented, $o = 1 \ldots O$:

   (a) place the object on the turntable;
   (b) acquire a set of n images by rotating the turntable by $\frac{360°}{n}$ each time;
   (c) in all images, make sure to adjust the background so that the object can be easily segmented from the background;
   (d) segment the object from the background (see Exercise 10.9);
   (e) normalize the images in scale and energy as stated in the assumptions;
   (f) represent the normalized images as vectors, $\mathbf{x}_p^o$, where p is the rotation index, $p = 1, \ldots, n$.

2. Compute the average image vector, $\bar{\mathbf{x}}$, of the complete database $\left\{ \mathbf{x}_1^1, \ldots \mathbf{x}_n^1, \mathbf{x}_1^2, \ldots \mathbf{x}_n^2, \ldots \mathbf{x}_n^O \right\}$.

3. Form the $N^2 \times N^2$ covariance matrix, $Q = XX^\top$, with $X = \left[\mathbf{x}_1^1 | \mathbf{x}_2^1 | \ldots \mathbf{x}_n^1 | \mathbf{x}_1^2 | \ldots \mathbf{x}_n^O \right]$.

4. Compute the eigenvalues of Q, keep the first k largest eigenvalues and the associated eigenvectors, $\mathbf{e}_1, \ldots \mathbf{e}_k$.

5. for each object, o:

   (a) compute the k-dimensional eigenspace points corresponding to the n images:

   $$\mathbf{g}_p^o = [\mathbf{e}_1 | \ldots \mathbf{e}_k] \, (\mathbf{x}_p^o - \bar{\mathbf{x}});$$

   (b) store the discrete eigenspace curve $\left\{ \mathbf{g}_1^o, \ldots, \mathbf{g}_p^o, \ldots, \mathbf{g}_n^o \right\}$, as the representation of object o.

The output is a set of O discrete curves in the k-dimensional eigenspace, each representing a 3-D object.

☞   As detailed in the Appendix, section A.6, you do not actually need to compute the eigenvalues and eigenvectors of $XX^\top$. Thanks to a fundamental property of the singular value

---

[8] If you have a few gigabytes to spare, that is.

decomposition, the eigenvalues of $XX^\top$ are the same as the eigenvalues of the $n \times n$ matrix $X^\top X$ (a matrix of much smaller size) and the eigenvectors of $XX^\top$ can be computed from the corresponding eigenvectors of $X^\top X$.

Notice that the dimensionality reduction is carried out on the global database. This ensures that the k important eigenvectors record visual information of all objects in the database. Conversely, as we see next, *recognition* is performed in the eigenspace of *individual* objects. To begin with, we suggest you try EIGENSPACE_LEARN with $N = 64$, $O = 5$, $n = 32$. This means that Q is 4096×4096, and X is 4096×32, all rather reasonable numbers.

☞   To implement EIGENSPACE_LEARN, you need a turntable to change the viewpoint by controlled rotations. The best would be to use a computer-controlled turntable, but the dish of an old record player, turning at constant speed, will do for the first attempts.

We now turn to the identification algorithm. Obviously enough, we assume that the learning and identification stage are run with the same illumination conditions and camera position.

Algorithm EIGENSPACE_IDENTIF

The input is a $N \times N$ image, I, of one of the objects in the database. The image I must satisfy the assumptions stated at the beginning of this section and acquired so that the object can be easily segmented from the background. We assume the same illumination conditions and camera position adopted in EIGENSPACE_LEARN.

1. Segment the object from the background.
2. Normalize I in scale and energy, and represent the normalized image as a vector, $\mathbf{i}$.
3. Compute the k-dimensional eigenspace point corresponding to $\mathbf{i}$:

$$\mathbf{g} = [\mathbf{e}_1| \ldots \mathbf{e}_k]\,(\mathbf{i} - \bar{\mathbf{x}}),$$

   where $\bar{\mathbf{x}}$ is the average image vector of the whole database.
4. Find the eigenspace point, $\hat{\mathbf{g}}$, created by EIGENSPACE_LEARN, closest to $\mathbf{g}$.

The output is the object associated to the curve on which $\hat{\mathbf{g}}$ lies; that is, the identity of the object in I.

Now for a few points of practical importance. First, finding the point of a curve or surface which is closest to a given point is not trivial; if the curve is represented by a high number of points (as in our case), brute force can prove too expensive. Second, it may not always be true that $k \ll n$. Third, finding the eigenvalues of very large matrices is computationally expensive, and special algorithms exist for this purpose. Finally, the figure-ground segmentation necessary to zero out background pixels is not trivial, and, in general, is simple only for certain classes of objects and with controlled scenes.

10.5   Concluding Remarks on Object Identification

How do the methods presented compare with each other? Although simple, INT_ TREE is a reasonable algorithm for real images. It copes with missing features, noisy features, and multiple object instances. The wild card inflates complexity, which becomes exponential in the number of model and data features; branch-and-bound and other methods (see Further Readings) alleviate this problem. INT_TREE performs grouping and identification *simultaneously* (see the four problems of section 10.1): It selects which image features are most likely to belong to the object, and performs the identity test. The inevitable price is a rather high complexity. *Alignment* or *hybrid methods* are another way to reduce the complexity of IT search. Such methods match only the number of data features strictly necessary for carrying out verification. More of this is discussed in the next chapter.

Invariants provide image measurements independent of viewpoint and intrinsic parameters, and suggest an easy strategy for model acquisition from real images. These are very valuable characteristics for practical recognition systems. However, their usability is subject to the possibility of defining invariants for the objects of interest, but not many invariants are known and easy to compute for 3-D shapes. Other points requiring attention include grouping, which is shared by all classifications methods based on local features, and the discriminational power of each invariant (how reliably can different objects be told apart given noisy images).

It can be more laborious to build models suitable for an interpretation-tree algorithm, as we presented it, than for invariants. Moreover, interpretation tree locate instances of a *given* object in an image, and require model search to recognize all objects present in an image; invariants, instead, support direct model indexing, and do not require model search. However, interpretation trees take care of feature grouping; invariant do not.

Invariants-based methods allow one to build model libraries from only one or two views per object, as opposed to the many views required to build a parametric eigenspace. However, eigenspaces can be built for *any* 3-D shape and do not require feature extraction, while invariants can cater only for special shape classes and depend on the performance of the feature extractor. Again, eigenspaces do not require feature grouping, invariants do. A disadvantage of parametric eigenspace methods is that they are vulnerable to occlusion and sensitive to segmentation.

10.6   3-D Object Modelling

As promised at the beginning of this chapter, we now bring together the hints to 3-D object modelling scattered throughout this chapter. By now you have certainly realized that designing adequate models is tremendously important, and, indeed, *3-D object modelling* is a much-investigated issue in computer vision. The aim of this section is *not* to list the many representations in existence (although some examples will be mentioned), but to make you reflect on the *necessity* and *design issues* of 3-D object modelling. We refer constantly to the identification methods we learnt to help you make practical sense of the discussion.[9]

Let us delimit the scope of this section, and give a definition. We consider only models of 3-D objects and their features, but models can be devised for image features (see for instance the edge models of Chapter 4) or even images (see the eigenspace representation of section 10.4). We limit ourselves to *geometric features*, as this book concentrates on the geometric properties of the world; the Further Readings point you to alternative models. Also, our treatment of object modelling is centered on identification and location, but 3-D models are used for various other purposes (e.g., inspection). And here is the definition promised.

Definition: Computer Vision Model of a 3-D Object

A computational representation of a sufficient number of geometric properties of the object to perform a desired vision tasks.

Our definition suggests that *computer vision models need not be exhaustive:* we want to represent only what we need to accomplish a given task. For instance, to tell X from Y, we would like the represent only the minimum information necessary to tell X and Y apart. This is different from computer graphics, where must be exhaustive: You cannot generate an image and leave blank areas!

The identification methods presented in this chapter suggest at least two pairs of alternatives for object modelling schemes: *feature-based* versus *appearance-based*, and *object-centered* versus *viewer-centered*.

10.6.1 Feature-based and Appearance-based Models

The first alternative concerns the basic elements of the representation: features or images.

Feature-based models represent 3-D objects through features, their type, and their spatial relations. The features can be those introduced in Chapters 4 and 5, or others, perhaps nongeometric, like color, reflectance properties, and polarization. We used feature-based models for interpretation trees (the features were surface patches) and invariants (the features were contours). With feature-based models, identification means finding a set of features which is uniquely distinctive for an object; location means, in essence, to match a number of image and object features, plug their positions in the projection equations, and solve for the position and orientation of the 3-D object.[10] The advantage of feature-based models is that they generate compact object descriptors, offer some robustness against occlusion (features are local), and some

---

[9] Incidentally, this is the reason why the section on object modelling comes *after* those on identification algorithms.

[10] Details about this idea are discussed in the next chapter.

invariance against illumination and pose variations. A disadvantage is that they cannot be compared directly with images and require feature extraction. Notice that feature-based models are further classified in terms of features type: For instance, *boundary representations*, or *B-reps*, describe an object by its boundaries (e.g., lines, surfaces); *volumetric representations* describe the position, size, and approximate shape of the main parts of an object, or just the shape of the volume of space occupied by the object.

Appearance-based models represent an object through one or more images, as in the eigenspace method. This works for 2-D objects, or for 3-D objects constrained to a limited number of poses (as assumed by EIGENSPACE_LEARN); however, as eigenspaces have shown, one needs *many* images to represent a 3-D object satisfactorily. Recognition (both identification and location) means to find the image in a model set which is most similar to the one to recognize. Similarity is quantified by a metric measuring the distance between images, for instance correlation or SSD. The advantage is that images and models can be compared directly, and objects with no obvious features can still be modelled. Disadvantages include the fact that illumination, pose, and location variations alter the images.

10.6.2 Object Versus Viewer-centered Representations

The second alternative concerns the fact that 3-D objects can be observed by different viewpoints.

Object-centered representations attach a reference frame to an object, and express the object geometry (points, lines, surfaces) in that frame. We did this with interpretation trees. The classic example of object-centered representations is the *generalized cone*, that we can figure as a section sweeping along and perpendicularly to an axis; the section may change as it moves. A special class of feature-based models is needed to describes *deformable objects* like faces, clothes, and human bodies. Snakes (Chapter 5) are examples of features at the heart of deformable models, and the idea can be extended to three dimensions. If the deformations allowed are limited, an interesting idea is to model a deformable object with a reference shape and the set of its deformations.

Viewer-centered representations model 3-D objects through a set of images or views, taken, ideally, in all possible conditions (viewpoint, illumination, sensor parameters). Views can be images or processed images (that is, visible features). Eigenspaces are an example of viewer-centered representations storing an image for each view; *aspect graphs*, instead, store a feature-based description for each view. An *aspect* is a set of viewpoints in space from which the same object features are simultaneously visible. When, moving the sensor or the object, some features appear or disappear (a situation called *visual event*), a new aspect is entered. Aspect graphs are computed by algorithms analyzing symbolic, feature-based object descriptions, in which the features are nearly invariably contours.[11] But a symbolic object description is not always available, and no feasible aspect graph algorithms exist for complex shapes. In this case, *approximate*

---

[11] Notice the differences between aspect graphs and eigenspaces: the former are learned from a symbolic description, the latter from images; the former are feature-based, the latter image-based. Moreover, visual events seldom cause drastic changes in correlation.

aspect graphs can be built. The set of possible viewpoints is restricted to a finite grid on a sphere centered on the object; the features visible from each viewpoint are found by raytracing a CAD model of the object. This method is applicable to any object for which a CAD model can be built, which are many more than those for which viable aspect graph algorithms exist. The problem is that not all the important features may be captured by the finite set of viewpoints on the sphere.

10.6.3 Concluding Remarks

How should we choose between feature and appearance-based representations? This depends largely on the object to be represented and the algorithms at hand. For instance, if we must represent a set of polyhedra, and have access to a line detector, a description in terms of lines and their relative position in space is reasonable: It describes well the objects' shape, and we can identify the features in input images. Feature-based models do not make sense for objects without detectable features; if the only features we can detect are straight lines, and we must represent a set of human faces, we would be better off with appearance-based methods.

Object-centered representations are intrinsically feature-based, so identification and location work exactly as explained for feature-based models, and advantages and disadvantages are similar. For instance, a disadvantage is that input images and models cannot be compared directly, as the sensor is not an element of the representation, and object appearance is not explicitly predicted. The features chosen should also create image features which are reasonably invariant to viewpoint and illumination (for instance, a line remains a line from all viewpoints, but a circle is turned into different ellipses). Finally, a shape may be too complex to be represented in terms of features.

One disadvantage of viewer-centered representations is their large size, so that efficient algorithms for view comparison are vital. The eigenspace method is basically an efficient way to perform image correlation, but it suffers from the problems of appearance-based matching; with aspect graphs, such problems are alleviated thanks to the use of feature-based views, but searching large aspect graphs is still a problem for applications. A great advantage of image-based methods is that *any* shape can be represented, no matter how complex, as long as we can take images of it.

10.7   Summary

After working through this chapter you should be able to:

- ❑ explain the nature of 3-D object recognition and identification, and describe and motivate the subproblems involved
- ❑ design a basic intrepretation tree algorithm and a few variations
- ❑ design a simple appearance-based system learning a small database of object models, and recognizing objects in unknown poses
- ❑ design a simple, invariants-based system for planar shapes
- ❑ discuss the main issues of representing 3-D objects for model-based recognition

10.8 Further Readings

Our discussion of interpretation trees, and the algorithm INT_TREE, is based largely on Grimson's book [14], which contains an extensive analysis of the method, details its use with both range and intensity images, and shows how geometric constraints can reduce the search. Fisher [12] reports an informative, experimental comparison of ten different variations on the basic intrepretation tree algorithms. Detailed introductions to state space methods, the historic roots of interpretation trees, are found in many books on artificial intelligence, for instance [25] and [16]. Haralick and Elliot [15] discuss efficient tree search for constraint satisfaction. Murray *et al.* [20] is an example of application of interpretation trees to intensity images.

Another approach to feature-based identification is *graph matching*, in which one builds graphs composed of features (nodes) and adjacency links (arcs) for both image data and model. Identification is cast as a *subgraph isomorphism problem*; that is, deciding whether the data graph is contained in the model graph. Solutions to this difficult problem are computationally complex, and for this reason no algorithm has been included here. The interested reader can consult [1, 24].

Methods for searching large databases of object models for identification purposes include model invocation [11].

Zisserman and Mundy's book [18] is the best starting point to investigate invariants; the first chapter is an excellent introduction to the field. INV_ACQ and INV_REC are based on the complete, invariant-based vision system for planar objects reported in Rothwell *et al.* [23], which includes interesting discussions of algebraic and nonalgebraic invariants, grouping, and recognition times. Invariance can be extended to image features other than contours; see for instance Nayar and Bolle [21].

The parametric eigenspace method for appearance-based identification is due to Murase and Nayar [19], who also suggest ways to address the problems pointed out at the end of section 10.4. Finding the closest point to a curve or surface is discussed in several articles on free-form object location, for instance [4].

Building automatically databases of 3-D models from multiple views of an object (*reverse engineering*) continues to attract substantial research. Accurate models can be obtained by fusing range views [4, 2, 8, 13]; see also Chapter 2 and references therein for model acquisition sensors. Notice that algorithms computing structure from stereo and motion (Chapters 7 and 8) can be regarded as the basis of a model acquisition system using multiple intensity views.

Object modelling and shape representation is a *pedigree* problem in computer vision, and the literature is vast. General discussions can be found in [1, 3, 7, 17, 24], to which we refer for pointers to the many representations devised by computer vision. The classic reference for generalized cones is Nevatia and Binford [22]. An increasingly popular type of deformable-object models based on a reference shape and its deformations is the *Point Distribution Model* [10] (see also the Further Readings on snakes in Chapter 5). For flexible models of 3-D objects, see for instance [9]. Bowyer and Dyer [6] give an overview of exact and approximate aspect graphs, and [5] is an instructive debate on their pros and cons. Useful articles are also found in computer graphics journals, for instance the *IEEE Transactions on Visualization and Computer Graphics* (see http://www.ieee.org/).

10.9 Review

Questions

❐ *10.1* Which nongeometric features could be used for 3-D object identification, and in which assumptions? Mention situations in which nongeometric features simplify visual identification, and others in which shape is the best or the only feature possible.

❐ *10.2* Think of real-world applications in which an identification system could be useful. Discuss potential advantages and problems for each application on your list, and the suitability of the methods presented in this chapter.

❐ *10.3* Consider the example of Figure 10.3. If the image features had been listed in a different order, the right path could have been found earlier. Can you think of any criteria for ordering image and model features so that the search is generally reduced?

❐ *10.4* Why did we not use conics for the estimation of the projective transformation in Step 4 of INV_REC?

❐ *10.5* Explain intuitively why, in the case that all the lines associated to an object hypothesis belong indeed to that object, the rank of the matrix A in (10.10) is 8.

❐ *10.6* Why do we need to assume that the illumination conditions are the same for EIGENSPACE_IDENTIF and EIGENSPACE_LEARN, if the grey levels are normalized ($\|\mathbf{x}\|^2 = 1$)? Does the normalization counteract the effects of changes in illuminant direction? And of illuminant intensity?

❐ *10.7* Explain the reasons which make each assumption of EIGENSPACE_IDENTIF necessary. (*Hint for Assumption 3:* Consider two different images, I_1 and I_2, such that the minimum value in I_2 is greater than the maximum value in I_1. The correlation of I_1 and I_2 is larger than the autocorrelation of I_1, yet I_2 cannot be more similar to I_1 than I_1 itself!)

❐ *10.8* What happens to algorithm EIGENSPACE_IDENTIF if the object in the input image is not one of the objects in the database?

❐ *10.9* Suppose two of the eigenspace curves built by EIGENSPACE_LEARN intersect in some points. What does this mean in terms of identification?

❐ *10.10* Compare the characteristics of the three identification techniques presented in this chapter. Identify classes of objects and applications for which some are better suited than others. Why is it the case?

❐ *10.11* We said that identification and location can be performed simultaneously. How could this be done?

❐ *10.12* Why did we say that object-centered representations are necessarily feature-based?

Exercises

○ *10.1* Estimate the complexity increase of the interpretation tree search caused by the introduction of a wild card.

○ *10.2* Expand the complete tree explored by the search illustrated by Figure 10.3. How many inconsistent solutions are generated?

○ *10.3* Modify INT_TREE into an *alignment* algorithm; that is, verification takes place as soon as enough features for computing the model-image transformation have been computed.

○ *10.4* What happens of an interpretation tree search without wild card, which comes across a spurious image feature as the first feature?

○ *10.5* Prove that, if no two of the three lines l_1, l_2, and l_3 are parallel, the determinant of M_{123} (in the same notation of (10.1)) equals the area of the triangle the vertices of which are the intersections of l_1, l_2, and l_3.

○ *10.6* Prove the invariance of I_1 or I_2 of (10.1). (*Hint:* Draw five lines on a sheet of paper and observe that all four triangles involved in the invariant have a side which lies on a same line. Now from the definition of cross-ratio . . .).

○ *10.7* Assume a set of objects characterized by a n-dimensional vector of invariants. Illustrate how an interpretation tree could be used to guide invariant-based recognition, and what particular problem of invariant-based recognition would the interpretation tree address.

○ *10.8* Give a criterion to estimate k automatically in EIGENSPACE_LEARN.

○ *10.9* Adapt algorithm CHANGE_DETECTION for segmenting the object from the background in Step 1 of EIGENSPACE_LEARN.

○ *10.10* Prove that the Euclidean distance between two images represented as vectors is the same as the SSD (sum of squared differences) distance used in Chapter 7 for stereo correspondence.

○ *10.11* Extend EIGENSPACE_LEARN and EIGENSPACE_IDENTIF to take into account a variable illumination direction.

Project

● *10.1* Implement two versions of INT_TREE, with and without wild card, and run both on a set of synthetic data (created, for instance, by writing files of symbolic patch descriptors by hand). Verify the complexity increase caused by the wild card.

References

[1] D.H. Ballard and C.M. Brown, *Computer Vision*, Prentice-Hall, Englewood Cliffs, NJ (1982).

[2] R. Bergevin, M. Soucy, H. Gagnon and D. Laurendeau, Towards a General Multiview Registration Technique, *IEEE Transactions on Pattern Analysis and Machine Intelligence*, Vol. PAMI-18, pp. 540-547 (1996).

[3] P.J. Besl, Geometric Modeling and Computer Vision, *Proceedings of the IEEE*, Vol. 76, pp. 936-958 (1988).

[4] P.J. Besl and N. McKay, A Method for Registration of 3-D Shapes, *IEEE Transactions on Pattern Analysis and Machine Intelligence*, Vol. PAMI-14, pp. 239-256 (1992).

[5]   K. Bowyer (ed.), Why Aspect Graphs Are Not (Yet) Practical for Computer Vision, *Computer Vision, Graphics and Image Processing: Image Understanding*, Vol. 55, pp. 212-218 (1992).

[6]   K. Bowyer and C.R. Dyer, Aspect Graphs: An Introduction and Survey of Recent Results, *International Journal of Imaging Systems and Technology*, Vol. 2, pp. 315-328 (1990).

[7]   M.J. Brady, Criteria for the Representation of Shape, in *Human and Machine Vision*, J. Beck, B. Hope and A. Rosenfeld eds., Academic Press (1983).

[8]   Y. Chen and G. Medioni, Object Modeling by Registration of Multiple Range Images, *Image and Vision Computing*, Vol. 10, pp. 145-155 (1992).

[9]   L.D. Cohen and I. Cohen, Finite Elements Methods for Active Contour Models in Balloons for 2-D and 3-D Images, *IEEE Transactions on Pattern Analysis and Machine Intelligence*, Vol. PAMI-15, pp. 1131–1147 (1993).

[10] T.F. Cootes, C.J. Taylor, D.H. Cooper and J. Graham, Active Shape Models – Their Training and Application, *Computer Vision and Image Understanding*, Vol. 61, pp. 38-59 (1995).

[11] R.B. Fisher, *From Surfaces to Objects*, John Wiley and Sons, Chichester (UK) (1989).

[12] R.B. Fisher, Performance Comparison of Ten Variations on the Interpretation-Tree Matching Algorithm, *Proc. European Conf. on Computer Vision*, Stockholm, pp. 507–512 (1994).

[13] A.W. Fitzgibbon, D.W. Eggert and R.B. Fisher, High-Level CAD Model Acquisition from Range Images, *Computer-Aided Design*, Vol. 29, pp. 321–330 (1997).

[14] W.E.L. Grimson, *Object Recognition by Computer: the Role of Geometric Constraints*, MIT Press, Cambridge (MA) (1990).

[15] R.M. Haralick and G.L. Elliott, Increasing Tree Search Efficiency for Constraint Satisfaction Problems, *Artificial Intelligence*, Vol. 14, pp. 263-313 (1980).

[16] G.E. Luger and W.A. Stubblefield, *Artificial Intelligence*, second edition, Benjamin/Cummings, Redwood City (CA) (1993).

[17] D. Marr, *Vision*, W. H. Freeman, San Francisco (1981).

[18] J.L. Mundy and A. Zisserman, *Geometric Invariance in Computer Vision*, MIT Press, Cambridge (MA) (1992).

[19] H. Murase and S.K. Nayar, Visual Learning and Recognition of 3-D Objects from Appearance, *International Journal of Computer Vision*, Vol. 14, pp. 5-24 (1995).

[20] D.W. Murray, D.A. Castelow and B.F. Buxton, From Images Sequences to Recognised Moving Polyhedral Objects, *International Journal of Computer Vision*, Vol. 3, pp. 181-209 (1989).

[21] S.K. Nayar and R.M. Bolle, Reflectance Ratio: A Photometric Invariant for Object Recognition, *Proc. IEEE International Conf. on Computer Vision*, pp. 280–285 (1993).

[22] R.Nevatia and T.Binford, Description and Recognition of Curved Objects, *Artificial Intelligence*, Vol. 38, pp. 77-98 (1977).

[23] C.A. Rothwell, A. Zisserman, D.A. Forsyth and J.L. Mundy, Planar Object Recognition Using Projective Shape Representation, *International Journal of Computer Vision*, Vol. 16, pp. 57–99 (1995).

[24] M. Sonka, V. Hlavac and R. Boyle, *Image Processing, Analysis and Machine Vision*, Chapman and Hall, London (1993).

[25] P.H. Winston, *Artificial Intelligence*, Third edition, Addison-Wesley, Reading (MA) (1993).

11

Locating Objects in Space

Ubi sunt pocula dulciora mellae?[1]

Horatio

This chapter introduces algorithms for *object location*; that is, determining the position and orientation of the objects in view, assuming their identity (model) is known.

Chapter Overview

Section 11.1 introduces the problem and the assumptions made for this chapter.

Section 11.2 solves object location given a full-perspective and weak-perspective intensity image.

Section 11.3 solves object location given a range image.

What You Need to Know to Understand this Chapter

- Basic feature extraction (Chapter 4) and object recognition (Chapter 10).
- Perspective and weak-perspective camera models (Chapter 2).
- Least squares, SVD (Appendix, section A.6).
- Rotation matrices and their parametrizations (Appendix, section A.9).

---

[1] Where are the chalices sweeter than honey?

11.1 Introduction

Once we have identified the objects in an image, how do we find their location in space? This problem is known in computer vision as *model-based object location*, but you are likely to come across other names, for instance *model matching, pose estimation,* and *optical jigging.* We begin with a general definition of the problem, which will be specialized for the intensity and range cases.

Problem Statement: Model-based Object Location

Given an image, a sensor model, and the geometric model of the object imaged, find which 3-D position and orientation of the model generated the image.

Position and *orientation* (or *pose*) refer to the 3-D translation and rotation, respectively, which bring the object model wherever it was observed by the sensor. The definition above hints at the assumptions we make in this chapter, which are summarized below.

Assumptions

1. Location uses a single image.
2. A model of geometric image formation (sensor model) is completely known.
3. Object models are object-centered, and based on geometric features.
4. The identification problem has already been solved.

Assumption 2 means, for instance in the case of intensity images, that a precise camera model is adopted (e.g., full perspective), and the camera's intrinsic parameters are known. Assumption 3 excludes appearance-based object models, and models formed by non-geometric features (see the Further Readings and Chapter 10 for more on such models). Assumption 4 means that we know that a precise part of the image (for instance, a group of features) corresponds to an object model, and, within that part of the image, which image feature corresponds to which model feature.

An important aspect of location, introduced in the previous chapter, is *verification.* Errors in the output of feature extraction and object identification can cause inaccurate or definitely wrong location estimates. We can verify the accuracy of the latter by creating an image of the model in the estimated position (*backprojecting* the model), and checking how close the model features are imaged to the corresponding features in the input image.[2]

---

[2] Notice that this is possible because the sensor model is known.

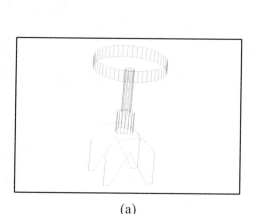

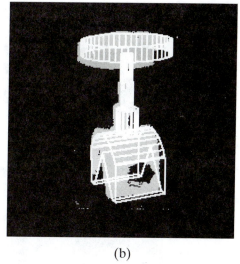

(a)   (b)

Figure 11.1   (a) Line-drawing rendering of a 3-D object model (a photographic stand). (b) The model's location has been estimated from a range image, and the model backprojected. Notice the imperfect alignment, mainly due to the incomplete range data. Courtesy of A. M. Wallace, Heriot-Watt University.

As an example, Figure 11.1 illustrates model-based matching and verification for a range image of a photographic stand. Another example is shown in Figure 11.2, in which intensity and range data are used together to compute location.

Before moving on to intensity-based location, we summarize the three basic modules of a recognition system,[3] as defined by the discussion and assumptions of this and the previous chapter.

Object Recognition: Basic Modules

1. *Object identification*, which selects object models from a database of models and establishes the correspondence of model and data features.

2. *Model-based location*, which positions the selected models in space.

3. *Verification*, which filters out the hypotheses (located models) that prove inconsistent with the input image.

☞   The type of images (range or intensity) makes a substantial difference for model-based matching. In the intensity case, models and data are different in nature (the model is 3-D, the image 2-D), and *must be transformed* to be compared; in the range case, both models and data are specified by 3-D coordinates, and can be compared *unaltered*.

---

[3] Image acquisition and feature extraction are taken for granted.

(a)

(b)

(c)

(d)

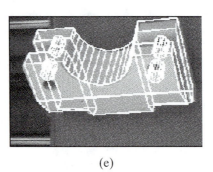

(e)

Figure 11.2 (a) Range image of a mechanical component. (b) Surface patches extracted (Chapter 5). (c) Intensity image of the same component. (d) Linear edges detected in the intensity image. (e) CAD model of the component located and backprojected onto the intensity image. Courtesy of W. Austin, Heriot-Watt University.

11.2 Matching from Intensity Data

This section discusses *model-based object location from a single intensity image* (more precisely, from a set of image features). All model points and vectors are expressed in the *model reference frame*, the frame associated with the object model (Chapter 10); all data points and vectors are expressed in the camera reference frame, as we assume the intrinsic parameters known.

We describe two methods. The first one adopts a full-perspective camera and is basically an application of Newton's iterative method for solving a system of simultaneous, nonlinear equations; the features can be either points or lines. The second one employs the more restrictive weak-perspective camera, but gives a closed-form solution (as opposed to an iterative one) using three corresponding pairs of image and model points. We start by stating the problem in the case of point features.

Problem Statement

Let $\mathbf{P}_1^m, \ldots, \mathbf{P}_n^m$, with $\mathbf{P}_i^m = \left[X_i^m, Y_i^m, Z_i^m \right]^\top$ and $n \geq 3$, expressed in the model reference frame, be n points of an object model. Let $\mathbf{P}_1, \ldots, \mathbf{P}_n$ with $\mathbf{P}_i = [X_i, Y_i, Z_i]^\top$, expressed in the camera reference frame, indicate the coordinates of the corresponding points on the object observed. Let $\bar{\mathbf{p}}_1, \ldots, \bar{\mathbf{p}}_n$, with $\bar{\mathbf{p}}_i = [\bar{x}_i, \bar{y}_i]^\top$, be the n image points, expressed in the camera frame, projections of the $\mathbf{P}_i$.

Determine the rigid transformation (rotation matrix $\bar{R}$ of entries r_{ij}, translation vector $\bar{\mathbf{T}} = \left[T_x, T_y, T_z \right]^\top$) aligning the camera and model reference frames; that is,

$$[X_i, Y_i, Z_i]^\top = \bar{R}\mathbf{P}_i^m + \bar{\mathbf{T}}. \tag{11.1}$$

☞　Notice that the problem is equivalent to determining the camera's extrinsic parameters with respect to the object reference frame. In this sense, solving the location problem means *locating the camera in space*, and consequently locating the vehicle or robot actuator on which the camera is possibly fitted.

11.2.1 3-D Location from a Perspective Image

The structure of a location algorithm depends on the camera model adopted, as the latter determines the equations of image formation. In the case of full perspective, the algorithm can be formulated as an application of Newton's iterative method for solving systems of nonlinear equations.[4]

Outline of the Algorithm.　The relation between an object point and an image point, both in camera coordinates, is given by the usual perspective projections:

---

[4] We derive all necessary equations without assuming *prior* knowledge of Newton's method. If you are familiar with it, you will soon realize that its simplicity is somewhat obscured by the fact that a 3×3 rotation matrix has nine entries, but these depend on three parameters only.

$$[x_i, y_i]^\top = \left[\frac{f X_i}{Z_i}, \frac{f Y_i}{Z_i} \right]^\top . \tag{11.2}$$

Plugging (11.1) into (11.2), we see that each point correspondence generates two non-linear equations,

$$x_i = f \frac{r_{11} X_i^m + r_{12} Y_i^m + r_{13} Z_i^m + T_1}{r_{31} X_i^m + r_{32} Y_i^m + r_{33} Z_i^m + T_3}$$
$$y_i = f \frac{r_{21} X_i^m + r_{22} Y_i^m + r_{23} Z_i^m + T_2}{r_{31} X_i^m + r_{32} Y_i^m + r_{33} Z_i^m + T_3}. \tag{11.3}$$

The unknown components of $\bar{R}$ and $\bar{T}$ can be determined from a sufficient number of correspondences, each bringing two equations like (11.3). The resulting system has six unknowns, as $\bar{R}$ depends only on three free parameters (Appendix, section A.9); in the following, these are the rotation angles about the three axes of the camera's frame, ϕ_1, ϕ_2, and ϕ_3. Our six unknowns are therefore $\phi_1, \phi_2, \phi_3, T_1, T_2, T_3$. Solving the system is where Newton's method comes into play.

Newton's Method. As any iterative technique, Newton's method starts off with an initial guess for $\bar{R}$ and $\bar{T}$, say R^0 and T^0, and computes the location of p_i through (11.3) with $R = R^0$ and $T = T^0$. If R^0 and T^0 are not too far from the true solution, $(\bar{R}, \bar{T})$, the residuals

$$\delta x_i = x_i(R^0, T^0) - \bar{x}_i$$
$$\delta y_i = y_i(R^0, T^0) - \bar{y}_i, \tag{11.4}$$

are small, and can be approximated by a first-order expansion of $x_i(\phi_1, \phi_2, \phi_3, T_1, T_2, T_3)$ and $y_i(\phi_1, \phi_2, \phi_3, T_1, T_2, T_3)$ given by (11.3), in a neighborhood of R^0 and T^0.

Let us compute the partial derivatives necessary for the expansion. The partial derivatives with respect to T_1, T_2, and T_3 are

$$\frac{\partial x_i}{\partial T_1} = \frac{f}{Z_i}, \quad \frac{\partial x_i}{\partial T_2} = 0, \quad \frac{\partial x_i}{\partial T_3} = -f \frac{X_i}{Z_i^2}$$

and

$$\frac{\partial y_i}{\partial T_1} = 0, \quad \frac{\partial y_i}{\partial T_2} = \frac{f}{Z_i}, \quad \frac{\partial y_i}{\partial T_3} = -f \frac{Y_i}{Z_i^2}.$$

For the partial derivatives with respect to the rotation angles, we must recall the definition of derivative of a vector with respect to a rotation angle, illustrated in Figure 11.3. Given our choice of angles, we have

$$\frac{\partial \mathbf{P}_i}{\partial \phi_1} = [1, 0, 0]^\top \times \mathbf{P}_i,$$

$$\frac{\partial \mathbf{P}_i}{\partial \phi_2} = [0, 1, 0]^\top \times \mathbf{P}_i,$$

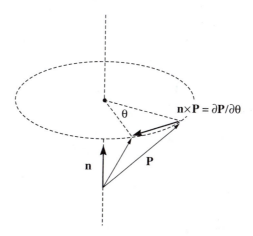

Figure 11.3 The derivative of a point vector, **P**, with respect to a rotation angle, θ, about a direction through the origin, **n** ($\|\mathbf{n} = 1\|$) is given by the vector product of **n** and **P**.

and

$$\frac{\partial \mathbf{P}_i}{\partial \phi_3} = [0, 0, 1]^\top \times \mathbf{P}_i,$$

from which we obtain

$$\frac{\partial x_i}{\partial \phi_1} = -\frac{f X_i Y_i}{Z_i^2}, \quad \frac{\partial x_i}{\partial \phi_2} = f \frac{X_i^2 + Z_i^2}{Z_i^2}, \quad \frac{\partial x_i}{\partial \phi_3} = -f \frac{Y_i}{Z_i}$$

and

$$\frac{\partial y_i}{\partial \phi_1} = -f \frac{Z_i^2 + Y_i^2}{Z_i^2}, \quad \frac{\partial y_i}{\partial \phi_2} = f \frac{X_i Y_i}{Z_i^2}, \quad \frac{\partial y_i}{\partial \phi_3} = f \frac{X_i}{Z_i}.$$

Therefore, for each image-model point correspondence, the expansion yields the following pair of linear equations:

$$\sum_{j=1}^{3} [\frac{\partial x_i}{\partial T_j} \Delta T_j + \frac{\partial x_i}{\partial \phi_j} \Delta \phi_j] = \delta x_i$$

$$\sum_{j=1}^{3} [\frac{\partial y_i}{\partial T_j} \Delta T_j + \frac{\partial y_i}{\partial \phi_j} \Delta \phi_j] = \delta y_i.$$

(11.5)

The six unknowns ΔT_j and $\Delta \phi_j$ can be determined if at least three point correspondences are known. To counteract the effect of inaccurate measurements or correspondences however, we try to use as many correspondences as possible.

The method proceeds by producing new estimates for the rotation matrix and translation vector, and iterating the procedure until the residuals δx_i and δy_i become small enough.

Some Necessary Details.   In order to arrive at the customary algorithm box, we must address three problems.

1. Given the current estimates, R_j and $\mathbf{T}_j$, and corrections, ΔT_j and $\Delta \phi_j$, how do we compute the new estimates, R_{j+1} and $\mathbf{T}_{j+1}$?
2. Is the solution unique?
3. How do we determine a satisfactory initial guess?

Updating the Estimates.   Updating $\mathbf{T}_j$ is straightforward:

$$\mathbf{T}_{j+1} = \mathbf{T}_j + \Delta T_j.$$

Updating R_j, instead, requires some care. The new rotation matrix, R_{j+1}, is obtained by repeated matrix multiplication of R_j with the three rotation matrices defined by the correction angles, $\Delta \phi_1$, $\Delta \phi_2$, and $\Delta \phi_3$. Appendix, section A.9 gives the structure of the necessary matrices.

☞   Upon convergence, the correction angles become sufficiently small to make the order of multiplication irrelevant (Exercise 11.1). However, this is not true at the beginning. In general, it is best to fix an order, and stick to your choice.

Uniqueness of Solution.   If a large number of feature points in nondegenerate configurations (e.g., not all coplanar) are employed, the issue of multiple solutions, with the exception of symmetrical objects, can be safely ignored. Multiple solutions do arise with small numbers of features (e.g., $n = 3$); in this case, one way of finding the correct solution is to run the method from several, different starting positions. For a thorough discussion of this issue see the Further Readings.

Determining the Initial Guess.   Fortunately, this problem is less critical than expected. The (11.3) are linear with respect to translation and scaling over the image plane, and approximately linear over a wide range of values of the rotational parameters. Hence, the method is likely to converge to the desired solution for a rather wide range of possible starting positions, and even rough estimates of $\mathbf{T}$ and R should be sufficient to ensure convergence to the true solution. Suggestions for this initialization stage are given in Exercise 11.2.

Summary of the Algorithm.   We are now ready to state a location algorithm.

Algorithm 3D_POSE

The input is formed by n corresponding image and model points, with $n \geq 3$, and the initial estimates R^0 and $\mathbf{T}^0$.

1. Use the current estimates of $\bar{R}$ and $\bar{\mathbf{T}}$ and (11.1) to compute the $\mathbf{P}_i$; that is, the predicted location of the model points $\mathbf{P}_i^m$ in the camera frame.
2. Project the $\mathbf{P}_i$ onto the image plane through (11.2).
3. Compute the residuals δx_i and δy_i $i = 1, \ldots, n$, from (11.4).
4. Solve the linear system formed by n instances of (11.5) for the unknown corrections, ΔT_j and $\Delta \phi_j$, $j = 1, 2, 3$.
5. Update the current estimates of the translation vector and the rotation matrix.
6. If the residuals are sufficiently small, exit; else go to step 1.

☞ During the first iterations, the residuals should decrease by about one order of magnitude per iteration. Thus a few iterations are usually sufficient to obtain a satisfactory solution. Remember that system (11.5) (step 4) ought to be overconstrained ($n > 3$) to counteract noise, and solved by least squares.

Extending the Method to Line Features. Since lines are often easier to detect and localize in images than points, line-based location has important practical applications. We just sketch the principles behind the extension of 3D_POSE to line features; the development of an algorithm is left as an exercise (Exercise 11.3).

We write the equation of a line in the image in the form

$$\frac{-m}{\sqrt{m^2 + 1}} x + \frac{1}{\sqrt{m^2 + 1}} y = d.$$

Observe that the distance between this line and a point, $[x_0, y_0]^\top$, is simply $|d - d_0|$, with

$$d_0 = \frac{-m}{\sqrt{m^2 + 1}} x_0 + \frac{1}{\sqrt{m^2 + 1}} y_0; \tag{11.6}$$

therefore, the derivatives of d_0 are simply a linear combination of the derivatives of x_0 and y_0, with the same weights of (11.6).

Now, given a set of pairs of corresponding image and model lines, we choose two points on each matched image line, and compute the distance between each point and the matched model line, as shown in Figure 11.4. Since each point gives one equation for the correction parameters, and two points are sufficient to uniquely identify the model line, a line-to-line correspondence brings home the same information (two equations) of a point-to-point correspondence, and the structure of algorithm 3D_POSE remains unchanged. The only changes are that the (11.6) are replaced by the two point-line distance equations, and the expressions of the two distances replace those of the residuals (11.5).

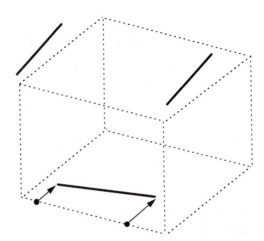

Figure 11.4 The dotted lines show the line-based model of a box backprojected onto the image. The solid lines show the lines detected in the input image of the box, each corresponding to the closest model line. The arrows show the two points and distances used by the method for one of the pairs of matched lines.

☞   In order to achieve stable estimates, the two points on the matched model line should be the endpoints of the matched image line (see Figure 11.4).

11.2.2 3-D Location from a Weak-perspective Image

This section presents an alternative method, easily illustrated in geometric terms. The method uses point correspondences and assumes a *weak-perspective camera*.[5] It also eases us into our next task, locating objects from range data.

Setting the Location Problem in Geometric Terms.   First of all, let us summarize the assumptions behind the method.

Assumptions

1. The viewing geometry can be approximated by the weak-perspective camera model.
2. Three image points, p_0, p_1, p_2, correspond to the model points P_0^m, P_1^m, P_2^m, respectively.

The most obvious difference between this method and 3D_POSE is the camera model. The simpler weak-perspective camera, although more restrictive, is mathematically simpler the than full-perspective one, and does not contain intrinsic parameters

---

[5] If you do not remember the weak-perspective camera model, review sections 2.2 and 2.4 of Chapter 2 before moving on.

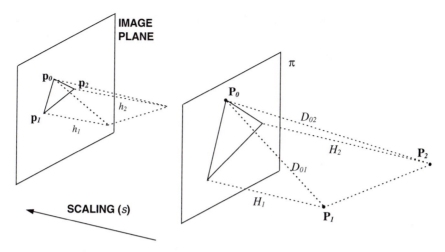

Figure 11.5 Three object points, $\mathbf{P}_0, \mathbf{P}_1, \mathbf{P}_2$, and their images under weak perspective, $\mathbf{p}_0, \mathbf{p}_1, \mathbf{p}_2$, generated by an orthographic projection followed by a scaling factor s. Notice that, as a consequence, the pyramid on the left is a scaled version of the pyramid on the right.

like principal point or focal length. The price to pay, as we shall see in detail, is that we cannot determine how far each point is as a weak-perspective camera does not give information on distances.

☞   There is no such thing like an optimal camera model *per se*. The optimal camera model is simply *the simplest geometric model compatible with the constraints of the problem at hand*. If you can guarantee that the depth of the scene observed is small with respect to its distance from the camera, a weak-perspective camera is perfectly adequate, and there is no need for more sophisticated models.

Under weak perspective, the geometry of image formation is described by an orthographic projection followed by scaling. This is illustrated by Figure 11.5, which also shows the geometric quantities used in the following. In a notation similar to the one used for 3D_POSE, we let $\mathbf{P}_0, \mathbf{P}_1, \mathbf{P}_2$ denote three points (in camera coordinates) on the object imaged, $\mathbf{P}_0^m, \mathbf{P}_1^m, \mathbf{P}_2^m$ the corresponding model points (in model coordinates), and $\mathbf{p}_0, \mathbf{p}_1, \mathbf{p}_2$ the image points projections of $\mathbf{P}_0, \mathbf{P}_1, \mathbf{P}_2$. First, we project $\mathbf{P}_1$ and $\mathbf{P}_2$ orthographically onto a plane, π, parallel to the image plane and passing through the third object point, $\mathbf{P}_0$. We then scale down the resulting triangle on π to the image plane triangle $\mathbf{p}_0\mathbf{p}_1\mathbf{p}_2$. The two triangles, as well as the pyramids shown in Figure 11.5, differ by an unknown scale factor, s. The distances from π to $\mathbf{P}_1$ and $\mathbf{P}_2$, which we consider *signed*, are denoted with H_1 and H_2, respectively.

The method works in two stages:

1. Compute the 3-D coordinates of the object points in the camera frame.
2. Find the rigid transformation bringing the model points onto the object points.

☞   Notice that stage 1 produces a set of range data; that is, a cloud of 3-D points. Stage 2, therefore, must solve the location problem from range data.

Stage 1: Solving for the Scene Points.   The 3-D coordinates of the scene points in the camera frame can be obtained in closed form as follows. We can derive three constraints from the three right triangles in Figure 11.5:

$$h_1^2 + d_{01}^2 = (s\,D_{01})^2$$
$$h_2^2 + d_{02}^2 = (s\,D_{02})^2 \qquad (11.7)$$
$$(h_1 - h_2)^2 + d_{12}^2 = (s\,D_{12})^2,$$

where the distances

$$d_{ij} = \|\mathbf{p}_i - \mathbf{p}_j\|$$

and

$$D_{ij} = \|\mathbf{P}_i - \mathbf{P}_j\|$$

are known: the d_{ij} can be computed from the image, the D_{ij} from the model. These distances, as well as the scale factor, s, are positive; h_1 and h_2 (likewise H_1 and H_2), instead, are signed. The "-" sign in $(h_1 - h_2)$ is simply due to the particular configuration of Figure 11.5.

Adding the first two equations in (11.7) and subtracting the third gives

$$2h_1 h_2 = s^2(D_{01}^2 + D_{02}^2 - D_{12}^2) - (d_{01}^2 + d_{02}^2 - d_{12}^2).$$

Squaring this last equation and using the first two in (11.7) yields a quartic equation in the scale factor:

$$as^4 - 2bs^2 + c = 0, \qquad (11.8)$$

with

$$a = (D_{01} + D_{02} + D_{12})(-D_{01} + D_{02} + D_{12})(D_{01} - D_{02} + D_{12})(D_{01} + D_{02} - D_{12})$$
$$b = D_{01}^2(-d_{01}^2 + d_{02}^2 + d_{12}^2) + D_{02}^2(d_{01}^2 - d_{02}^2 + d_{12}^2) + D_{12}^2(d_{01}^2 + d_{02}^2 - d_{12}^2) \qquad (11.9)$$
$$c = (d_{01} + d_{02} + d_{12})(-d_{01} + d_{02} + d_{12})(d_{01} - d_{02} + d_{12})(d_{01} + d_{02} - d_{12})$$

Assuming that the model triangle is not degenerate, simple but rather lengthy calculations (see Exercise 11.6 for some hints, and the Further Readings for the complete story) show that (11.8) has a unique solution given by

$$s = \sqrt{\frac{b + \sqrt{b^2 - ac}}{a}}, \qquad (11.10)$$

and that the coordinates of the scene points in the camera frame are

$$\mathbf{P}_0 = \tfrac{1}{s}(x_0, y_0, w)\,, \; \mathbf{P}_1 = \tfrac{1}{s}(x_1, y_1, w + h_1)\,, \; \mathbf{P}_2 = \tfrac{1}{s}(x_2, y_2, w + h_2) \qquad (11.11)$$

with

$$(h_1, h_2) = \pm(\sqrt{(s D_{01})^2 - d_{01}^2}, \sigma(\sqrt{(s D_{02})^2 - d_{02}^2})),$$

$$(H_1, H_2) = \frac{1}{s}(h_1, h_2),$$

$$\sigma = \begin{cases} 1 & \text{if } d_{01}^2 + d_{02}^2 - d_{12}^2 \leq s^2(D_{01}^2 + D_{02}^2 - D_{12}^2) \\ -1 & \text{otherwise.} \end{cases}$$

☞   Notice that w, an unknown depth offset, cannot be eliminated because the weak-perspective projection does not carry information on distances.

The ambiguity in the overall sign of h_1 and h_2 reflects the ambiguity of the location of the points $\mathbf{P}_1$ and $\mathbf{P}_2$ with respect to the plane π: it corresponds to a flip of the plane through $\mathbf{P}_0$, $\mathbf{P}_1$ and $\mathbf{P}_2$ with respect to that plane. This ambiguity can only be resolved by checking the reconstructed location of further model points, which must not lie on the plane defined by $\mathbf{P}_0$, $\mathbf{P}_1$ and $\mathbf{P}_2$.

☞   Since $s \to \infty$ for $a \to 0$, (11.11) tells us that a *3-D pose cannot be estimated if the three model points are collinear*. In practice, the numerical solution will be ill-conditioned whenever the model points are nearly collinear; this situation should be avoided.

Here is a summary of the algorithm that recovers the coordinates of the scene points in the camera frame.

Algorithm WEAK_PERSP_INV

The input is formed by three model points, $\mathbf{P}_0^m$, $\mathbf{P}_1^m$, $\mathbf{P}_2^m$, and the three corresponding image points, $\mathbf{p}_0$, $\mathbf{p}_1$, $\mathbf{p}_2$.

1. Compute the scale factor using (11.10) and (11.9).
2. Compute the coordinates of the scene points using (11.11).

The output is formed by the coordinates of the scene points imaged, $\mathbf{P}_0$, $\mathbf{P}_1$, $\mathbf{P}_2$, expressed in the camera reference frame.

Stage 2: Solving for the Pose.   We now turn briefly to the second stage of our method, i.e., how to compute the rigid transformation bringing the model points onto the scene points expressed in the camera reference frame. We pointed out already that, as the input is now a set of 3-D points expressed in the camera frame, this corresponds exactly to solving location from range data. Several solutions to this problem are discussed in section 11.3, and any of them can be coupled to WEAK_PERSP_INV to complete the solution to the location problem from a weak-perspective image.

☞   You can regard WEAK_PERSP_INV as an algorithm reconstructing *shape (structure) from point features*.

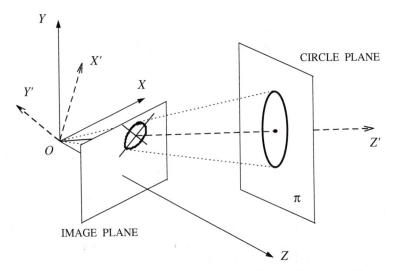

Figure 11.6   The cone and two of the three reference frames used by POSE_FROM_ELLIPSE. Notice that the axis Z' forms a *generic* angle with π.

11.2.3 Pose from Ellipses

We conclude the section on intensity-based location with an algorithm estimating the pose of a circle in space from a single intensity image. Given that many man-made objects contain circles, which are imaged as ellipses, this is a useful tool in many situations; e.g., estimating the orientation of a vehicle relative to a circular target for vision-guided docking.

The geometry of the algorithm, POSE_FROM_ELLIPSE, is illustrated in Figure 11.6. The image ellipse defines a cone with vertex in the center of projection of the pinhole camera. We can find the orientation of the circle's plane, π, by rotating the camera so that the intersection of the cone with the image plane becomes a circle, which happens when the image plane is parallel to the circle. This rotation is estimated as the composition of two successive rotations: The first put the Z axis through the center of the circle, and aligns the X and Y axes with the axes of the image ellipse; the second rotates the reference frame around the new Y axis until the image plane becomes parallel to π. In general this problem has two distinct solution, due to a twofold ambiguity in the determination of the second rotation's angle.

How do we estimate the two rotations? Let $aX^2 + bXY + cY^2 + dX + eY + f = 0$ be the equation of the ellipse in the image plane. We assume all distances expressed in multiples of the focal length, which is therefore set to 1. The equation of the cone is then

$$aX^2 + bXY + cY^2 + dXZ + eYZ + fZ^2 = \mathbf{P}^\top C \mathbf{P} = 0, \qquad (11.12)$$

where $\mathbf{P} = [X, Y, Z]^\top$ and C is the real, symmetric matrix of the ellipse. The first rotation, transforming the frame $OXYZ$ into $OX'Y'Z'$ and sending a point $\mathbf{P}$ to $\mathbf{P}'$, is determined simply by diagonalizing C. If $\lambda_1, \lambda_2, \lambda_3$ are the eigenvalues of C, with $\lambda_1 < \lambda_2 < \lambda_3$, and $\mathbf{e}_1, \mathbf{e}_2, \mathbf{e}_3$ the corresponding eigenvalues, we have

$$\mathbf{P}' = R_1^\top \mathbf{P} = [\mathbf{e}_1|\mathbf{e}_2|\mathbf{e}_3]^\top \mathbf{P}.$$

The second rotation is determined by imposing the equality of the coefficients of X^2 and Y^2, resulting in a rotation around the Y' axis by an angle

$$\theta = \pm \arctan \sqrt{\frac{\lambda_2 - \lambda_1}{\lambda_3 - \lambda_2}},$$

which sets $a = c = \lambda_2$. Notice the twofold ambiguity in θ, which cannot be resolved in the absence of further constraints. In the $X'Y'Z'$ frame, the second rotation is therefore identified by a matrix

$$R_2 = \begin{bmatrix} \cos\theta & 0 & \sin\theta \\ 0 & 1 & 0 \\ -\sin\theta & 0 & \cos\theta \end{bmatrix}$$

which brings $OX'Y'Z'$ onto a new frame $OX''Y''Z''$. The global rotation matrix, R, is the composition of R_1 and R_2; that is, $R = R_1 R_2$. The normal to plane π is therefore

$$\mathbf{n} = R \begin{bmatrix} 0 \\ 0 \\ -1 \end{bmatrix} = \begin{bmatrix} -R_{13} \\ -R_{23} \\ -R_{33} \end{bmatrix}.$$

Following is a description of the algorithm.

Algorithm POSE_FROM_ELLIPSE

The input is the implicit equation of an image ellipse, (11.12), perspective projection of a circle in space on plane π. The focal length is set to 1; all distances are expressed in focal length units.

1. Compute the eigenvalues $\lambda_1, \lambda_2, \lambda_3$ of C ($\lambda_1 < \lambda_2 < \lambda_3$), and the corresponding eigenvectors $\mathbf{e}_1, \mathbf{e}_2, \mathbf{e}_3$.

2. Compute the two values $\theta = \pm \arctan \sqrt{\frac{\lambda_2-\lambda_1}{\lambda_3-\lambda_2}}$.

3. Compute the rotation matrix

$$R = [\mathbf{e}_1|\mathbf{e}_2|\mathbf{e}_3] \begin{bmatrix} \cos\theta & 0 & \sin\theta \\ 0 & 1 & 0 \\ -\sin\theta & 0 & \cos\theta \end{bmatrix}$$

4. Set $\mathbf{n} = [-R_{13}, -R_{23}, -R_{33}]^\top$.

The output is $\mathbf{n}$, the normal to the plane π.

☞   Some ellipses may be partially visible because of surface discontinuities or occlusion from other objects. In such cases, errors may creep into the ellipse fit, and consequently into the orientation estimates. Notice that POSE_FROM_ELLIPSE can be regarded as a *deterministic shape-from-texture algorithm* for surfaces covered by circles. The corresponding ellipses in the images must be large enough to be extracted reliably by an ellipse detector (Chapter 5).

11.2.4 Concluding Remarks

We conclude this section with some comments on the strengths and weaknesses of the two methods described, and a word of caution. First of all, which method should we use, and when? The answer depends on a number of factors. In short, 3D_POSE is considerably more complicated from the computational viewpoint, but can be employed with points, lines, or arbitrary combinations thereof. Further strengths of 3D_POSE include the ability to adjust automatically to a variable number of data features, and the possibility to include unknown model parameters in the estimation process. A limit is the need for an initial guess of the solution. WEAK_PERSP_INV is computationally simpler, but is likely to be less effective in the presence of large amounts of noise, as practically every method based on the minimum amount of information needed to compute a unique solution. For details on the important issue of robustness against noise, see the Further Readings (as well as Appendix, section A.7).

We reiterate that any answers should be *verified* against the data. Ideally, verification could exploit *all* the features extracted from the image, which in general are many more than those used for solving location. However, occlusions, missing features, false correspondences, and inaccuracies of numerical estimates contribute to make verification a difficult operation in itself. As discussed in the context of invariants-based identification (Chapter 10), verification can be used not only to check the accuracy of the estimated location of an object, but also the hypothesized object identity (the model selected). The complexity of efficient procedures for pose verification are beyond the scope of this introductory chapter; the interested reader is referred to the Further Readings.

11.3   Matching from Range Data

We now turn to *model-based object location from a single range image*. As done throughout this chapter, we assume that the identification problem has been solved; we must now match a given object-centered, feature-based model to a given subset of range data, knowing the corresponding pairs of model and data features. All model points and vectors are expressed in the *model reference frame* associated with the model; all data points and vectors are expressed in the *sensor reference frame*, defined by the range sensor used to acquire the image, and known from calibration.

As model and data are now directly comparable (both are specified by 3-D coordinates), we are better off than in the intensity case: We just have to find the rigid transformation which aligns the model with the data, and no projection is involved.

☞   This is actually an assumption that does not reflect the way in which some range sensors operate (for instance, a scale factor could be involved).

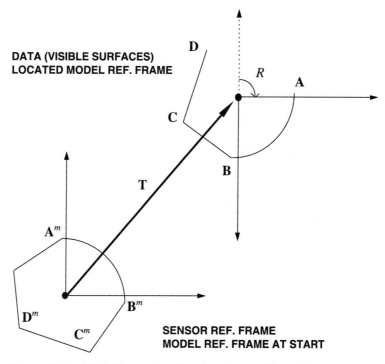

Figure 11.7 A 2-D illustration of the location problem with range data. Starting with the model reference frame superimposed on the sensor reference frame, we must estimate the transformation R, $\mathbf{T}$ bringing the model onto the data.

You can visualize a model reference frame initially coincident with the sensor reference frame, then rotated and translated so that the model points coincide with their corresponding data points; Figure 11.7 illustrates this idea with a 2-D example.

In the following, we assume that the features forming both model and data are surface patches, extracted as discussed in Chapter 4; for simplicity, we consider only *planar patches*.

☞   Planar patches are easy to represent, and their normals can be estimated reliably from range data provided a large enough portion of the patch is visible. A sufficiently high number of planar patches lead to accurate location estimates.

Following are the problem statement and our explicit assumptions.

Assumptions

The range image contains a known object in unknown position and orientation.

A set of feature descriptors, $\mathbf{d}_1, \ldots, \mathbf{d}_N$, describing the visible planar patches of the object is available; each patch descriptor contains estimates of the patch's centroid, $\mathbf{p}_i^d$, and the patch's normal, $\mathbf{n}_i^d$, expressed in the sensor reference frame, as well as the distance of the plane of the patch from the origin, d_i^d.

An object-centered, surface-based model of the object is available. The origin of the model reference frame is the object centroid. The model descriptors of the visible patches, $\mathbf{m}_1, \ldots, \mathbf{m}_N$, include the patch centroid, $\mathbf{p}_i^m$, and the patch's normal, $\mathbf{n}_i^m$, both expressed in the model reference frame, as well as the distance of the plane of the patch from the origin, d_i^m.

Model descriptor $\mathbf{m}_k$ corresponds to data descriptor $\mathbf{f}_k$, for all k between 1 and N.

Problem Statement

Compute the rigid transformation (rotation and translation of the model reference frame) aligning model and data.

The data are supplied by segmentation and identification algorithms, and typically corrupted by noise of at least three kinds. First, all data measurements (normals, centroids, distances) are estimates and carry numerical errors. Second, some correspondences may be false. Third, some data patches may be partially occluded (so that their centroid does not coincide with that of the complete model patch), and some model patches completely invisible (so that only a subset of the model patches can be used for location). All these factors contribute to the uncertainty of location estimates.

We solve the location problem in two ways, differing for the order in which we estimate rotation and translation, and for the data used:

- estimating translation first, and representing the planar patches by their normals and centroids;

- estimating rotation first, and representing the planar patches by their normals as well as the equations of the associated planes.

11.3.1 Estimating Translation First

This method computes the translation of the object's centroid first, then uses the result to compute the rotation, *in the translated reference frame*, bringing the translated model onto the data.

Estimating Translation. A first idea is suggested by the fact that, at the beginning, the model reference frame coincides with the sensor reference frame (Figure 11.7); therefore, an estimate of the translation of the object centroid is obtained simply by averaging the centroids of all the data patches. The resulting algorithm is extremely simple.

Algorithm TRANSL_RANGE

The input is formed by the observed centroids of the data patches, $\mathbf{p}_1^d, \ldots, \mathbf{p}_N^d$.

The output is an estimate of the object centroid:

$$\hat{\mathbf{T}} = \frac{1}{N} \sum_{j=1}^{N} \mathbf{p}_j^d \tag{11.13}$$

This simple algorithm does not need rotation estimates, but is not free from problems. How reliable is $\hat{\mathbf{T}}$? In general, the average attenuates the errors on the centroid data; however, if some of the patches imaged are partially occluded, their centroids will be wrong, and this error will affect $\hat{\mathbf{T}}$ (Exercise 11.7). Moreover, the average of the *visible* centroids is generally different than the average of *all* the patch centroids in the model, and some views of the object can make this problem particularly severe. The conclusion is that this algorithm is only a first idea and must be improved upon; we do so in the next section.

Unfortunately, 3-D rotations are trickier than translations, as they can be represented in several different ways (Appendix, section A.9). We must therefore choose a representation first, then use it to estimate the unknown rotation. We present two simple methods to estimate rotations, based on two different representations. For more on the representation of 3-D rotations, you are referred to the Further Readings.

Estimating the Rotation Matrix.   A first idea is to estimate the entries of the rotation matrix directly, then ensure that the resulting matrix is numerically orthogonal.[6] We need a minimum of three pairs of corresponding data-model normals; the nine simultaneous equations resulting,

$$
\begin{aligned}
\mathbf{n}_1^d &= R\mathbf{n}_1^m \\
\mathbf{n}_2^d &= R\mathbf{n}_2^m \\
\mathbf{n}_3^d &= R\mathbf{n}_3^m,
\end{aligned}
\tag{11.14}
$$

constrain completely the nine entries of the rotation matrix, R, which brings the model onto the data.[7]

☞   It is a good idea to discard corresponding pairs of normals in which the two normals form different angles (allowing for a reasonable numerical tolerance).

To counteract the effects of noise, we use *all* the available pairs of corresponding normals, and solve the least-squares problem

$$
\hat{R} = \arg\min_R \{ \sum_{i=1}^{N} \| \mathbf{n}_i^d - R\mathbf{n}_i^m \|^2 \}
\tag{11.15}
$$

To solve problem (11.15), we introduce a vector, $\mathbf{r} = [r_{11}, r_{12}, r_{13}, r_{21} \ldots r_{33}]^\top$, formed by the entries of the unknown rotation matrix, and write the associated, overconstrained system of equations

$$
M\mathbf{r} = \mathbf{f},
\tag{11.16}
$$

where M is a $3N \times 9$ block-diagonal matrix, and $\mathbf{f}$ is a $3N$-element vector.

---

[6] This is very similar to what we did, for instance, in the derivation of the calibration algorithm EXPL_PARS_CAL (Chapter 6).

[7] Notice that the three vectors needed could be defined by the centroids of four patches, but this would be more sensitive to occlusions.

☞ Notice that, thanks to the structure of M, system (11.16) is formed by three independent systems, each yielding one line of R (Exercise 11.8).

Here is the usual algorithm box.

Algorithm ROT_MATRIX_RANGE

The input is formed by N pairs of corresponding data and model normals, $\mathbf{n}_1^d, \ldots, \mathbf{n}_N^d$ and $\mathbf{n}_1^m, \ldots, \mathbf{n}_N^m$, respectively.

1. Form the overconstrained system (11.16), and compute the least-squares solution, $\hat{\mathbf{r}}$, corresponding to a matrix R'.

2. To enforce orthogonality on R', compute its SVD decomposition $R' = U\Lambda V^\top$, where

$$\Lambda = V \begin{pmatrix} 1 & 0 & 0 \\ 0 & 1 & 0 \\ 0 & 0 & r \end{pmatrix} U^\top, \tag{11.17}$$

and set r to 1 or -1, whatever is closest to $\det(VU^\top)$.

3. Compute $\hat{R} = U\Lambda V^\top$ using the corrected Λ.

The output is $\hat{R}$, the best least-squares estimate of the rotation matrix.

☞ As usual, the least-squares solution suffers from the presence of outliers, created by false correspondences and occasional large errors in the data normal. In this case, you can run a robust estimator (Appendix, section A.7) to identify the outliers, eliminate them from the data set, and reapply ROT_MATRIX_RANGE.

Estimating Axis and Angle. The second method is a geometric algorithm based on the *axis-angle parametrization* (Appendix, section A.9). Any 3-D rotation can be expressed as a rotation around an axis,[8] represented by a 3-D unit vector $\mathbf{n}$, by a certain angle, ϕ, so that $\mathbf{n}$ and ϕ satisfy the right-hand rule. We assume $\phi \in [0, \pi]$, and represent rotations by angles between π and 2π by using $-\mathbf{n}$ instead of $\mathbf{n}$.

Two corresponding pairs of directions (two model normals, two data normals) are sufficient to identify uniquely $\mathbf{n}$ and ϕ, which can be computed by a simple geometric algorithm. The algorithm considers three possible cases: First, the axis of rotation and the model and data normals are all independent; second, two of the normals coincide, and therefore define the axis of rotation; third, the model normals, the data normals, and the axis of rotation are all coplanar. As usual, we shall use all the available correspondences. Following is the detailed description of the algorithm.

---

[8] This is known as *Euler's theorem* (but how many theorems are known by this name?).

Algorithm AXIS_ANGLE_RANGE

The input is formed by N pairs of corresponding data and model normals, $(\mathbf{n}_{i1}^d, \mathbf{n}_{i1}^m)$, $(\mathbf{n}_{i2}^d, \mathbf{n}_{i2}^m)$, with $i = 1, \ldots, N$. We assume no pair is trivial; that is, $\mathbf{n}_{i1}^d \neq \mathbf{n}_{i1}^m$ or $\mathbf{n}_{i2}^d \neq \mathbf{n}_{i2}^m$ for all pairs.

1. For each i-th pair of corresponding normals:

   (a) IF $\mathbf{d} = (\mathbf{n}_{i1}^d - \mathbf{n}_{i1}^m) \times (\mathbf{n}_{i2}^d - \mathbf{n}_{i2}^m) \neq [0, 0, 0]^\top$,

   $$\mathbf{n}_i = \frac{\mathbf{d}}{\|\mathbf{d}\|}.$$

   (i) IF $(\mathbf{n}_i^\top \mathbf{n}_{ij}^d \times \mathbf{n}_{ij}^m) < 0)$ for $j = 1$ or $j = 2$, let $\mathbf{n}_i = -\mathbf{n}_i$.

   (ii) The rotation angle $\phi \in [0, \pi]$ is given by

   $$\phi_i = \arccos \frac{(\mathbf{n}_1^m)^\top \mathbf{n}_1^d - (\mathbf{n}_i^\top \mathbf{n}_1^d)^2}{1 - (\mathbf{n}_i^\top \mathbf{n}_1^d)^2} = \arccos \frac{(\mathbf{n}_2^m)^\top \mathbf{n}_2^d - (\mathbf{n}_i^\top \mathbf{n}_2^d)^2}{1 - (\mathbf{n}_i^\top \mathbf{n}_2^d)^2}. \quad (11.18)$$

   (b) ELSE IF $\mathbf{n}_{i1}^d = \mathbf{n}_{i1}^m$,

   (i) $\mathbf{n}_i = \mathbf{n}_{i1}^m$;

   (ii) IF $(\mathbf{n}_i^\top \mathbf{n}_{i2}^d \times \mathbf{n}_{i2}^m < 0$, let $\mathbf{n}_i = -\mathbf{n}_i$;

   (iii)

   $$\phi_i = \arccos \frac{(\mathbf{n}_2^m)^\top \mathbf{n}_2^d - (\mathbf{n}_i^\top \mathbf{n}_2^d)^2}{1 - (\mathbf{n}_i^\top \mathbf{n}_2^d)^2}.$$

   (c) ELSE IF $\mathbf{n}_{i2}^d = \mathbf{n}_{i2}^m$,

   (i) $\mathbf{n}_i = \mathbf{n}_{i2}^m$;

   (ii) IF $(\mathbf{n}_i^\top \mathbf{n}_{i1}^d \times \mathbf{n}_{i1}^m < 0)$, let $\mathbf{n}_i = -\mathbf{n}_i$;

   (iii)

   $$\phi_i = \arccos \frac{(\mathbf{n}_1^m)^\top \mathbf{n}_1^d - (\mathbf{n}_i^\top \mathbf{n}_1^d)^2}{1 - (\mathbf{n}_i^\top \mathbf{n}_1^d)^2}.$$

   (d) ELSE (axis of rotation coplanar with the data normals, and consequently with the model normals)

   (i) $\mathbf{n}_i = (\lambda \mathbf{n}_{i1}^m - \mathbf{n}_{i2}^m)/(\|\lambda \mathbf{n}_{i1}^m - \mathbf{n}_{i2}^m\|)$, with $\lambda = (\mathbf{n}_{j1}^d - \mathbf{n}_{j1}^m)^\top (\mathbf{n}_{j2}^d - \mathbf{n}_{j2}^m)/2(1 - (\mathbf{n}_{j1}^d)^\top \mathbf{n}_{j1}^d)$

   (ii) IF $(\mathbf{n}_i^\top \mathbf{n}_{i1}^d \times \mathbf{n}_{i1}^m < 0)$ or $((\mathbf{n}_i)^\top \mathbf{n}_{i2}^d \times \mathbf{n}_{i2}^m < 0)$, let $\mathbf{n}_i = -\mathbf{n}_i$;

   (iii)

   $$\phi_i = \arccos \frac{(\mathbf{n}_1^m)^\top \mathbf{n}_1^d - (\mathbf{n}_i^\top \mathbf{n}_1^d)^2}{1 - (\mathbf{n}_i^\top \mathbf{n}_1^d)^2} = \arccos \frac{(\mathbf{n}_2^m)^\top \mathbf{n}_2^d - (\mathbf{n}_i^\top \mathbf{n}_2^d)^2}{1 - (\mathbf{n}_i^\top \mathbf{n}_2^d)^2}. \quad (11.19)$$

2. Let $\hat{\mathbf{n}} = \frac{1}{N} \sum_{i=1}^N \mathbf{n}_i$ and $\hat{\phi} = \frac{1}{N} \sum_{i=1}^N \phi_i$.

The output are the estimates of the rotation axis, $\hat{\mathbf{n}}$, and the rotation angle, $\hat{\phi}$.

☞   The algorithm can be made more robust in various ways. You can consider only the pairs in which the data and model normals form the same angles, within a reasonable numerical tolerance. You can try to identify outliers (severely wrong estimates) in the $\mathbf{n}_i$ and ϕ_i (Exercise 11.10), discard them, and recompute the averages. You can take the median instead of the mean in step 2.

11.3.2   Estimating Rotation First

The second solution of the location problem computes the rotation of the object first, then computes the translation of the *rotated* model reference frame which brings the model onto the data.

Estimating Rotation.   The rotation matrix is computed essentially as in ROT_MATRIX_RANGE, but we incorporate the *reliability of the data normals* possibly supplied by the acquisition stage, and solve the least-squares problem in a different way.

Algorithm ROT_MATRIX_RANGE2

The input is formed by N pairs of corresponding data and model normals, $\mathbf{n}_1^d, \ldots, \mathbf{n}_N^d$ and $\mathbf{n}_1^m, \ldots, \mathbf{n}_N^m$, respectively, as well as a set of positive weights $w_1, \ldots, w_N$ expressing the *reliability* associated with the data normals (the larger w_i, the more reliable the estimate $\mathbf{n}_i^d$).

1. Build the *correlation matrix* $K = \sum_{i=1}^{N} w_i \mathbf{n}_i^d (\mathbf{n}_i^m)^\top$.
2. Compute the SVD $K = V \Lambda U^\top$, Λ being the diagonal matrix of the singular values.
3. Compute $\delta = \det(V U^\top)$, and set δ to 1 or -1, whatever is closest.
4. The best least-squares estimate of the rotation matrix is

$$\hat{R} = V \begin{pmatrix} 1 & 0 & 0 \\ 0 & 1 & 0 \\ 0 & 0 & \delta \end{pmatrix} U^\top \tag{11.20}$$

The output is the estimate of the rotation matrix, $\hat{R}$.

☞   If no information on data reliability is available, set $w_1 = w_2 = \ldots = w_N = 1$.

☞   ROT_MATRIX_RANGE2 and ROT_MATRIX_RANGE are essentially two ways of solving the *same* least squares problem. ROT_MATRIX_RANGE too could be modified to take into account data uncertainties (Exercise 11.9).

Estimating Translation.   We now rotate the model reference frame using $\hat{R}$, and proceed to estimate the translation bringing the *rotated* model onto the data. We make use of the equations of the patch planes,

$$\mathbf{P}^\top \mathbf{n}_j^d = d_j^d \qquad \mathbf{P}^\top \mathbf{n}_j^m = d_j^m, \tag{11.21}$$

for data and model respectively, instead of the patch centroids used by TRANSL_RANGE (we shall comment on this choice in the next section). In the equations above, d_j^d and d_j^m are the distances of the planes from the origin, and **P** is a generic point in space. Three independent planes constrain the translation completely.

In the ideal case, the projection of the unknown translation along the normal of any plane would be exactly the distance between the untranslated and translated planes, $(d_j^d - d_j^m)$. Therefore, we can determine **T** as the solution of the least squares problem

$$\hat{\mathbf{T}} = \arg \min_{\mathbf{T}} \{ \sum_{j=1}^{N} (d_j^d - d_j^m - \mathbf{T}^\top \mathbf{n}_j^{m,r})^2 \}. \tag{11.22}$$

where $\mathbf{n}_j^{m,r}$ is the normal of the *rotated* j-th model plane. Problem (11.22) corresponds to the overconstrained system of N equations

$$\mathbf{T}^\top \mathbf{n}_j^{m,r} = d_j^d - d_j^m,$$

with $j = 1, \ldots, N$, which can be rewritten in the familiar matrix form as

$$L\mathbf{T} = \mathbf{d}, \tag{11.23}$$

with L a $3N \times 3$ matrix, and **d** a $3N$-dimensional vector. Here is the usual summary.

Algorithm TRANSL_RANGE2

The input is formed by the estimated rotation matrix, $\hat{R}$, and by the equations of N corresponding planar patches, $\mathbf{P}^\top \mathbf{n}_j^d = d_j^d$ and $\mathbf{P}^\top \mathbf{n}_j^m = d_j^m$ for data and model respectively.

1. Rotate the normals of all the model patches:

$$\mathbf{n}_j^{m,r} = \hat{R}\mathbf{n}_j^m.$$

2. Solve the overconstrained system (11.23) by least squares.

The output is the estimate of the translation vector, $\hat{\mathbf{T}}$.

☞   The considerations on outliers made for ROT_MATRIX_RANGE apply here too.

11.3.3 Concluding Remarks

From a mathematical point of view, range-based location appears as little more than an exercise on least squares: One way or another, we always set up a least squares problem, then just turned the crank. The real challenges are *robustness* and *accuracy*, which depend on how well we can cope with the errors and occlusions corrupting the data, and how well we can identify and discard outliers in sets of input data or numerical estimates. The Further Readings and Appendix, section A.7 point to suggestions on the latter point. Related challenges include segmenting and pairing features, and estimating feature geometry accurately.

Using plane equations leads to more reliable translation estimates than using centroids, because the latter are more sensitive to partial occlusion. Nonplanar (e.g., cylindrical, spherical) patches can also support range matching, at the price of more complex descriptors and procedures.

11.4   Summary

After working through this chapter you should be able to:

- ❏ explain the problem of 3-D location from intensity and range images
- ❏ explain the different problems involved by the use of intensity and range data
- ❏ design algorithms for 3-D model matching from intensity and range images, given a set of corresponding feature pairs
- ❏ find the pose of a circle in space from its image

11.5   Further Readings

The literature on model-based location from intensity images is very rich. [14] contains an interesting collection of papers on 3-D recognition and location. For a review on the recovery of 3-D pose from point triplets in the general perspective case, see Haralick *et al.* [8]. A closed-form solution based on quaternions is given by Horn [9]. Murase and Nayar [15] is an example of location system adopting viewer-centered object models and models not based on features.

3D_POSE is based on work by Lowe [11], who deals with the uniqueness of the solution and other issues in his account of the SCERPO system. Lowe has also reported an extension for the location of arbitrary 3-D surfaces and articulated objects [12]. WEAK_PERSP_INV is based on Alter's thesis [1] (see also [2]), which is an excellent source of references. Algorithms ROT_MATRIX_RANGE and RANGE_TRANSL were used by Bhanu [3] in an early system for automatic model acquisition from multiple range views. Grimson and Lozano-Perez [7] use the axis-angle parameterization and show an application of AXIS_ANGLE_RANGE. Our algorithm is adapted from Orr [16]. ROT_MATRIX_RANGE2 is based on the solution given in Kanatani's book ([10], Chapter 5), which includes proofs and a detailed discussion of the algorithm. The same chapter gives a complete mathematical treatment of rotation estimation, least squares and SVD.

Among the several influential papers on model-based matching, Faugeras and Hebert [5] apply the pseudoinverse method to obtain the best least-squares transformation; Marshall and Martin [13] discuss and refine this method. Bolles and Horaud describe the *local feature focus* method, and its implementation in the 3DPO system [4], designed to direct a robot arm to grasp industrial parts. RANSAC [6] (for *random sample consensus*) is an example of robust algorithm, in the sense of good tolerance to outliers in the data (see the Appendix, section A.7 for more on robustness).

11.6 Review

Questions

❐ *11.1*   What is the location problem? How does it relate to identification?

❐ *11.2*   How would you modify 3D_POSE if the focal length is unknown?

❐ *11.3*   In the case of line-to-line correspondence, 3D_POSE requires the choice of two points on each matched model line. What is the best possible choice?

❐ *11.4*   In what situations is the weak-perspective camera model suitable for model-based matching?

❐ *11.5*   Explain why a model triangle parallel to the image plane (Figure 11.5) leads to instabilities in algorithm WEAK_PERSP_INV (*Hint:* Look at what happens if the location of one of the three points on plane π is perturbed in the direction orthogonal to the line identified by the other two points.)

❐ *11.6*   The two methods for range-based location suggested in section 11.3 estimate rotation and translation in different orders. Does this mean that the estimated rotation and translation should be applied in different orders to align model and data? Why?

❐ *11.7*   What is the exact difference between applying the rigid transformations $R(\mathbf{x} + \mathbf{T})$ and $R\mathbf{x} + \mathbf{T}$ to a pair of independent vectors $\mathbf{u}$, $\mathbf{v}$? Visualize your results by moving a real object in space. Can you imagine cases in which the result is the same independent of the order of application of rotation and translation?

❐ *11.8*   Under which hypotheses is it sensible to take the average of several estimates to find an approximate answer, as done in TRANSL_RANGE?

❐ *11.9*   We suggested that algorithm AXIS_ANGLE_RANGE can be made more robust by using the median, not the mean, in Step 2. Under what assumptions would this work? When would it fail?

❐ *11.10*   How would you assign the reliability weights to the data input to a location algorithm like ROT_MATRIX_RANGE2? In other words, which elements in the processing of the previous modules contribute measurable uncertainty?

❐ *11.11*   We suggested planar patches as the main features for solving the 3-D location problem with range data. Can you imagine what problems would arise in our solutions if curved patches were adopted instead?

Exercises

○ *11.1*   Devise numerical and geometric examples showing that the order of multiplication of rotation matrices matters in general (Appendix, section A.9). Is this still the case if the rotation angles are very small? Check the results by rotating a real object by two large and two small rotation angles around the same axes, respectively, and altering the order of matrix multiplication.

○ *11.2*   Think of simple heuristics for the estimation of the initial values of R and $\mathbf{T}'$ in 3D_POSE. (*Hint:* Try to make use of information about the center of gravity of the data (for T_1 and T_2), the apparent size (for f), and orientation of the matched image lines (for two of the rotation angles).)

○ *11.3* Write out a complete algorithm extending 3D_POSE to the case of lines, as suggested in section 11.2.1.

○ *11.4* The extension suggested in Exercise 11.3 is based on the explicit line equation (11.6), but this representation is inadequate for vertical and near-vertical lines. Adapt your solution to the line representation $x \cos \theta + y \sin \theta = d$, which does not have this problem.

○ *11.5* Devise an algorithm for backprojecting model features onto data, and a procedure based on the RMS error for estimating the fit between backprojected model features and the corresponding image features.

○ *11.6* Show that the quantities a and c in (11.8) are proportional to the areas of the model and image triangle, respectively. Show that $a > 0$ implies $b \geq 0$ and $b^2 - ac \geq 0$. (*Hint:* Make use of the angles $\mathbf{p}_1\mathbf{p}_0\mathbf{p}_2$ and $\mathbf{P}_1\mathbf{P}_0\mathbf{P}_2$.)

○ *11.7* If only a subset of the model patches are visible, algorithm TRANSL_RANGE produces, in general, wrong translations (centroids). Assuming no false correspondences, explain how the model can be altered to counteract this problem. (*Hint:* Compute the model centroid using only the matched patches.)

○ *11.8* Write out the components of M and $\mathbf{f}$ in algorithm ROT_MATRIX_RANGE. Use the result to identify three independent systems within (11.16).

○ *11.9* How would you modify ROT_MATRIX_RANGE so that the solution incorporates uncertainties on the data normals?

○ *11.10* How would you identify outliers in the $\mathbf{n}_i$ and ϕ_i of AXIS_ANGLE_RANGE? If you need help, check out Appendix, section A.7.

○ *11.11* Using the equations in the Appendix, section A.9, write a rotation matrix R which expresses a rotation around $[1, 1, 1]^\top$ by 45° anticlockwise. Assume a right-handed reference frame.

○ *11.12* Write a program which (a) rotates two synthetic vectors $\mathbf{u}, \mathbf{v}$ into $\mathbf{u}_r, \mathbf{v}_r$, using the rotation matrix defined in the previous exercise; (b) computes an estimate $\hat{R}$ of R using algorithm ROT_MATRIX_RANGE. Is $\hat{R}$ a rotation matrix within reasonable numerical accuracy?

○ *11.13* Using the $\hat{R}$ estimated in the previous exercise, write a program that rotates back $\mathbf{u}_r, \mathbf{v}_r$ using $R^\top$, producing $\hat{\mathbf{u}}, \hat{\mathbf{v}}$. Compute the distance between the tips of $\mathbf{u}, \mathbf{u}_r$, and $\mathbf{v}, \mathbf{v}_r$. How large are the errors? How do they change if you perturb $\mathbf{u}_r, \mathbf{v}_r$ with Gaussian noise of increasing standard deviation?

Projects

● *11.1* Implement the algorithm 3D_POSE for point-to-point and line-to-line correspondences respectively. Using the solution to Exercise 11.3, study the convergence properties of the algorithm as a function of the number of features. Test your code on real images of a polyhedral object using the solutions to Exercise 11.5.

- *11.2* Implement WEAK_PERPS_INV, and test the stability of the results (in term of reconstruction errors) with several data corrupted by varying amounts of Gaussian noise. Run experiments with several triplets of object points, so that the angle formed by the image plane with the plane defined by a triplet varies between about 50° and 0°.

- *11.3* Implement algorithms ROT_MATRIX_RANGE2 and TRANSL_ RANGE2 using your favorite numerical package. Test your code on synthetic data, obtained by generating sets of points, rotating and translating them, and corrupting the results with noise of various intensity (see the Appendix, section A.1 for guidelines on experiments) to check robustness. If you have implemented the patch extractor proposed in the project of Chapter 2, or if you have access to a program extracting adequate features from range data, you can test your code with real data as well.

References

[1] T.D. Alter, Robust and Efficient 3-D Recognition by Alignment, *Tech. Report AI-TR-1410*, Massachusetts Institute of Technology, Cambridge (MA) (1992).

[2] T.D. Alter and E.W.L. Grimson, Fast and Robust 3-D Recognition by Alignment, *Proceedings. 3rd IEEE International Conference on Computer Vision* (1993).

[3] B. Bhanu, Representation and Shape Matching of 3-D Objects, *IEEE Transactions on Pattern Analysis and Machine Intelligence*, Vol PAMI-6, pp. 340-351 (1984).

[4] R.C. Bolles and P. Horaud, 3DPO: A Three-Dimensional Part Orientation System, *IEEE International Journal of Robotics Research*, Vol 5, pp. 3–26 (1986).

[5] O.D. Faugeras and M. Hebert, The Representation, Recognition, and Location of 3D Shapes from Range Data, *International Journal of Robotics Research*, Vol. 5, pp. 27-52 (1986).

[6] M.A. Fischler and R.C. Bolles, Random Sample Consensus: A Paradigm for Model Fitting with Applications to Image Analysis and Automated Cartography, *Communications of the ACM*, Vol. 24, pp. 381-395 (1981).

[7] W.E.L. Grimson and T. Lozano-Pérez, Model-Based Recognition and Localization from Sparse Range or Tactile Data, *International Journal of Robotics Research*, Vol. 3, pp. 3-33 (1984).

[8] R.M. Haralick, C. Lee, K. Ottenberg and M. Nolle, Analysis of Solutions of the Three Point Perspective Pose Estimation Problem, *Proc. IEEE Int. Conf. on Computer Vision and Pattern Recognition*, pp. 592-598 (1991).

[9] B.K.P. Horn, Closed-Form Solution of Absolute Orientation Using Unit Quaternions, *Journal of the Optical Society of America A*, Vol. 4 (1987).

[10] K. Kanatani, *Geometric Computation for Machine Vision*, Oxford University Press, Oxford (1993).

[11] D. Lowe, Three-Dimensional Object Recognition from Two-Dimensional Images, *Artificial Intelligence*, Vol. 31, pp. 355-395 (1987).

[12] D. Lowe, Fitting Parametrized Three-Dimensional Models to Images, *IEEE Transactions on Pattern Analysis and Machine Intelligence*, Vol. PAMI-13, pp. 441-450, (1991).

[13] A.D. Marshall and R.R. Martin, *Computer Vision, Models and Inspection*, World Scientific, London (1992).

[14] J.E.W. Mayhew and J.P. Frisby, *3D Model Recognition from Stereoscopic Cues*, MIT Press, Cambridge (MA) (1991).

[15] H. Murase and S.K. Nayar, Visual Learning and Recognition of 3-D Objects from Appearance, *International Journal of Computer Vision*, Vol. 14, pp. 5-24 (1995).

[16] M.J L. Orr, On Estimating Rotations, DAI Working Paper 233, Dept. of Artificial Intelligence, University of Edinburgh, (1992).

A

Appendix

In this appendix, we give you some details on a number of facts needed to read and make profitable use of this book.

A.1 Experiments: Good Practice Hints

This section is a concise introduction to the *performance assessment* of computer vision programs, and is concerned mainly with the *design of experimental tests*. The purpose is only to provide a list of good-practice principles; be warned that the foundations of *sound* experimental testing are in statistics! Some references are provided at the end; notice the special issue of Machine Vision and Application journal on performance characterization of vision algorithms. In addition to the Web sites quoted in Chapter 1, check out the informative site on performance assessment techniques for computer vision systems and components, `http://pandora.imag.fr/ECVNet/benchmarking.html`, maintained by the European Computer Vision Network (ECV Net). Another interesting site describes HATE, a tool for generating scripts for testing computer vision programs (`http://peipa.essex.ac.uk/hate/`).

Why an Appendix on Experiments?

The reason for an appendix on experiments in a textbook on computer vision is twofold. First and foremost, computer vision is an experimental discipline: the ultimate proof of its theories is whether or not a system can be built which performs as predicted. It is therefore essential to know how to design experiments assessing whether an implementation does what expected, and how well. Second, the literature of computer vision is punctuated by algorithms supported by unconvincing or even scanty experimental evidence; it is important to recognize and avoid such bad practice.

Experimental Performance Assessment

Testing is not just about taking separate measurements, but *discovering how a system behaves as its parameters and experimental conditions vary*. This behavior can be investigated theoretically and experimentally; here we are interested in the latter method. Our objective is therefore *to assess experimentally and quantitatively how well a computer vision program performs its task*.

To achieve this objective, we must

1. identify the key variables and parameters of the algorithm;
2. predict the behavior of the output variables as functions of the parameters and input variables;
3. devise a systematic testing procedure, checking the results of the tests against expected values.

We discuss briefly each point in the next sections. In essence, we run the program undergoing testing in a large number of conditions, and record the value of target variables, their discrepancies from the expected true values, or *errors*, and the error statistics. If the true values are not known in advance, we can still measure the difference between each measure and its average, and its statistics.

Identifying the Key Variables

The key variables of an algorithm are:

- its *input variables*;
- its *output* or *target variables*;
- its *parameters*, for instance constants and thresholds.

We must also consider quantifiable experimental conditions which can affect the results of tests (e.g., illumination direction, uncertainties on input variables), that we shall call *experimental parameters*.

The nature of the target variables and their associated errors depends on the algorithm. For the purposes of this discussion, we identify two broad classes of algorithms, aimed respectively at *measurement* or *detection*.

Measurement Algorithms.   If an algorithm outputs measurements like lengths or areas, such measurements are the target variables. In this case, typical errors used are

- the *mean absolute error*,

$$e_{ma} = \frac{1}{N} \sum_{i=0}^{N} |\hat{x}_i - x_T|,$$

where N is the number of tests, $\hat{x}_i$ is the value of the target variable, x, measured at the i-th test, and x_T the variable's true value;

- the *mean error*,

$$e_m = \frac{1}{N} \sum_{i=0}^{N} (\hat{x}_i - x_T);$$

- the *RMS (root mean square) error*,

$$e_{RMS} = \sqrt{\frac{1}{N} \sum_{i=0}^{N} (\hat{x}_i - x_T)^2}.$$

Detection Algorithms.   If an algorithm is meant to detect instances of a particular entity, the errors above do not make sense; instead, we observe

- the numbers of *false positives*; that is, spurious responses not corresponding to real instances.
- the number of *false negatives*; that is, real instances missed by the algorithm.

☞   Notice that the quantities above are estimates of the *a posteriori* probability of false positives and false negatives.

Predicting the Behavior of the Output Variables

There are two basic cases, depending on the nature of the input data:

1. the test data are *synthetic*
2. the test data are *real*.

Synthetic data are simulated data sets. They allow us to run our program in perfectly controlled situations; as we design the data, we know the value of *all* input variables and parameters (sometimes called *ground truth*), as well as the exact value of the corresponding output variables.

Real data are obtained from real images. When testing with real data, you should endeavor to control the input variables, parameters and experimental parameters, so that results can be compared with the predictions of synthetic tests.

☞   Synthetic data are useful as long as they mirror realistic situations as closely as possible. It is good practice to begin testing with such synthetic data, then use the results to guide the choice of parameters and input variables to be used in tests with real data. It is *definitely bad practice* to claim that "the program works well" because "the results look good" with a few, uncharacterized real images.

Designing Systematic Tests

The next step is to design systematic experiments to investigate the behavior of the program as input variables, parameters and experimental parameters vary. Again, the procedure depends on the nature of data; we begin with synthetic data.

Synthetic Data.

Reference Procedure with Synthetic Data

1. Choose a realistic range of variation for each input variable, algorithm parameter, and experimental parameters. Discretize each range using sampling steps small enough to guarantee a significant sampling of the behavior of the output variables, but large enough to make the total number of runs in step 4 (below) feasible in reasonable times, given the computational resources available.

2. Select a model of random, additive noise to be added to the ideal data. The model (statistical distribution and its parameters) should reflect the real noise measured in the application; in the absence of such information, use Gaussian noise plus outliers. Choose a realistic range for the amount (standard deviation) of the noise, and discretize such range according to the criteria of point 1.

3. For each possible combination of values of input parameters, experimental parameters and amounts of noise:

   (a) generate the data set corresponding to the current values, and corrupt it with a number of different realizations of noise of the current amount;
   (b) run the program on all the noisy data sets obtained;
   (c) store the average values of the target variables together with the current values of noise amount, input variables, algorithm parameters, and experimental parameters.

4. Compute global statistics (over all tests) for the errors of the target variables.

It is convenient to show the results as graphs plotting the error statistics against variations of input variables, program and experimental parameters, and noise levels.

☞   As we cannot test the program in *all* possible situations, a practical question arises: What is the minimum number of tests guaranteeing that the results are representative for the general behavior of the program, within a given level of confidence? An exhaustive answer comes from statistics, and the considerations required are beyond the scope of this appendix. Refer to the Further Readings (for instance Lapin's book) for the whole story.

Real Data.   Ultimately, the algorithm must work with real data from the target application. Hence, you should test your program with real, controlled images. What and how much you can control in real data depends on the application; for the purposes of performance assessment, you should know in advance reliable estimates of input and output variables, and experimental parameters. This includes the statistical properties of the image noise (Chapter 3) and the uncertainty on any quantity, which is input to the program.

In these assumptions, the reference algorithm for synthetic data can be adapted and applied. You cannot vary values as you please any more, but you can certainly use the results of the synthetic tests to identify key values for each controllable range, and run real tests for the selected combinations of input variables and parameters.

References

R.M. Haralick, C. Lee, K. Ottenberg, and M. Nolle, Analysis of Solutions of the Three Point Perspective Pose Estimation Problem, *Proc. IEEE Int. Conf. on Computer Vision and Pattern Recognition*, pp. 592-598 (1991).

L.W. Lapin, *Probability and Statistics for Modern Engineering*, 2nd edition, PWS-Kent, Boston (MA) (1990).

Machine Vision and Applications, Special Issue on Performance Characteristics of Vision Algorithms, Vol. 9, no. 5/6 (1997).

A.2 Numerical Differentiation

This section reminds you of some basic formulae for numerical derivatives, and how to apply them to digital images. Given our definition of digital image and the needs of the algorithms in this book, we limit ourselves to *first and second derivatives* in the case of *equally spaced data*. To dispel any potential impression that the problem is solved by a few, straightforward formulae, we list in the last section a few of the many dangers lurking in the perilous land of numerical differentiation.

Deriving Formulae for Numerical Derivatives

Consider a real-valued function, $f(x)$, of which we know only N samples taken at equidistant points, $x_1, \ldots, x_N$, such that $x_i = x_{i-1} + h$ for some $h > 0$. We want to compute finite-difference estimates of f' and f'', the first and second derivatives of f, as well as the error on these estimates[1]. To solve the problem we write Taylor's polynomial approximation of $f(x)$ at $x + h$ and $x - h$:

$$f(x + h) = f(x) + hf'(x) + \frac{1}{2}h^2 f''(x) + O(h^3) \qquad (A.1)$$

$$f(x - h) = f(x) - hf'(x) + \frac{1}{2}h^2 f''(x) + O(h^3) \qquad (A.2)$$

Subtracting (A.2) from (A.1) and solving for $f'(x_0)$ we get the desired formula for the first derivative:

$$f'(x_0) = \frac{f(x_0 + h) - f(x_0 - h)}{2h} + O(h^2).$$

The quantity $O(h^2)$ means that the *truncation error*, caused by stopping the polynomial approximation to second order, tends to zero as h^2 as h tends to zero. Similarly, summing up (A.1) and (A.2) and solving for $f''(x_0)$ we get the desired formula for the second derivative:

---

[1] Notice that we *assume* that f' and f'' exist, but this may not be true at some points; the samples do not really tell us!

$$f''(x_0) = \frac{f(x_0+h) - 2f(x_0) + f(x_0-h)}{h^2} + O(h).$$

We can derive finite-difference formulae in which the truncation error vanishes more rapidly if we begin with higher-order polynomial approximations (which involve more samples).[2] Here is a summary of the useful formulae and their truncation errors, in the notation introduced at the beginning of the section.

First Derivatives: Central-difference Approximations

$$f'_i = \frac{f_{i+1} - f_{i-1}}{2h} + O(h^2)$$

$$f'_i = \frac{-f_{i+2} + 8f_{i+1} - 8f_{i-1} + f_{i-2}}{12h} + O(h^4)$$

Second Derivatives: Central-difference Approximations

$$f''_i = \frac{f_{i+1} - 2f_i + f_{i-1}}{h^2} + O(h)$$

$$f''_i = \frac{-f_{i+2} + 16f_{i+1} - 30f_i + 16f_{i-1} - f_{i-2}}{12h^2} + O(h^3)$$

☞   With digital images, one nearly invariably sets $h = 1$.

The equations above are called *central-difference approximations*, as they estimate the derivatives at x_i using samples from a symmetric interval centered on x_i. One can use asymmetric intervals instead, leading to the so-called *forward* and *backward approximations*; for instance, for the first derivatives,

$$f'(x_0) = \frac{f_{i+1} - f_i}{h} + O(h)$$

$$f'(x_0) = \frac{f_i - f_{i-1}}{h} + O(h)$$

are the forward and backward approximation respectively. Notice that these formulae carry larger truncation errors; therefore, *central differences should be your choice whenever derivatives must be estimated with simple formulae* (see the next subsection for hints to more sophisticated methods).

---

[2] Again, this implies the *assumption* that higher-order derivatives of f exist.

Computing Image Derivatives

One of the nice features of the formulae above is that *they allow us to compute image derivatives by convolution* (see algorithm LINEAR_FILTER in Chapter 3). For instance, suppose you want to estimate the image gradient, $\nabla I = [\partial I/\partial x, \partial I/\partial y]^\top$, at all pixels (i, j). You just convolve the rows and columns with the mask

$$[\,1 \quad 0 \quad -1\,]$$

and this implements the first formula in the box at all image points. Such masks are sometimes called *stencils*.

☞   Notice that the result can be represented as two images, one for each gradient component, assuming we allow noninteger pixel values (and do not consider peripheral pixels).

Building the masks implementing the other formulae in the box is left as an exercise. You may also want to try and show that the stencil for the Laplacian of $f = f(x, y)$,

$$\Delta f = \frac{\partial^2 f}{\partial x^2} + \frac{\partial^2 f}{\partial y^2},$$

is

$$\begin{bmatrix} 0 & 1 & 0 \\ 1 & -4 & 1 \\ 0 & 1 & 0 \end{bmatrix}.$$

A Word of Caution

What we said so far allows you to take numerical derivatives with simple formulae, but is just the tip of the iceberg; a word of caution must be spent on *errors*. The total error of a numerical approximation is due to two sources, *roundoff errors* and *truncation errors*.

This section has considered only the latter. The former is due to the fact that most numbers are not represented exactly by binary codes in the computer; this can lead to significant errors in the result of even simple computations. Moreover, it turns out that there is an optimal value of h minimising the total error of a given derivative of f at x_i; to make things worse, this value varies, in general, with x_i. Since we do not control the spatial sampling of a digital image, we must expect that the errors associated with numerical image derivatives vary across the image.

☞   A useful take-home message is: *the fractional accuracy of the simple finite-difference approximations in the box is always worse than the fractional accuracy with which the function can be computed* (which, in turn, is generally worse than the machine accuracy).

Better accuracies can be obtained with more sophisticated methods. For instance, one can fit a spline to the data, and compute the derivatives from the spline's coefficient. A good, concise introduction to these methods (including C code) is given by the Numerical Recipes.

References

C.F. Gerald and P.O. Wheatley, *Applied Numerical Analysis*, Fourth edition, Addison-Wesley, Reading (MA) (1970).

W.H. Press, S.A. Teulosky, W.T. Vetterling, and B.P. Flannery, *Numerical Recipes in C*, 2nd edition, Cambridge University Press, Cambridge (UK) (1992).

A.3 The Sampling Theorem

This section states the celebrated *Sampling Theorem* and discusses a few related concepts. This should give you the minimum knowledge necessary to go through Chapters 2 and 3. But if this is the first time you hear about the Sampling Theorem, you are strongly encouraged to read more about it.

The Theorem

Let $F(\omega)$ be the Fourier transform of a function $f(t)$, with $t \in (-\infty, \infty)$. We assume that f is *band-limited*; that is, $F(\omega) = 0$ for $|\omega| > \omega_c > 0$. Then the following theorem holds true.

Sampling Theorem

The function f can be *exactly* reconstructed for every $t \in (-\infty, +\infty)$ from a sequence of equidistant samples, $f_n = f(n\pi/\omega_c)$, according to the formula

$$f(t) = \sum_{n=-\infty}^{+\infty} f_n \frac{\sin(\omega_c t - n\pi)}{\omega_c t - n\pi} = \sum_{n=-\infty}^{+\infty} f_n sinc(\omega_c t - n\pi). \tag{A.3}$$

☞   The Sampling Theorem is nontrivial in at least three respects. First, it tells you that any bandlimited function can be reconstructed *exactly* at locations where it has not been observed. Second, the reconstruction relies upon discrete observations. Third, the information content of a function is not "local": as shown by (A.3), the reconstruction of any value $f(t)$ receives finite contributions from *all* samples.

A few important remarks now follow.

Aliasing and the Nyquist Frequency

The frequency $\nu_c = \omega_c/\pi$, inverse of the sampling interval $T_c = \pi/\omega_c$, is named *Nyquist frequency* and is typical of the signal. It is *the minimal sampling frequency necessary to reconstruct the signal*. This means that, for $\omega < \omega_c$, the series

$$f_\omega(t) = \sum_{n=-\infty}^{+\infty} f_n \frac{\sin(\omega t - n\pi)}{\omega t - n\pi}$$

does *not* converge to $f(t)$. The difference between f and f_ω, that is, the *reconstruction error*, is due to the fact that the sampling distance is too coarse to capture the higher frequencies of the signal. This phenomenon is called *aliasing* because f_ω, the reconstruction of the original signal f, is corrupted by higher frequencies which behave as if they *were* lower frequencies.

If the Function is Not Band-limited

How strong is the assumption that f is band-limited? In practice, not very: After all, because of the integrability conditions, the Fourier transform of any function (if it exists) must be very close to zero outside some finite frequency range. The problem is, one does not necessarily know in advance the value of ω_c, and it may well be the case that π/ω_c is too small with respect to the finest sampling distance achievable in practice. The consequence of this fact can be appreciated by means of the following example.

Assume you are given a sequence of equidistant samples $\ldots, g(-T), g(0), g(T),$ $g(2T), \ldots$ of a function $g(t)$. You don't know whether or not g is band-limited. If you let $\omega = \pi/T$, the series

$$\sum_{n=-\infty}^{\infty} g_n \frac{\sin(\omega t - n\pi)}{\omega t - n\pi}$$

converges to a band-limited function $\tilde{g}$ with $\omega_c = \omega$ (the proof of this fact is left to you as an exercise). If g is not band-limited, or perhaps its band is larger than 2ω, g and $\tilde{g}$ differ, and the difference increases with the amplitude of G, the Fourier transform of g, outside the band $[-\omega, \omega]$.

Function Reconstruction

Perhaps a weakness of the Sampling Theorem is that the reconstruction given by (A.3) converges rather slowly. The function $\sin x/x$ goes to zero as $1/x$ for $x \to \infty$. This means that samples far away from the location where the reconstruction takes place might still give important contributions.

It is instructive to evaluate the derivative of a band-limited function $f(t)$ making use of (A.3). To compute $f'(0)$, the derivative of $f(t)$ at $t = 0$, we take the derivative of both sides of (A.3), set $t = 0$, and obtain

$$f'(0) = \frac{\omega_c}{\pi} \sum_{n \neq 0} (-1)^{n+1} \frac{f_n}{n},$$

where n ranges from $-\infty$ to $+\infty$ (with the exception of $n = 0$, where the derivative of $sinc(x)$ vanishes). Thus for $\omega_c = \pi$ (the typical computer vision setting in which the pixel width is one) we have

$$f'_i = f_{i+1} - f_{i-1} + \frac{f_{i+2} - f_{i-2}}{2} + \ldots \frac{f_{i+n} - f_{i-n}}{n} + \ldots$$

This formula should be compared with the stencils proposed in section A.2.

References

A. Papoulis, *The Fourier Integral and Its Applications*, McGraw-Hill, New York (1962).

A.4 Projective Geometry

In this section, which is not meant to be a rigorous introduction to projective geometry, we give you the minimum information necessary to go through the projective material of the book. We first define projective transformations and standard bases, then discuss briefly the most important projective invariant, the cross-ratio.

Definitions

The projective geometry immediately relevant for computer vision deals with points, lines and their relations in 2-D and 3-D. In this section, we discuss the main concepts in the *planar case*; the extension to the 3-D case is straightforward.

The Projective Plane

We begin by defining the *projective plane*.

Definition: Projective Plane

The *projective plane*, P^2, is the set of equivalence classes of triplets of real numbers (not all zero), where two triplets, $\mathbf{p} = [x, y, z]^\top$ and $\mathbf{p}' = [x', y', z']^\top$, are equivalent if and only if

$$[x, y, z]^\top = \lambda[x', y', z']^\top,$$

where λ is a real number.

A point $\mathbf{p} \in P^2$ is thus identified by three numbers, called *homogeneous coordinates*, defined up to an undetermined factor. This redundant representation allows us to develop a more general geometry than Euclidean geometry. We retain only the elementary concepts of *point*, *line*, and *incidence*, but we do not talk of *angles* and *lengths*.

A Useful Model

A useful model of the projective plane can be obtained in 3-D space: each projective point $\mathbf{p}$ is put in correspondence with a 3-D line through the origin. The proof that this is a *faithful* model of the projective plane is left to you as an exercise.

In this setting, all 3-D lines (or, equivalently, all points of P^2) stand on an equal footing. Instead, if we cut the bundle of 3-D lines in our model with a plane, π, not going through the origin (say the plane of equation $z = 1$), we can distinguish between *proper* and *improper* points:

- each point of the projective plane with $z \neq 0$ is a *proper* point, identified by the coordinates $[x/z, y/z, 1]^\top$;

- each point with $z = 0$ is an *improper* point, identified by the coordinates $[x, y, 0]^\top$.

The same reasoning can be applied to both the 1-D and 3-D case, for the definitions of P^1 and P^3 respectively). *Mutatis mutandis*, the picture is identical. You add one coordinate for the description of a n-dimensional point, subject to the condition that the $(n + 1)$-tuple of numbers (not all zero) are unique up to an undetermined factor. Hence a point in the projective line is identified by two numbers, whereas a point in the projective space by four numbers. In both cases, the homogeneous coordinates are unique up to an undetermined factor.

The Projective Line

We now close this preliminary section by introducing the notion of *projective line*. This can be easily done through the model above, since *collinear points in P^2 correspond to coplanar lines in the 3-D model*.

Definition: Projective Line

A *projective line*, $\mathbf{u}$, is represented by a 3-D plane going through the origin, or

$$\mathbf{u}^\top \mathbf{p} = 0. \tag{A.4}$$

☞   In the projective plane, points and lines are *dual*. In (A.4), one can alternatively think of $\mathbf{p}$ as (*a*) a point lying on the line $\mathbf{u}$, or (*b*) a line going through the point $\mathbf{u}$.

Projective Transformations

A projective transformation is *a linear transformation between projective spaces*. In computer vision, there are at least two important classes of projective transformations:

- linear invertible transformations of P^n, $n = 1, 2, 3$, into themselves.
- transformations between P^3 and P^2, which model image formation.

In what follows, we are interested in the first class. In particular, we want to establish that *a projective transformation of P^n onto itself is completely determined by its action on $n + 2$ points*. For the sake of simplicity, we prove this general result in the particular case of $n = 2$; the extension to the case of a generic $n > 0$ does not pose any problem and is left for an exercise.

Determining a Projective Transformation

A projective transformation of the projective plane onto itself is completely determined once the transformation is known on four points, of which no three are collinear.

As a projective transformation is a linear, invertible transformation, we can represent it in matrix form and write

$$T\mathbf{p}' = \mathbf{p}.$$

Since the coordinates of both $\mathbf{p}$ and $\mathbf{p}'$ are known up to an undetermined factor, the entries of T are also known up to an undetermined factor.

We want to show that T can be written in terms of the four points $\mathbf{p}'_i = [x'_i, y'_i, z'_i]^\top$, $i = 1, \ldots, 4$, image of $\mathbf{p}_1 = [1, 0, 0]^\top$, $\mathbf{p}_2 = [0, 1, 0]^\top$, $\mathbf{p}_3 = [0, 0, 1]^\top$, and $\mathbf{p}_4 = [1, 1, 1]^\top$, respectively. Thanks to the fact that T is a 3×3 invertible matrix, our statement is proven if it can be proven for T^{-1}.

We start by writing

$$T^{-1}\mathbf{p}_1 = \mathbf{p}'_1.$$

From this equation we find that the first column of T^{-1} can be written as $\lambda[x'_1, y'_1, z'_1]^\top$, with λ undetermined. By using the knowledge of $\mathbf{p}'_2$ and $\mathbf{p}'_3$, we then find that T^{-1} can be written as

$$T^{-1} = \begin{pmatrix} \lambda x'_1 & \mu x'_2 & \nu x'_3 \\ \lambda y'_1 & \mu y'_2 & \nu y'_3 \\ \lambda z'_1 & \mu z'_2 & \nu z'_3 \end{pmatrix}.$$

Since no three of the four points $\mathbf{p}'_i$ are collinear, we can now determine λ, μ, and ν, up to an unknown factor. To do this, we use the last available point, $\mathbf{p}_4$, to find

$$\lambda x'_1 + \mu x'_2 + \nu x'_3 = \rho x'_4$$
$$\lambda y'_1 + \mu y'_2 + \nu y'_3 = \rho y'_4$$
$$\lambda z'_1 + \mu z'_2 + \nu z'_3 = \rho z'_4.$$

The nine entries of T^{-1} are therefore known up to an undetermined factor.

In summary, we have shown that a projective transformation of P^2 onto itself is characterized by its action on four points, no three of which are collinear. The four points, $\mathbf{p}_1 \ldots \mathbf{p}_4$, are called the *standard basis of P^2*.

By means of similar arguments, you should be able to show that one can pick $[1, 0]^\top, [0, 1]^\top, [1, 1]^\top$ as the standard basis for the case of line-to-line projective transformations, and $[1, 0, 0, 0]^\top, [0, 1, 0, 0]^\top, [0, 0, 1, 0]^\top, [0, 0, 0, 1]^\top, [1, 1, 1, 1]^\top$ as the standard basis for the case of space-to-space projective transformations.

The Cross-ratio

We close this brief appendix on projective geometry by touching upon the vast subject of *invariants*. We consider the most important and simplest invariant, the *cross-ratio*.

Definition: Cross-ratio

Given four distinct points of P^1, described in homogeneous coordinates as $\mathbf{p}_i = [x_i, y_i]^\top$ ($i = 1, \ldots, 4$), the *cross-ratio* c is defined as

$$c(1, 2, 3, 4) = \frac{d(1, 2)d(3, 4)}{d(1, 3)d(2, 4)}$$

with

$$d(i, j) = x_i y_j - x_j y_i$$

the *determinant* between $\mathbf{p}_i$ and $\mathbf{p}_j$.

Cross-ratio Invariance

The cross-ratio is invariant to projective transformations of P^1 onto itself.

☞   Notice that the order of the points matters in the definition of cross-ratio, and each point appears once in the numerator and once in the denominator. In addition, the cross-ratio can also take on the improper values $\pm\infty$.

You should be able to prove that, given four points, you can define six different cross-ratios. A slightly more difficult exercise is to show that, if λ is one of the six cross-ratios, the other five are $1/\lambda$, $1 - \lambda$, $1/(1 - \lambda)$, $(\lambda - 1)/\lambda$, and $\lambda/(\lambda - 1)$. This tells you also that if, for some labelling, the cross-ratio goes to infinity, it is always possible to find a different labeling which evaluates to a finite value.

It is instructive to go through the proof of the cross-ratio invariance step by step. If we form the determinant between two of the four points, say $\mathbf{p}_1$ and $\mathbf{p}_2$, and assume y_1 and y_2 are not zero, we can write

$$d(1, 2) = x_1 y_2 - x_2 y_1 = y_1 y_2 \left(\frac{x_1}{y_1} - \frac{x_2}{y_2} \right). \tag{A.5}$$

As $[x, y]^\top$ are homogeneous coordinates of a line point, $d(1, 2)$ can be interpreted as the Euclidean distance between $\mathbf{p}_1$ and $\mathbf{p}_2$ times an undetermined factor, $y_1 y_2$, depending on both $\mathbf{p}_1$ and $\mathbf{p}_2$. Now let T, a 2×2 invertible matrix, represent a generic projective transformation,

$$T\mathbf{p}_i = \mathbf{p}'_i.$$

If the determinant between $\mathbf{p}'_1$ and $\mathbf{p}'_2$ is denoted by $d'(1, 2)$, we have

$$d'(1, 2) = d(1, 2)|T|,$$

where $|T|$ denotes the determinant of T. We can see that the determinant of the transformation cancels out in any ratio of determinants like (A.5), but not the factors $y_i y_j$; such factors cancel out, however, in the definition of cross-ratio, for which invariance is achieved.

References

J.L. Mundy and A. Zisserman, Appendix—Projective Geometry for Machine Vision. In *Geometric Invariants in Computer Vision*, Mundy, J.L. and Zisserman, A., eds, MIT Press, Cambridge (MA) (1992).

C.E. Springer, *Geometry and Analysis of Projective Spaces*, Freeman (1964).

A.5 Differential Geometry

This section complements our discussion of range image segmentation (Chapter 4) by recalling a few concepts from the differential geometry of surfaces.

Surface Tangent, Normal, and Area

Consider a parametric surface, $\mathbf{S}(u, v) = [x(u, v), y(u, v), z(u, v)]^\top$, and a point P on $\mathbf{S}$. Assume that all first and second derivatives of $\mathbf{S}$ with respect to u and v exist at P. The *tangent plane* at P is identified by the two vectors $\mathbf{S}_u = \partial \mathbf{S}/\partial u$ and $\mathbf{S}_v = \partial \mathbf{S}/\partial v$.

The *surface normal* is the unit normal vector to the tangent plane of S at P; that is,

$$\mathbf{n}_S(P) = \frac{\mathbf{S}_u \times \mathbf{S}_v}{\|\mathbf{S}_u \times \mathbf{S}_v\|}.$$

Range images in r_{ij} form (Chapter 4) correspond to the parametrization $S(x, y) = [x, y, h(x, y)]^\top$. In this case,

$$\mathbf{n}_S(P) = \frac{(-h_x, -h_y, 1)}{\sqrt{1 + h_x^2 + h_y^2}},$$

where $h_s = \partial h / \partial s$.

It is also useful to know how to compute the area of a surface patch, $\mathbf{Q}$, from a generic parametrization, $\mathbf{Q}(u, v)$, as well as from the parametrization $\mathbf{Q}_{xy} = (x, y, h(x, y))$. If we call A_Q the former and A_{xy} the latter, we have

$$A_Q = \int \int_Q \|\mathbf{Q}_u \times \mathbf{Q}_v\| \, du \, dv$$

$$A_{xy} = \int \int_Q \sqrt{1 + h_x^2 + h_y^2} \, dx \, dy.$$

Surface Curvatures

Curvatures make useful shape descriptors as they are invariant to viewpoint and parametrization. We now want to *extend the notion of curvature of a curve to define the curvature of a surface*, in order to derive the quantities used in Chapter 4.

To begin with, recall that the curvature of a parametric curve $\alpha(t) = (x(t), y(t))$, with t a parameter, is given at each point by

$$k(t) = \frac{x'y'' - x''y'}{((x')^2 + (y')^2)^{\frac{3}{2}}},$$

where $\alpha' = d\alpha/dt$. For the common parametrization $\alpha(x) = [x, y(x)]^\top$, the curvature becomes

$$k(t) = \frac{y''}{(1 + (y')^2)^{\frac{3}{2}}}.$$

Consider a parametric surface, $S(u, v)$, and a point P on S. Assume that all first and second derivatives of S with respect to u, v exist at P. We define surface curvatures in four steps.

Step 1: Normal Curvature of a Curve on S. Consider a curve C on S going through P. We define the *normal curvature of C at P* as

$$k_n = k \cos \theta,$$

where k is the curvature of C at P, and θ is the angle formed by the surface normal at P, $\mathbf{n}_S(P)$, with the curve normal, $\mathbf{n}_C(P)$.[3]

Step 2: Normal Curvature along a Direction. It can be proven that k_n *does not depend on the particular curve C chosen*, but only on the tangent of C at P, identified by the unit vector $\mathbf{d}$. This enables us to speak of *normal curvature along a direction*. For the sake of visualization, we can choose C as the planar curve obtained by intersecting S with a plane through P and containing both $\mathbf{d}$ and $\mathbf{n}_S(P)$. Obviously, C is a cross-section of S along $\mathbf{d}$, and describes the surface shape along that direction. Notice that, in this case, $k_n = k$.

Step 3: Principal Curvatures and Directions. We could now describe the local shape of S at P by taking the normal curvatures at P in *all* directions. This is totally impractical, but fortunately unnecessary. Assume we know the maximum and minimum normal curvatures at P, k_1 and k_2 respectively, called *principal curvatures*, and the corresponding directions, $\mathbf{d}_1$ and $\mathbf{d}_2$, called *principal directions*. It can be proven that

- the principal directions are always orthogonal;
- the normal curvature along *any* direction, $\mathbf{v} = (\cos \beta, \sin \beta)$, where β is the angle from $\mathbf{d}_1$ to $\mathbf{d}$, can be computed through *Euler's formula:*

$$k_n = k_1 \cos^2 \beta + k_2 \sin^2 \beta;$$

- consequently, the local shape of the surface is completely specified by the principal curvatures and directions.

---

[3] You can think of k_n as the projection of $k\mathbf{n}_C(P)$ along the surface normal, $\mathbf{n}_S(P)$.

Step 4: Classifying Local Shape.   Finally, the shape classification given in Chapter 4 is achieved by defining two further quantities, the *mean curvature, H,* and the *Gaussian curvature, K:*

$$H = -\frac{k_1 + k_2}{2} \quad K = k_1 k_2.$$

One can show that *the Gaussian curvature measures how fast the surface moves away from the tangent plane around P*, and in this sense is an extension of the 1-D curvature k. The formulae giving H and K for a range surface in r_{ij} form, $(x, y, h(x, y))$ are given in Chapter 4.

References

M.P. Do Carmo, *Differential Geometry of Curves and Surfaces*, Prentice-Hall, Englewood Cliffs (NJ) (1976).

A.6   Singular Value Decomposition

The aim of this section is to collect the basic information needed to understand the *Singular Value Decomposition* (SVD) as used throughout this book. We start giving the definition of SVD for a generic, rectangular matrix A and discussing some related concepts. We then illustrate three important applications of the SVD:

- solving systems of nonhomogeneous linear equations;
- solving rank-deficient systems of homogeneous linear equations;
- guaranteeing that the entries of a matrix estimated numerically satisfy some given constraints (e.g., orthogonality).

Definition

Singular Value Decomposition

Any $m \times n$ matrix A can be written as the product of three matrices:

$$A = UDV^\top. \tag{A.6}$$

The columns of the $m \times m$ matrix U are mutually orthogonal unit vectors, as are the columns of the $n \times n$ matrix V. The $m \times n$ matrix D is diagonal; its diagonal elements, σ_i, called *singular values*, are such that $\sigma_1 \geq \sigma_2 \geq \ldots \sigma_n \geq 0$.

☞   While both U and V are not unique, the singular values σ_i are fully determined by A.

Some important properties now follow.

Properties of the SVD

Property 1. The singular values give you valuable information on the singularity of a square matrix, *A square matrix, A, is nonsingular if and only if all its singular values are different from zero*. Most importantly, the σ_i also tell you how close A is to be singular: the ratio

$$C = \frac{\sigma_1}{\sigma_n},$$

called *condition number*, measures the *degree of singularity of A*. When $1/C$ is comparable with the arithmetic precision of your machine, the matrix A is *ill-conditioned* and, for all practical purposes, can be considered singular.

Property 2. *If A is a rectangular matrix, the number of nonzero σ_i equals the rank of A.* Thus, given a fixed tolerance, ϵ (typically of the order of 10^{-6}), the number of singular values greater than ϵ equals the *effective rank* of A.

Property 3. *If A is a square, nonsingular matrix, its inverse can be written as*

$$A^{-1} = VD^{-1}U^{\top}.$$

Be A singular or not, the *pseudoinverse* of A, A^+, can be written as

$$A^+ = VD_0^{-1}U^{\top},$$

with D_0^{-1} equal to D^{-1} for all nonzero singular values and zero otherwise. If A is nonsingular, then $D_0^{-1} = D^{-1}$ and $A^+ = A^{-1}$.

Property 4. The columns of U corresponding to the nonzero singular values span the range of A, the columns of V corresponding to the zero singular value the null space of A.

Property 5. The squares of the nonzero singular values are the nonzero eigenvalues of both the $n \times n$ matrix $A^{\top}A$ and $m \times m$ matrix $AA^{\top}$. The columns of U are eigenvectors of $AA^{\top}$, the columns of V eigenvectors of $A^{\top}A$. Morevoer, $A\mathbf{u}_k = \sigma_k \mathbf{v}_k$ and $A^{\top}\mathbf{v}_k = \sigma_k \mathbf{u}_k$, where $\mathbf{u}_k$ and $\mathbf{v}_k$ are the columns of U and V corresponding to σ_k.

Property 6. One possible distance measure between matrices can use the *Frobenius norm*. The Frobenius norm of a matrix A is simply the sum of the squares of the entries a_{ij} of A, or

$$\|A\|_F = \sum_{i,j} a_{ij}^2. \tag{A.7}$$

By plugging (A.6) in (A.7), it follows that

$$\|A\|_F = \sum_i \sigma_i^2.$$

We are now ready to summarize the applications of the SVD used throughout this book.

Least Squares

Assume you have to solve a system of m linear equations,

$$A\mathbf{x} = \mathbf{b},$$

for the unknown n-dimensional vector $\mathbf{x}$. The $m \times n$ matrix A contains the coefficients of the equations, the m-dimensional vector $\mathbf{b}$ the data. If not all the components of $\mathbf{b}$ are null, the solution can be found by multiplying both sides of the above equation for $A^\top$ to obtain

$$A^\top A\mathbf{x} = A^\top \mathbf{b}.$$

It follows that the solution is given by

$$\mathbf{x} = (A^\top A)^+ A^\top \mathbf{b}.$$

This solution is known to be optimal in the least square sense.

It is usually a good idea to compute the pseudoinverse of $A^\top A$ through SVD. In the case of more equations than unknowns the pseudoinverse is more likely to coincide with the *inverse* of $A^\top A$, but keeping an eye on the condition number of $A^\top A$ (**Property 1**) won't hurt.

☞   Notice that linear fitting amounts to solve exactly the same equation. Consequently, you can use the same strategy!

Homogeneous Systems

Assume you are given the problem of solving a homogeneous system of m linear equations in n unknowns

$$A\mathbf{x} = 0,$$

with $m \geq n - 1$ and $rank(A) = n - 1$. Disregarding the trivial solution $\mathbf{x} = 0$, a solution unique up to a scale factor can easily be found through SVD. *This solution is simply proportional to the eigenvector corresponding to the only zero eigenvalue of $A^\top A$ (all other eigenvalues being strictly positive because $rank(A) = n - 1$).* This can be proven as follows.

Since the norm of the solution of a homogeneous system of equations is arbitrary, we look for a solution of unit norm in the least square sense. Therefore we want to minimize

$$\|A\mathbf{x}\|^2 = (A\mathbf{x})^\top A\mathbf{x} = \mathbf{x}^\top A^\top A\mathbf{x},$$

subject to the constraint

$$\mathbf{x}^\top \mathbf{x} = 1.$$

Introducing the Lagrange multiplier λ this is equivalent to minimize the *Lagrangian*

$$\mathcal{L}(\mathbf{x}) = \mathbf{x}^\top A^\top A \mathbf{x} - \lambda(\mathbf{x}^\top \mathbf{x} - 1).$$

Equating to zero the derivative of the Lagrangian with respect to $\mathbf{x}$ gives

$$A^\top A \mathbf{x} - \lambda \mathbf{x} = 0.$$

This equation tells you that λ is an eigenvalue of $A^\top A$, and the solution, $\mathbf{x} = \mathbf{e}_\lambda$, the corresponding eigenvector. Replacing $\mathbf{x}$ with $\mathbf{e}_\lambda$, and $A^\top A \mathbf{e}_\lambda$ with $\lambda \mathbf{e}_\lambda$ in the Lagrangian yields

$$\mathcal{L}(\mathbf{e}_\lambda) = \lambda.$$

Therefore, the minimum is reached at $\lambda = 0$, the least eigenvalue of $A^\top A$. But from **Properties 4** and **5**, it follows that this solution could have been equivalently established as *the column of V corresponding to the only null singular value of A* (the kernel of A). This is the reason why, throughout this book, we have not distinguished between these two *seemingly* different solutions of the same problem.

Enforcing Constraints

One often generates numerical estimates of a matrix, A, whose entries are not all independent, but satisfy some algebraic constraints. This is the case, for example, of orthogonal matrices, or the fundamental matrix we met in Chapter 7. What is bound to happen is that the errors introduced by noise and numerical computations alter the estimated matrix, call it $\hat{A}$, so that its entries no longer satisfy the given constraints. This may cause serious problems if subsequent algorithms assume that $\hat{A}$ satisfies *exactly* the constraints.

Once again, SVD comes to the rescue, and allows us to *find the closest matrix to $\hat{A}$, in the sense of the Frobenius norm (**Property 6**), which satisfies the constraints exactly*. This is achieved by computing the SVD of the estimated matrix, $\hat{A} = UDV^\top$, and estimating A as $UD'V^\top$, with D' obtained by changing the singular values of D to those expected when the constraints are satisfied exactly.[4] Then, the entries of $A = UD'V^\top$ satisfy the desired constraints by construction.

References

G. Strang, *Linear Algebra and its Applications*, Harcourt Brace Jovanovich, Orlando (FL) (1988).

---

[4] If $\hat{A}$ is a good numerical estimate, its singular values should not be too far from the expected ones.

A.7   Robust Estimators and Model Fitting

This section sketches a few, introductory concepts behind *robust estimators*, in particular the so called *M-estimators*. Our limited aim is to support the discussion of Chapter 5 by explaining

- why least squares are a *maximum likelihood estimator* from the point of view of statistics,
- why least squares is skewed significantly by outliers, and
- why an estimator based on the least absolute value, as used in Chapter 5, tolerates outliers better than conventional least squares.

The subject has a vast literature; an initial set of further readings is provided at the end of this appendix. Several robust estimators not detailed here have become popular in computer vision. If interested, you should check out at least the *least median of squares*, discussed in the review by Meer *et al.* and detailed in Rousseeuw and Leroy's book, and the *RANSAC (Random Sample Consensus)* algorithm, introduced by Fischler and Bolles and also discussed by Meer *et al.*

Least Squares as Maximum Likelihood Estimator

Consider N data points $\mathbf{p}_i = [x_i, y_i]^\top$, $i = 1, \ldots, N$, and a *model*, $y = f(x, \mathbf{a})$, where $\mathbf{a}$ is a vector of parameters, and f is a known function. Assume that the data points are observations corrupted by noise (to be characterized better in the following). The well-known estimate of the parameter vector, $\mathbf{a}_o$, such that $f(x, \mathbf{a}_o)$ interpolates the data best in the least squares sense is

$$a_o = \min_{\mathbf{a}} \sum_{i=1}^{N} \mid y_i - f(x_i, \mathbf{a}) \mid^2 . \tag{A.8}$$

We want to show briefly that $\mathbf{a}_o$ *is the parameter vector maximizing the probability that the data are a noisy version of* $f(x, \mathbf{a})$, given appropriate assumptions on the noise.

Notice that we should estimate $\mathbf{a}$ by maximizing the probability that $\mathbf{a}$ *is correct given the data*, but we cannot estimate this probability (why?). However, if we assume that the noise corrupting each data point is *additive* and *Gaussian*, with *zero mean* and standard deviation σ, and that the amounts of noise at different data points are all independent, we *can* express the probability P that, for a given $\mathbf{a}$, all data points fall within Δy of the true value:

$$P \propto \prod_{i=1}^{N} (e^{-\frac{(y_i - f(x_i, \mathbf{a}))^2}{2\sigma^2}} \Delta y). \tag{A.9}$$

In essence, *maximum likelihood estimation* follows from the assumption that *the parameter vector which maximizes P is also the most likely one to occur given the ob-*

served data. To obtain (A.8) from (A.9) is now easy: we just notice that P is maximized by minimizing the negative of its logarithm; that is,

$$\left[\sum_{i=1}^{N} \frac{(y_i - f(x_i, \mathbf{a}))^2}{2\sigma^2} \right] - N \log \Delta y,$$

and the constants σ, N, Δy can be ignored in the minimization.

Why Least Squares Fits are Skewed by Outliers

Since we assumed that the noise corrupting the data points is Gaussian, the probability that a noisy point lies within a distance d of its corresponding true point decreases very rapidly with d. Consequently, the least squares estimator believes that most points lie within a few standard deviations of the true (unknown) model. Now suppose that the data contain even a small percentage of points that are just *way off* and presumably not consistent with the Gaussian hypothesis.[5] Points like these are called *outliers*. In the absence of further information, a least squares estimator believes that outliers too are close to the model, as the probability of an outlier being as far as it really is from the true model is practically zero. The result is that the outliers "pull" the best fit away from the true model much more than they should.

Why Absolute Value Estimators Tolerate Outliers Better

The essence of the problem with least squares is that the Gaussian distribution decreases very rapidly as d becomes larger than σ. Therefore, a solution is to adopt a noise distribution which does not vanish as quickly as a Gaussian; that is, *which considers outliers more likely to occur.* An example of such a distribution is the *double exponential,*

$$Pr\{d\} \propto e^{-\left| \frac{y_i - f(x_i, \mathbf{a})}{\sigma} \right|}.$$

In this case, the probability P becomes

$$P \propto \prod_{i=1}^{N} (e^{-\left| \frac{y_i - f(x_i, \mathbf{a})}{\sigma} \right|} \Delta y). \tag{A.10}$$

The same reasoning which took us from (A.9) to (A.8) takes us from (A.10) to the *robust maximum likelihood estimator*

$$\min_{\mathbf{a}} \sum_{i=1}^{N} |y_i - f(x_i, \mathbf{a})|.$$

The price we pay is that, unlike (A.8), this problem cannot be solved in closed form in most cases, and numerical methods must be employed.

---

[5] The reasons for these points being in the data set can be diverse and, for the purpose of this brief introduction, not quite relevant.

References

M.A. Fischler and R.C. Bolles, Random Sample Consensus: A Paradigm for Model Fitting with Applications to Image Analysis and Automated Cartography, *Communications of the ACM*, Vol. 24, pp. 381-395 (1981).

P. Meer, D. Mintz and A. Rosenfeld, Robust Regression Methods for Computer Vision: a Review, *International Journal of Computer Vision*, Vol. 6, pp. 59-70 (1991).

P.J. Rousseeuw and A.M. Leroy, *Robust Regression and Outlier Detection*, Wiley, New York (1987).

A.8   Kalman Filtering

This section is a brief account of a classic tool of optimal estimation theory, the *linear Kalman filter*. It is meant only to enable understanding of our treatment of feature-based tracking in Chapter 8. For an extensive discussion of Kalman filtering, see Maybeck's classic book.

The purpose of optimal estimation algorithms is to *produce optimal estimates of the state of a dynamic system, on the basis of noisy measurements and an uncertain model of the system's dynamics*. The theory is based on two ingredients, the notions of *system model* and *measurement model*.

The System Model

A physical system is modelled by a *state vector*, $\mathbf{x}$, often called simply the *state*, and a set of equations, called the *system model*. The state is a time-dependent vector, $\mathbf{x}(t)$, the components of which are system variables, in sufficient number to capture the dynamic properties of the system. The system model is a vector equation describing the evolution of the state in time. We indicate discrete equally spaced time instants with $t_k = t_0 + k\Delta T$, with $k = 0, 1, \ldots$ and ΔT a sampling interval, and denote with $\mathbf{x}_k$ the state $\mathbf{x}(t_k)$. We also assume that ΔT is small enough to capture the system's dynamics, so that *the state does not change much between consecutive time instants*, and a *linear* system model is an adequate approximation of the state change within ΔT. The linear system model is usually written in vector form as

$$\mathbf{x}_k = \Phi_{k-1}\mathbf{x}_{k-1} + \xi_{k-1},$$

where ξ_{k-1} is a random vector modelling additive system noise. The subscript $k-1$ indicates that the *state transition matrix*, Φ, is a function of time, and hence accounts for more complicated dynamics than the one we considered for feature tracking in Chapter 8.

The Measurement Model

The second ingredient of estimation theory is the *measurement model*. We assume that, at any time instant t_k, a noisy measurement of the state vector (or at least of some components) is taken, and that the following, linear, relation holds between the measurements and the true system state:

$$\mathbf{z}_k = H_k \mathbf{x}_k + \boldsymbol{\mu}_k.$$

In this equation, $\mathbf{z}_k$ is the vector of measurements taken at time t_k, H_k the so-called *measurement matrix*, and $\boldsymbol{\mu}_k$ a random vector modelling additive noise, which accounts for the uncertainty associated with the measurements.

The key points and assumptions are summarized as follows:

Optimal Linear Estimation Theory (Kalman Filtering)

Assumptions

The state of a dynamic system at time t_k ($k = 1, 2 \ldots$) is described by an n-dimensional vector $\mathbf{x}_k$, the *state vector*. The evolution of the system's state is modelled by the linear equation

$$\mathbf{x}_k = \Phi_{k-1} \mathbf{x}_{k-1} + \boldsymbol{\xi}_{k-1}$$

with Φ_{k-1} a time-dependent $n \times n$ matrix called *state transition matrix*, and $\boldsymbol{\xi}_k$ a n-vector that accounts for the noise associated with the system model.

At any instant t_k, a m-dimensional vector of measurements, $\mathbf{z}_k$, is obtained. The relation between state and measurements is linear:

$$\mathbf{z}_k = H_k \mathbf{x}_k + \boldsymbol{\mu}_k$$

with H_k a time-dependent $m \times n$ matrix, and $\boldsymbol{\mu}_k$ a m-vector that accounts for the noise associated with the measurement process.

The noise terms, $\boldsymbol{\xi}_k$ and $\boldsymbol{\mu}_k$, are assumed to be white, zero-mean, Gaussian processes, with covariance matrices Q_k and R_k, respectively.

Problem Statement

Compute the best estimate of the system's state at t_k, $\hat{\mathbf{x}}_k$, taking into account the state estimate predicted by the system model at t_{k-1} *and* the measurements, $\mathbf{z}_k$, taken at t_k.

The Kalman Filter Algorithm

Optimal linear estimation theory gives you an algorithm based on four recursive equations, the celebrated *Kalman filter equations*, that returns the best estimate of the system state and its covariance at time t_k in the assumptions stated above. The equations integrate the measurements taken at t_k with the prediction of the state formulated at t_{k-1}. The resulting estimate is *optimal in a statistical sense*: over a large number of experiments, the Kalman filter's estimates would be better, on average, than those of any other filter.

The Kalman filter equations are characterized by the *state covariance matrices*, P_k and P'_k, and the *gain matrix*, K_k. P'_k is the covariance matrix of the k-th state estimate, $\hat{\mathbf{x}}'_k = \Phi_{k-1} \hat{\mathbf{x}}_{k-1}$, predicted by the filter *immediately before* obtaining the measurement $\mathbf{z}_k$. P_k is the covariance matrix of the k-th state estimate, $\hat{\mathbf{x}}_k$, computed by the filter *after* integrating the measurement, $\mathbf{z}_k$, with the prediction, $\mathbf{x}'_k$. The covariance matrices are a quantitative model of the *uncertainty* of $\mathbf{x}'_k$ and $\mathbf{x}_k$. Finally, K_k establishes the relative importance of the prediction, $\hat{\mathbf{x}}'_k$, and the state measurement, $\hat{\mathbf{x}}_k$.

The Kalman Filter Algorithm

Let Q_k and R_k be the covariance matrices of the noise processes ξ_k and μ_k. The Kalman filter equations are

$$P'_k = \Phi_{k-1} P_{k-1} \Phi_{k-1}^\top + Q_{k-1} \tag{A.11}$$

$$K_k = P'_k H_k^\top (H_k P'_k H_k^\top + R_k)^{-1} \tag{A.12}$$

$$\hat{\mathbf{x}}_k = \Phi_{k-1} \hat{\mathbf{x}}_{k-1} + K_k(\mathbf{z}_k - H_k \Phi_{k-1} \hat{\mathbf{x}}_{k-1}) \tag{A.13}$$

$$P_k = (I - K_k) P'_k (I - K_k)^\top + K_k R_k K_k^\top \tag{A.14}$$

In A.13, the term $\Phi_{k-1} \hat{\mathbf{x}}_{k-1} = \hat{\mathbf{x}}'_k$ is often called *prediction*, the term $(\mathbf{z}_k - H_k \Phi_{k-1} \hat{\mathbf{x}}_{k-1})$ *innovation*.

☞   Notice that the term $(H_k P'_k H_k^\top + R_k)$ in A.12 is the covariance of the innovation.

The (A.11–A.14) estimate the state and its covariance recursively: they assume known initial estimates of the state covariance matrix, P_0, and of the state, $\hat{\mathbf{x}}_0$. The entries of P_0 are usually set to high, arbitrary values. The value of $\hat{\mathbf{x}}_0$ depends of the initial knowledge of the state; if none is available, an arbitrary value is used.

Let us briefly go through the four steps. First, the state covariance matrix at t_k, P'_k, is estimated according to the system model (A.11). Second, the *gain* of the Kalman filter is computed by (A.12), before reading the measurements. Third, the *optimal state estimate* at time t_k, $\hat{\mathbf{x}}_k$, is formed by (A.13), which integrates the state predicted by the system model ($\Phi_{k-1} \hat{\mathbf{x}}_{k-1}$) with the discrepancy of prediction and observation ($\mathbf{z}_k - H_k \Phi_{k-1} \hat{\mathbf{x}}_{k-1}$) in a sum weighted by the gain matrix, K_k. Finally, the *new state covariance matrix*, P_k, is evaluated through (A.14).

A simpler expression for P_k (A.14) is

$$P_k = (I - K_k H_k) P'_k,$$

but this can lead to serious numerical problems: if the entries of both $(I - K_k H_k)$ and P'_k are small, the accumulation of round-off errors in time can destroy the symmetry and positive definiteness of the covariance matrices, two essential assumptions of the Kalman filter. Violating these assumptions results in unstable state estimates. For this reason, one prefers (A.14), which guarantees both properties for P_k (why?).

☞   Notice that *the state covariance matrix does not depend on the measurements*. Therefore, if the time dependence of Φ_k, H_k, Q_k and R_k is known, P_k can be computed off-line and then approximated by a stepwise function. This property can be crucial for real-time implementations of the Kalman filter.

It can be proven that the Kalman filter is the optimal estimator in the assumptions of linear system and white, Gaussian noise. The filter's assumptions may seem restrictive, but are actually justified for many applications. Interestingly, if the noise is *not* Gaussian, *the Kalman filter can still be proven to be the best linear unbiased filter*. Check the classic book by Maybeck for more on this subject.

Uncertainty

What is the uncertainty on the state vector estimates computed by the Kalman filter? The answer is embedded in the state covariance matrix, which tells us *which region of the state space contains the true state with a given probability*. We illustrate briefly this important idea for the 2-D case.

Let $\mathbf{x} = [x_1, x_2]^\top$ be a 2-D state vector. The Kalman filter computes the optimal state estimate, $\hat{\mathbf{x}}_k$, as the maximum of the conditional probability density of $\mathbf{x}_k$ given the past state estimates, $\hat{\mathbf{x}}_1, \dots \hat{\mathbf{x}}_{k-1}$, the past measurements, $\mathbf{z}_1, \dots \mathbf{z}_{k-1}$, and the current measurement, $\mathbf{z}_k$. This density function is assumed Gaussian, so its maximum coincides with its mean. The important consequences in practice are:

- the region of the plane centered around $\hat{\mathbf{x}}_k$ which contains the *true* state with a given probability c^2 is the ellipse

$$(\mathbf{x} - \hat{\mathbf{x}}_k)(P_k)^{-1}(\mathbf{x} - \hat{\mathbf{x}}_k)^\top \le c^2; \tag{A.15}$$

- the axes of this ellipse are $\pm c\sqrt{\lambda_i}\mathbf{e}_i$, $i = 1, 2$, where λ_i and $\mathbf{e}_i$ are the eigenvalues and eigenvectors, respectively, of P_k;
- the variable $(\mathbf{x} - \hat{\mathbf{x}}_k)(P_k)^{-1}(\mathbf{x} - \hat{\mathbf{x}}_k)^\top$ has a chi-square distribution, so that the probability of the region described by the ellipse is $(1 - \alpha)$, where c^2 is the upper (100α)-th percentile of a chi-square distribution with two degrees of freedom (in general, with a number of degrees of freedom equal to the dimension of the state vector).

In feature tracking, you use these *uncertainty ellipses* to reduce the search for the feature in the next frame, as follows. At t_{k-1}, the filter's prediction of the state at t_k is $\hat{\mathbf{x}}'_k = \Phi_{k-1}\hat{\mathbf{x}}_{k-1}$, with covariance P'_k. You diagonalize P'_k to get its eigenvectors,[6] consider the components of the state vector giving the feature's position $[x_1, x_2]^\top$, build an uncertainty ellipse (A.15) (centered in $\hat{\mathbf{x}}'_k$ and with a desired probability $(1 - \alpha)$), and look for the feature in the k-th frame *only* within that ellipse. Once your feature detector measures the feature's position at t_k, you compute $\hat{\mathbf{x}}_k$ and P_k, which allow you to build the uncertainty ellipse containing the true feature at t_k with the probability level chosen.

---

[6] This can always be done, because covariance matrices are always symmetric.

References

W.S. Cooper, Use of Optimal Estimation Theory, in Particular the Kalman Filter, in Data Analysis and Signal Processing, *Review of Scientific Instrumentation*, Vol. 57, pp. 2862-2869 (1986).

P.S. Maybeck, *Stochastic Models, Estimation, and Control*, Vol. I, Academic Press, New York (1979).

R. A. Johnson and D. W. Wichern, *Applied Multivariate Statistical Analysis*, Third edition, Prentice-Hall, Englewood Cliffs (1992).

A.9 Three-dimensional Rotations

This section collects some useful facts about *3-D rotations* and their representation. We limit ourselves to a few formulae involved by the discussions in this book, most importantly Chapter 11.

General Properties of Rotation Matrices

A rigid rotation of a 3-D vector $\mathbf{v}$ onto a 3-D vector $\mathbf{v}'$ can be represented by a linear transformation, defined by a 3×3 matrix R:

$$\mathbf{v}' = R\mathbf{v},$$

subject to the constraints

$$RR^\top = R^\top R = I \qquad \det(R) = 1,$$

with I the identity matrix. The first constraint tells you that the *inverse* of R equal its *transpose*. The second that *the transformation preserves the relative orientation of the reference frame*, and therefore its right- or left-handedness.

A matrix R for which $RR^\top = I$ is called *orthogonal*. The orthogonality property can be better appreciated by assuming that $\mathbf{v}$ and $\mathbf{v}'$ are expressed in two orthogonal reference frames, defined by the unit vectors $\mathbf{e}_1$, $\mathbf{e}_2$, $\mathbf{e}_3$, and $\mathbf{e}'_1$, $\mathbf{e}'_2$, $\mathbf{e}'_3$ respectively. It can easily be seen that the generic entry r_{ij} of R is *the cosine of the angle formed by the base vector*, $\mathbf{e}_i$, *with the rotated base vector* $\mathbf{e}'_j$:

$$r_{ij} = \mathbf{e}_i^\top \mathbf{e}'_j.$$

Therefore we have that

$$\sum_{j=1}^{3} r_{ij} r_{kj} = \sum_{j=1}^{3} r_{ji} r_{jk} = \begin{cases} 0 & i \neq k \\ 1 & i = k \end{cases} \tag{A.16}$$

that is, *the rows (and columns) of R are mutually orthogonal unit vectors*.

As discussed in section A.6, this property makes the numerically estimation of a rotation matrix a rather tricky task. Even a slight perturbation of the entries of R, due to noise or small errors, destroys the orthogonality property and affects the estimation of the rotation parameters.

Two Useful Parametrizations

A 3×3 orthogonal matrix has nine elements which have to satisfy the six orthogonality constraints (A.16). This reduces the number of *degrees of freedom* of a 3-D rotation to $9 - 6 = 3$ and tells you that describing a rotation matrix through its nine entries is redundant and not really natural; in most cases it is useful and simpler to represent a rotation by means of more *natural* parametrization. In what follows we limit ourselves to the two parametrizations used in the book: *rotations around the coordinate axes* and *axis and angle.* Further parametrizations, well-known in computer vision, include *Euler angles, quaternions,* and *pitch,roll,yaw angles* (see References).

Rotations Around the Coordinate Axes. We can express a 3-D rotation as the result of three consecutive rotations around the coordinate axes, $\mathbf{e}_1, \mathbf{e}_2$, and $\mathbf{e}_3$, by angles α, β, and γ respectively. The angles are then the three free parameters of R, and each rotation is expressed as a rotation matrix, R_j, rotating vectors around $\mathbf{e}_j$, that is,

$$R_1(\alpha) = \begin{bmatrix} 1 & 0 & 0 \\ 0 & \cos \alpha & - \sin \alpha \\ 0 & \sin \alpha & \cos \alpha \end{bmatrix}$$

$$R_2(\beta) = \begin{bmatrix} \cos \beta & 0 & \sin \beta \\ 0 & 1 & 0 \\ - \sin \beta & 0 & \cos \beta \end{bmatrix}$$

$$R_3(\gamma) = \begin{bmatrix} \cos \gamma & - \sin \gamma & 0 \\ \sin \gamma & \cos \gamma & 0 \\ 0 & 0 & 1 \end{bmatrix}$$

The matrix R describing the overall rotation is the product of the R_j:

$$R = R_1 R_2 R_3 =$$

$$\begin{bmatrix} \cos \beta \cos \gamma & - \cos \beta \sin \gamma & \sin \beta \\ \sin \alpha \sin \beta \cos \gamma + \cos \alpha \sin \gamma & - \sin \alpha \sin \beta \sin \gamma + \cos \alpha \cos \gamma & - \sin \alpha \cos \beta \\ - \cos \alpha \sin \beta \cos \gamma + \sin \alpha \sin \gamma & \cos \alpha \sin \beta \sin \gamma + \sin \alpha \cos \gamma & \cos \alpha \cos \gamma \end{bmatrix}$$

☞   The order of multiplication matters. Different sequences give different results, with the same triplet of angles. Notice that there are six ways to represent a rotation matrix as a product of rotations around three different coordinate axes.

The recovery of α, β, and γ from R is easy (and is left to you as an exercise). However, if the matrix R has been obtained as the output of some numerical computation, you should always make sure that the estimated R is *really* orthogonal; section A.6 gives you the recipe that we employed throughout the book.

Axis and Angle. According to *Euler's theorem*, any 3-D rotation can be described as a rotation by an angle, θ, around an axis identified by a unit vector $\mathbf{n} = [n_1, n_2, n_3]^\top$. The corresponding rotation matrix, R, can then be obtained in terms of θ and the components of $\mathbf{n}$, which gives you a total of four parameters. The redundancy of this parametrization (four parameters for three degrees of freedom) is eliminated by adding the constraint that $\mathbf{n}$ has unit norm, that is, by dividing each n_i by $\sqrt{n_1^2 + n_2^2 + n_3^2}$.

The matrix R in terms of θ and $\mathbf{n}$ is given by

$$R = I\cos\theta + (1 - \cos\theta)\begin{bmatrix} n_1^2 & n_1n_2 & n_1n_3 \\ n_2n_1 & n_2^2 & n_2n_3 \\ n_3n_1 & n_3n_2 & n_3^2 \end{bmatrix} + \sin\theta\begin{bmatrix} 0 & -n_3 & n_2 \\ n_3 & 0 & -n_1 \\ -n_2 & n_1 & 0 \end{bmatrix} \quad \text{(A.17)}$$

Conversely, both θ and $\mathbf{n}$ can be obtained from the eigenvalues and eigenvectors of R. The three eigenvalues of R are 1, $\cos\theta + i\sin\theta$, and $\cos\theta - i\sin\theta$, where i is the imaginary unit. The unit vector, $\mathbf{n}$, is proportional to the eigenvector of R corresponding to the eigenvalue 1; the angle θ can be obtained from either of the two complex eigenvalues. To resolve the ambiguity in the sign of both θ and $\mathbf{n}$, you can check the consistency of (A.17).

References

S. L. Altmann, *Rotations, Quaternions, and Double Groups*, Oxford University Press, Oxford (1986).

O. D. Faugeras and M. Hebert, The Representation, Recognition, and Locating of 3-D Shapes from Range Data, *International Journal of Robotic Research*, Vol. 5, pp. 27-52 (1986).

G. A. Korn and T. M. Korn, *Mathematical Handbook for Scientists and Engineers*, second edition, McGraw-Hill, New York (1968).

Index